T0264924

Deep Learning in Computer Vision

Digital Imaging and Computer Vision Series

Series Editor
Rastislav Lukac
Foveon, Inc./Sigma Corporation San Jose, California, U.S.A.

Deep Learning in Computer Vision

Principles and Applications

Edited by
Mahmoud Hassaballah and Ali Ismail Awad

CRC Press
Taylor & Francis Group
Boca Raton London New York

CRC Press is an imprint of the
Taylor & Francis Group, an **Informa** business

CRC Press
Taylor & Francis Group
6000 Broken Sound Parkway NW, Suite 300
Boca Raton, FL 33487-2742

© 2020 by Taylor & Francis Group, LLC
CRC Press is an imprint of Taylor & Francis Group, an Informa business

No claim to original U.S. Government works

Printed on acid-free paper

International Standard Book Number-13: 978-1-138-54442-0 (Hardback)

Library of Congress Cataloging-in-Publication Data

Names: Hassaballah, Mahmoud, editor. | Awad, Ali Ismail, editor.
Title: Deep learning in computer vision : principles and applications / edited by M. Hassaballah and Ali Ismail Awad.
Description: First edition. | Boca Raton, FL : CRC Press/Taylor and Francis, 2020. | Series: Digital imaging and computer vision | Includes bibliographical references and index.
Identifiers: LCCN 2019057832 (print) | LCCN 2019057833 (ebook) | ISBN 9781138544420 (hardback ; acid-free paper) | ISBN 9781351003827 (ebook)
Subjects: LCSH: Computer vision. | Machine learning.
Classification: LCC TA1634 .D437 2020 (print) | LCC TA1634 (ebook) | DDC 006.3/7--dc23
LC record available at https://lccn.loc.gov/2019057832
LC ebook record available at https://lccn.loc.gov/2019057833

Visit the Taylor & Francis Web site at
http://www.taylorandfrancis.com

and the CRC Press Web site at
http://www.crcpress.com

Contents

Foreword

Deep learning, while it has multiple definitions in the literature, can be defined as "inference of model parameters for decision making in a process mimicking the understanding process in the human brain"; or, in short: "brain-like model identification". We can say that deep learning is a way of data inference in machine learning, and the two together are among the main tools of modern artificial intelligence. Novel technologies away from traditional academic research have fueled R&D in convolutional neural networks (CNNs); companies like Google, Microsoft, and Facebook ignited the "art" of data manipulation, and the term "deep learning" became almost synonymous with decision making.

Various CNN structures have been introduced and invoked in many computer vision-related applications, with greatest success in face recognition, autonomous driving, and text processing. The reality is: deep learning is an art, not a science. This state of affairs will remain until its developers develop the theory behind its functionality, which would lead to "cracking its code" and explaining why it works, and how it can be structured as a function of the information gained with data. In fact, with deep learning, there is good and bad news. The good news is that the industry—not necessarily academia—has adopted it and is pushing its envelope. The bad news is that the industry does not share its secrets. Indeed, industries are never interested in procedural and textbook-style descriptions of knowledge.

This book, *Deep Learning in Computer Vision: Principles and Applications*—as a journey in the progress made through deep learning by academia—confines itself to deep learning for computer vision, a domain that studies sensory information used by computers for decision making, and has had its impacts and drawbacks for nearly 60 years. Computer vision has been and continues to be a system: sensors, computer, analysis, decision making, and action. This system takes various forms and the flow of information within its components, not necessarily in tandem. The linkages between computer vision and machine learning, and between it and artificial intelligence, are very fuzzy, as is the linkage between computer vision and deep learning. Computer vision has moved forward, showing amazing progress in its short history. During the sixties and seventies, computer vision dealt mainly with capturing and interpreting optical data. In the eighties and nineties, geometric computer vision added science (geometry plus algorithms) to computer vision. During the first decade of the new millennium, modern computing contributed to the evolution of object modeling using multimodality and multiple imaging. By the end of that decade, a lot of data became available, and so the term "deep learning" crept into computer vision, as it did into machine learning, artificial intelligence, and other domains.

This book shows that traditional applications in computer vision can be solved through invoking deep learning. The applications addressed and described in the eleven different chapters have been selected in order to demonstrate the capabilities of deep learning algorithms to solve various issues in computer vision. The content of this book has been organized such that each chapter can be read independently of the others. Chapters of the book cover the following topics: accelerating the CNN inference on field-programmable gate arrays, fire detection in surveillance applications, face recognition, action and activity recognition, semantic segmentation for autonomous driving, aerial imagery registration, robot vision, tumor detection, and skin lesion segmentation as well as skin melanoma classification.

From the assortment of approaches and applications in the eleven chapters, the common thread is that deep learning for identification of CNN provides accuracy over traditional approaches. This accuracy is attributed to the flexibility of CNN and the availability of large data to enable identification through the deep learning strategy. I would expect that the content of this book to be welcomed worldwide by graduate and postgraduate students and workers in computer vision, including practitioners in academia and industry. Additionally, professionals who want to explore the advances in concepts and implementation of deep learning algorithms applied to computer vision may find in this book an excellent guide for such purpose. Finally, I hope that readers would find the presented chapters in the book interesting and inspiring to future research, from both theoretical and practical viewpoints, to spur further advances in discovering the secrets of deep learning.

Prof Aly Farag, PhD, Life Fellow, IEEE, Fellow, IAPR
Professor of Electrical and Computer Engineering
University of Louisville, Kentucky

Preface

Simply put, computer vision is an interdisciplinary field of artificial intelligence that aims to guide computers and machines toward understanding the contents of digital data (i.e., images or video). According to computer vision achievements, the future generation of computers may understand human actions, behaviors, and languages similarly to humans, carry out some missions on their behalf, or even communicate with them in an intelligent manner. One aspect of computer vision that makes it such an interesting topic of study and active research field is the amazing diversity of daily-life applications such as pedestrian protection systems, autonomous driving, biometric systems, the movie industry, driver assistance systems, video surveillance, and robotics as well as medical diagnostics and other healthcare applications. For instance, in healthcare, computer vision algorithms may assist healthcare professionals to precisely classify illnesses and cases; this can potentially save patients' lives through excluding inaccurate medical diagnoses and avoiding erroneous treatment. With this wide variety of applications, there is a significant overlap between computer vision and other fields such as machine vision and image processing. Scarcely a month passes where we do not hear from the research and industry communities with an announcement of some new technological breakthrough in the areas of intelligent systems related to the computer vision field.

With the recent rapid progress on deep convolutional neural networks, deep learning has achieved remarkable performance in various fields. In particular, it has brought a revolution to the computer vision community, introducing non-traditional and efficient solutions to several problems that had long remained unsolved. Due to this promising performance, it is gaining more and more attention and is being applied widely in computer vision for several tasks such as object detection and recognition, object segmentation, pedestrian detection, aerial imagery registration, video processing, scene classification, autonomous driving, and robot localization as well as medical image-related applications. If the phrase "deep learning for computer vision" is searched in Google, millions of search results will be obtained. Under these circumstances, a book entitled *Deep Learning in Computer Vision* that covers recent progress and achievements in utilizing deep learning for computer vision tasks will be extremely useful.

The purpose of this contributed volume is to fill the existing gap in the literature for the applications of deep learning in computer vision and to provide a bird's eye view of recent state-of-the-art models designed for practical problems in computer vision. The book presents a collection of eleven high-quality chapters written by renowned experts in the field. Each chapter provides the principles and fundamentals of a specific topic, introduces reviews of up-to-date techniques, presents outcomes, and points out challenges and future directions. In each chapter, figures, tables, and examples are used to improve the presentation and analysis of covered topics. Furthermore, bibliographic references are included in each chapter, providing a good starting point for deeper research and further exploration of the topics considered in this book. Further, this book is structured such that each chapter can be read independently from the others as follows:

Chapter 1 presents a state-of-the-art of CNN inference accelerators over FPGAs. Computational workloads, parallelism opportunities, and the involved memory accesses are analyzed. At the level of neurons, optimizations of the convolutional and fully connected layers are explained and the performances of the different methods compared, while at the network level, approximate computing and data-path optimization methods are covered and state-of-the-art approaches compared. The methods and tools investigated in this chapter represent the recent trends in FPGA CNN inference accelerators and will fuel future advances in efficient hardware deep learning.

Chapter 2 concentrates on object detection problem using deep CNN (DCNN): the recent developments of several classical CNN-based object detectors are discussed. These detectors significantly improve detection performance either through employing new architectures or through solving practical issues like degradation, gradient vanishing, and class imbalance. Detailed background information is provided to show the progress and improvements of different models. Some evaluation results and comparisons are reported on three datasets with distinctive characteristics.

Chapter 3 proposes three methods for fire detection using CNNs. The first method focuses on early fire detection with an adaptive prioritization mechanism for surveillance cameras. The second CNN-assisted method improves fire detection accuracy with a main focus on reducing false alarms. The third method uses an efficient deep CNN for fire detection. For localization of fire regions, a feature map selection algorithm that intelligently selects appropriate feature maps sensitive to fire areas is proposed.

Chapter 4 presents an accurate and real-time multi-biometric system for identifying a person's identity using a combination of two discriminative deep learning approaches to address the problem of unconstrained face recognition: CNN and deep belief network (DBN). The proposed system is tested on four large-scale challenging datasets with high diversity in the facial expressions—SDUMLA-HMT, FRGC V 2.0, UFI, and LFW—and new state-of-the-art recognition rates on all the employed datasets are achieved.

Chapter 5 introduces a study of the concept of sequence learning using RNN, LSTM, and its variants such as multilayer LSTM and bidirectional LSTM for action and activity recognition problems. The chapter concludes with major issues of sequence learning for action and activity recognition and highlights recommendations for future research.

Chapter 6 discusses semantic segmentation in autonomous driving applications, where it focuses on constructing efficient and simple architectures to demonstrate the benefit of flow and depth augmentation to CNN-based semantic segmentation networks. The impact of both motion and depth information on semantic segmentation is experimentally studied using four simple network architectures. Results of experiments on two public datasets—Virtual-KITTI and CityScapes—show reasonable improvement in overall accuracy.

Chapter 7 presents a method based on deep learning for geolocalizing drones using only onboard cameras. A pipeline has been implemented that makes use of the availability of satellite imagery and traditional computer vision feature detectors and descriptors, along with renowned deep learning methods (semantic segmentation), to be able to locate the aerial image captured from the drone within the satellite imagery. The method enables the drone to be autonomously aware of its surroundings and navigate without using GPS.

Chapter 8 is intended to be a guide for the developers of robot vision systems, focusing on the practical aspects of the use of deep neural networks rather than on theoretical issues.

The last three chapters are devoted to deep learning in medical applications. Chapter 9 covers basic information about CNNs in medical applications. CNN developments are discussed from different perspectives, specifically, CNN design, activation function, loss function, regularization, optimization, normalization, and network depth. Also, a deep convolutional neural network (DCNN) is designed for brain tumor detection using MRI images. The proposed DCNN architecture is evaluated on the RIDER dataset, achieving accurate detection accuracy within a time of 0.24 seconds per MRI image.

Chapter 10 discusses automatic segmentation of skin lesion boundaries from surrounding tissue and presents a novel deep learning segmentation methodology via full-resolution convolutional network (FrCN). Experimental results show the great promise of the FrCN method compared to state-of-the-art deep learning segmentation approaches such as fully convolutional networks (FCN), U-Net, and SegNet with overall segmentation.

Chapter 11 is about the automatic classification of color skin images, where a highly accurate method is proposed for skin melanoma classification utilizing two modified deep convolutional neural networks and consisting of three main steps. The proposed method is tested using the well-known MED-NODE and DermIS & DermQuest datasets.

It is very necessary to mention here that the book is a small piece in the puzzle of computer vision and its applications. We hope that our readers find the presented chapters in the book interesting and that the chapters will inspire future research both from theoretical and practical viewpoints to spur further advances in the computer vision field.

The editors would like to take this opportunity to express their sincere gratitude to the contributors for extending their wholehearted support in sharing some of their latest results and findings. Without their significant contribution, this book could not have fulfilled its mission. The reviewers deserve our thanks for their constructive and timely input. Special profound thanks go to Prof Aly Farag, Professor of Electrical and Computer Engineering, University of Louisville, Kentucky for writing the Foreword for this book. Finally, the editors acknowledge the efforts of the CRC Press Taylor & Francis for giving us the opportunity to edit a book on deep learning for computer vision. In particular, we would like to thank Dr Rastislav Lukac, the editor of the Digital Imaging and Computer Vision book series, and Nora Konopka for initiating this project. Really, the editorial staff at CRC Press has done a meticulous job, and working with them was a pleasant experience.

Mahmoud Hassaballah
Qena, Egypt

Ali Ismail Awad
Luleå, Sweden

Editors Bio

Mahmoud Hassaballah was born in 1974, Qena, Egypt. He received his BSc degree in Mathematics in 1997 and his MSc degree in Computer Science in 2003, both from South Valley University, Egypt, and his Doctor of Engineering (D Eng) in computer science from Ehime University, Japan in 2011. He was a visiting scholar with the department of computer & communication science, Wakayama University, Japan in 2013 and GREAH laboratory, Le Havre Normandie University, France in 2019. He is currently an associate professor of computer science at the faculty of computers and information, South Valley University, Egypt. He served as a reviewer for several journals such as *IEEE Transactions on Image Processing, IEEE Transactions on Fuzzy Systems, Pattern Recognition, Pattern Recognition Letters, IET Image Processing, IET Computer Vision, IET Biometrics, Journal of Real-Time Image Processing, and Journal of Electronic Imaging.* He has published over 50 research papers in refereed international journals and conferences. His research interests include feature extraction, object detection/recognition, artificial intelligence, biometrics, image processing, computer vision, machine learning, and data hiding.

Ali Ismail Awad (SMIEEE, PhD, PhD, MSc, BSc) is currently an Associate Professor (Docent) with the Department of Computer Science, Electrical, and Space Engineering, Luleå University of Technology, Luleå, Sweden, where he also serves as a Coordinator of the Master Programme in Information Security. He is a Visiting Researcher with the University of Plymouth, United Kingdom. He is also an Associate Professor with the Electrical Engineering Department, Faculty of Engineering, Al-Azhar University at Qena, Qena, Egypt. His research interests include information security, Internet-of-Things security, image analysis with applications in biometrics and medical imaging, and network security. He has edited or co-edited five books and authored or co-authored several journal articles and conference papers in these areas. He is an Editorial Board Member of the following journals: *Future Generation Computer Systems, Computers & Security, Internet of Things: Engineering Cyber Physical Human Systems,* and *Health Information Science and Systems.* Dr Awad is currently an IEEE senior member.

Contributors

Ahmad El Sallab
Valeo Company
Cairo, Egypt

Ahmed Nassar
IRISA Institute
Rennes, France

Alaa S. Al-Waisy
University of Bradford
Bradford, UK

Ali Ismail Awad
Luleå University of Technology
Luleå, Sweden
and
Al-Azhar University
Qena, Egypt

Amin Ullah
Sejong University
Seoul, South Korea

Ashraf A. M. Khalaf
Minia University
Minia, Egypt

François Berry
University Clermont Auvergne
Clermont-Ferrand, France

Guanghui Wang
University of Kansas
Kansas City, Kansas

Hazem Rashed
Valeo Company
Cairo, Egypt

Hesham F.A. Hamed
Egyptian Russian University
Cairo, Egypt
and
Minia University
Minia, Egypt

Javier Ruiz-Del-Solar
University of Chile
Santiago, Chile

Kaidong Li
University of Kansas
Kansas City, Kansas

Kamel Abdelouahab
Clermont Auvergne University
Clermont-Ferrand, France

Khalid M. Hosny
Zagazig University
Zagazig, Egypt

Khan Muhammad
Sejong University
Seoul, South Korea

Mahmoud Hassaballah
South Valley University
Qena, Egypt

Mahmoud Khaled Abd-Ellah
Al-Madina Higher Institute for
    Engineering and Technology
Giza, Egypt

Maxime Pelcat
University of Rennes
Rennes, France

Miyoung Lee
Sejong University
Seoul, South Korea

Mohamed A. Kassem
Kafr El Sheikh University
Kafr El Sheikh, Egypt

Mohamed Elhelw
Nile University
Giza, Egypt

Mohamed M. Foaud
Zagazig University
Zagazig, Egypt

Mohammed A. Al-Masni
Kyung Hee University
Seoul, South Korea
and
Yonsei University
Seoul, South Korea

Mugahed A. Al-Antari
Kyung Hee University
Seoul, South Korea
and
Sana'a Community College
Sana'a, Republic of Yemen

Patricio Loncomilla
University of Chile
Santiago, Chile

Rami Qahwaji
University of Bradford
Bradford, UK

Salman Khan
Sejong University
Seoul, South Korea

Senthil Yogamani
Valeo Company
Galway, Ireland

Shumoos Al-Fahdawi
University of Bradford
Bradford, UK

Sung Wook Baik
Sejong University
Seoul, South Korea

Tae-Seong Kim
Kyung Hee University
Seoul, South Korea

Tanveer Hussain
Sejong University
Seoul, South Korea

Usman Sajid
University of Kansas
Kansas City, Kansas

Wenchi Ma
University of Kansas
Kansas City, Kansas

Yuanwei Wu
University of Kansas
Kansas City, Kansas

1 Accelerating the CNN Inference on FPGAs

*Kamel Abdelouahab, Maxime Pelcat,
and François Berry*

CONTENTS

1.1   INTRODUCTION

The exponential growth of big data during the last decade motivates for innovative methods to extract high semantic information from raw sensor data such as videos, images, and speech sequences. Among the proposed methods, convolutional neural networks (CNNs) [1] have become the de facto standard by delivering near-human accuracy in many applications related to machine vision (e.g., classification [2], detection [3], segmentation [4]) and speech recognition [5].

This performance comes at the price of a large computational cost as CNNs require up to 38 GOPs to classify a single frame [6]. As a result, dedicated hardware is required to accelerate their execution. Graphics processing units GPUs are the most widely used platform to implement CNNs as they offer the best performance in terms of pure computational throughput, reaching up 11 TFLOPs [7]. Nevertheless, in terms of power consumption, field-programmable gate array (FPGA) solutions are known to be more energy efficient (vs. GPU). While GPU implementations have demonstrated state-of-the-art computational performance, CNN acceleration will soon be moving towards FPGAs for two reasons. First, recent improvements in FPGA technology put FPGA performance within striking distance of GPUs with a reported performance of 9.2 TFLOPs for the latter [8]. Second, recent trends in CNN development increase the sparsity of CNNs and use extremely compact data types. These trends favor FPGA devices, which are designed to handle irregular parallelism and custom data types. As a result, next-generation CNN accelerators are expected to deliver up to 5.4× better computational throughput than GPUs [7].

As an inflection point in the development of CNN accelerators might be near, we conduct a survey on FPGA-based CNN accelerators. While a similar survey can be found in [9], we focus in this chapter on the recent techniques that were not covered in the previous works. In addition to this chapter, we refer the reader to the works of Venieris et al. [10], which review the toolflows automating the CNN mapping process, and to the works of Sze et al., which focus on ASICs for deep learning acceleration.

The amount and diversity of research on the subject of CNN FPGA acceleration within the last 3 years demonstrate the tremendous industrial and academic interest. This chapter presents a state-of-the-art review of CNN inference accelerators over FPGAs. The computational workloads, their parallelism, and the involved memory accesses are analyzed. At the level of neurons, optimizations of the convolutional and fully connected (FC) layers are explained and the performances of the different methods compared. At the network level, approximate computing and data-path optimization methods are covered and state-of-the-art approaches compared. The methods and tools investigated in this survey represent the recent trends in FPGA CNN inference accelerators and will fuel the future advances on efficient hardware deep learning.

1.2 BACKGROUND ON CNNS AND THEIR COMPUTATIONAL WORKLOAD

In this first section, we overview the main features of CNNs, mainly focusing on the computations and parallelism patterns involved during their inference.

1.2.1 GENERAL OVERVIEW

Deep* CNNs are feed-forward[†], sparsely connected[‡] neural networks. A typical CNN structure consists of a pipeline of layers. Each layer inputs a set of data, known as a feature map (FM), and produces a new set of FMs with *higher-level semantics*.

1.2.2 INFERENCE VERSUS TRAINING

As typical machine learning algorithms, CNNs are deployed in two phases. First, the *training* stage works on a known set of annotated data samples to create a model with a *modeling* power (which semantics extrapolates to natural data outside the training set). This phase implements the *back-propagation* algorithm [11], which iteratively updates CNN parameters such as convolution weights to improve the predictive power of the model. A special case of CNN training is *fine-tuning*. When *fine-tuning* a model, weights of a previously trained network are used to initialize the parameters of a new training. These weights are then adjusted for a new constraint, such as a different dataset or a reduced precision.

The second phase, known as *inference*, uses the learned model to classify new data samples (i.e., inputs that were not previously seen by the model). In a typical setup, CNNs are trained/fine-tuned only once, on large clusters of GPUs. By contrast, the inference is implemented each time a new data sample has to be classified. As a consequence, the literature mostly focuses on accelerating the inference phase. As a result, our discussion overviews the main methods employed to accelerate the inference. Moreover, since most of the CNN accelerators benchmark their performance on models trained for image classification, we focus our chapter on this application. Nonetheless, the methods detailed in this survey can be employed to accelerate CNNs for other applications such object detection, image segmentation, and speech recognition.

1.2.3 INFERENCE, LAYERS, AND CNN MODELS

CNN inference refers to the *feed-forward* propagation of B input images across L layers. This section details the computations involved in the major types of these layers. A common practice is to manipulate layers, parameters, and FMs as multidimensional arrays, as listed in Table 1.1. Note that when it will be relevant, the type of the layer will be denoted with superscript, and the position of the layer will be denoted with subscript.

---

* Includes a large number of layer, typically above three.
[†] The information flows from the neurons of a layer ℓ towards the neurons of a layer. $\ell + 1$
[‡] CNNs implement the weight sharing technique, applying a small number of weights across all the input pixels (i.e., image convolution).

TABLE 1.1

Tensors Involved in the Inference of a Given Layer ℓ with Their Dimensions

X	Input FMs	$B \times C \times H \times W$	B	Batch size (Number of input frames)
Y	Output FMs	$B \times N \times V \times U$	$W/H/C$	Width/Height/Depth of Input FMs
Θ	Learned Filters	$N \times C \times J \times K$	$U/V/N$	Width/Height/Depth of Output FMs
β	Learned biases	N	K/J	Horizontal/Vertical Kernel size

A convolutional layer (*conv*) carries out the feature extraction process by applying – as illustrated in Figure 1.1 – a set of three-dimensional convolution filters Θ^{conv} to a set of B input volumes $\mathbf{X}^{conv}$. Each input volume has a depth C and can be a color image (in the case of the first *conv* layer), or an output generated by previous layers in the network. Applying a three-dimensional filter to three-dimensional input results in a 2D *(FM)*. Thus, applying N three-dimensional filters in a layer results in a three-dimensional output with a depth N.

In some CNN models, a learned offset β^{conv} – called a *bias* – is added to processed feature maps. However, this practice has been discarded in recent models [6]. The computations involved in feed-forward propagation of *conv* layers are detailed in Equation 1.1.

$$\forall \{b,n,u,v\} \in [1,B] \times [1,N] \times [1,V] \times [1,U]$$

$$\mathbf{Y}^{conv}[b,n,v,u] = \beta^{conv}[n]$$

$$+ \sum_{c=1}^{C}\sum_{j=1}^{J}\sum_{k=1}^{K}\mathbf{X}^{conv}[b,c,v+j,u+k] \cdot \Theta^{conv}[n,c,j,k] \tag{1.1}$$

One may note that applying a depth convolution to a 3D input boils down to applying a mainstream 2D convolution to each of the 2D channels of the input, then, at each point, summing the results across all the channels, as shown in Equation 1.2.

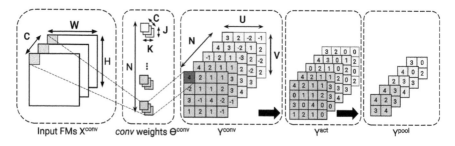

Input FMs $\mathbf{X}^{conv}$ *conv* weights Θ^{conv} $\mathbf{Y}^{conv}$ $\mathbf{Y}^{act}$ $\mathbf{Y}^{pool}$

FIGURE 1.1 Feed-forward propagation in *conv*, *act*, and *pool* layers (batch size $B=1$, bias β omitted).

$$\forall n \in [1,N]$$

$$Y[n]^{\mathrm{conv}} = \beta^{\mathrm{conv}}[n] + \sum_{c=1}^{C} \mathrm{conv2D}\Big(\mathbf{X}[c]^{\mathrm{conv}}, \Theta[c]^{\mathrm{conv}}\Big) \qquad (1.2)$$

Each *conv* layer of a CNN is usually followed by an activation layer that applies a *nonlinear* function to all the values of FMs. Early CNNs were trained with TanH or Sigmoid functions, but recent models employ the rectified linear unit (ReLU) function, which grants faster training times and less computational complexity, as highlighted in Krizhevsky et al. [12].

$$\forall \{b,n,u,v\} \in [1,B] \times [1,N] \times [1,V] \times [1,U]$$

$$Y^{\mathrm{act}}[b,n,h,w] = \mathrm{act}(X^{\mathrm{act}}[b,n,h,w]) \mid \mathrm{act} := \mathrm{TanH, Sigmoid, ReLU} \ldots \qquad (1.3)$$

The convolutional and activation parts of a CNN are directly inspired by the cells of visual cortex in neuroscience [13]. This is also the case with *pooling* layers, which are periodically inserted in between successive *conv* layers. As shown in Equation 1.4, *pooling* sub-samples each channel of the input FM by selecting either the *average*, or, more commonly, the *maximum* of a given neighborhood **K**. As a result, the dimensionality of an FM is reduced, as illustrated in Figure 1.1.

$$\forall \{b,n,u,v\} \in [1,B] \times [1,N] \times [1,V] \times [1,U]$$

$$Y^{\mathrm{pool}}[b,n,v,u] = \max_{p,q \in [1:K]} \Big(X^{\mathrm{pool}}[b,n,v+p,u+q]\Big) \qquad (1.4)$$

When deployed for classification purposes, the CNN pipeline is often terminated by FC layers. In contrast with convolutional layers, FC layers do not implement weight sharing and involve as much weight as input data (i.e., $W = K$, $H = J, U = V = 1$). Moreover, in a similar way as *conv* layers, a nonlinear function is applied to the outputs of FC layers.

$$\forall \{b,n\} \in [1,B] \times [1,N]$$

$$Y^{\mathrm{fc}}[b,n] = \beta^{\mathrm{fc}}[n] + \sum_{c=1}^{C} \sum_{h=1}^{H} \sum_{w=1}^{W} X^{\mathrm{fc}}[b,c,h,w] \cdot \Theta^{\mathrm{fc}}[n,c,h,w] \qquad (1.5)$$

The Softmax function is a generalization of the Sigmoid function, and "squashes" a N-dimensional vector X to $Sigmoid(X)$ where each output is in the range [0,1]. The Softmax function is used in various multi-class classification methods, especially in CNNs. In this case, the Softmax layer is placed at the end of the network and the dimension of vector it operates on (i.e., N) represents the number of classes in the considered dataset. Thus, the input of the Softmax is the data generated by the last fully connected layer, and the output is the probability predicted for each class.

$$\forall \{b,n\} \in [1,B] \times [1,N]$$

$$\text{Softmax}(\mathbf{X}[b,n]) = \frac{\exp(\mathbf{X}[b,n])}{\sum_{c=1}^{N} \exp(\mathbf{X}[b,c])} \tag{1.6}$$

Batch normalization was introduced [14] to speed up training by linearly shifting and scaling the distribution of a given batch of inputs B to have zero mean and unit variance. These layers find also their interest when implementing binary neural networks (BNNs) as they reduce the quantization error compared to an arbitrary input distribution, as highlighted in Hubara et al. [15]. Equation 1.7 details the processing of *batch norm* layers, where the mean μ and the variance σ are statistics collected during the training, α and γ are parameters learned during the training, and ϵ is a hyper-parameter set empirically for numerical stability purposes (i.e., avoiding division by zero).

$$\forall \{b,n,u,v\} \in [1,B] \times [1,N] \times [1,V] \times [1,U]$$

$$\mathbf{Y}^{\text{BN}}[b,n,v,u] = \frac{\mathbf{X}^{\text{BN}}[b,n,u,v] - \mu}{\sqrt{\sigma^2 + \epsilon}} \gamma + \alpha \tag{1.7}$$

1.2.4 WORKLOADS AND COMPUTATIONS

The accuracy of CNN models has been increasing since their breakthrough in 2012 [12]. However, this accuracy comes at a high computational cost. The main challenge that faces CNN developers is to improve classification accuracy while maintaining a tolerable computational workload. As shown in Table 1.2, this challenge was successfully addressed by Inception [16] and ResNet models [17], with their use of bottleneck 1×1 convolutions that reduce both model size and computations while increasing depth and accuracy.

1.2.4.1 Computational Workload

As shown in Equations 1.1 and 1.5, the processing of CNN involves an intensive use of Multiply Accumulate (MAC) operation. All these MAC operations take place at *conv* and FC layers, while the remaining parts of network are element-wise transformations that can be generally implemented with low-complexity computational requirements.

TABLE 1.2

Popular CNN Models with Their Computational Workload*

Model	AlexNet [12]	GoogleNet [16]	VGG16 [6]	VGG19 [6]	ResNet101 [17]	ResNet-152 [17]
Top1 err (%)	42.9%	31.3%	28.1%	27.3%	23.6% %	23.0%
Top5 err (%)	19.80%	10.07%	9.90%	9.00%	7.1%	6.7%
L_c	5	57	13	16	104	155
$\sum_{\ell=1}^{L_c} C_\ell^{conv}$	666 M	1.58 G	15.3 G	19.5 G	7.57 G	11.3 G
$\sum_{\ell=1}^{L_c} W_\ell^{conv}$	2.33 M	5.97 M	14.7 M	20 M	42.4 M	58 M
Act			ReLU			
Pool	3	14	5	5	2	2
L_f	3	1	3	3	1	1
$\sum_{\ell=1}^{L_f} C_\ell^{fc}$	58.6 M	1.02 M	124 M	124 M	2.05 M	2.05 M
$\sum_{\ell=1}^{L_f} W_\ell^{fc}$	58.6 M	1.02 M	124 M	124 M	2.05 M	2.05 M
C	724 M	1.58 G	15.5 G	19.6 G	7.57 G	11.3 G
W	61 M	6.99 M	138 M	144 M	44.4 M	60 M

* Accuracy Measured on Single-Crops of ImageNet Test-Set

In this chapter, the computational workload C of a given CNN corresponds to the number of MACs it involves during inference*. The number of these MACs mainly depends on the topology of the network, and more particularly on the number of *conv* and FC layers and their dimensions. Thus, the computational workload can be expressed as in Equation 1.8, where L_c is the number of *conv* (fully connected) layers, and C_ℓ^{conv} (C_ℓ^{fc}) is the number of MACs occurring on a given convolution (fully connected) layer ℓ.

$$C = \sum_{\ell=1}^{L_c} C_\ell^{conv} + \sum_{\ell=1}^{L_f} C_\ell^{fc} \tag{1.8}$$

$$C_\ell^{conv} = N_\ell \times C_\ell \times J_\ell \times K_\ell \times U_\ell \times V_\ell \tag{1.9}$$

$$C_\ell^{fc} = N_\ell \times C_\ell \times W_\ell \times H_\ell \tag{1.10}$$

* Batch size is set to 1 for clarity purposes.

In a similar way, the number of weights, and consequently the size of a given CNN model, can be expressed as follows:

$$\mathcal{W} = \sum_{\ell=1}^{L_c} \mathcal{W}_\ell^{\text{conv}} + \sum_{\ell=1}^{L_f} \mathcal{W}_\ell^{\text{fc}} \tag{1.11}$$

$$\mathcal{W}_\ell^{\text{conv}} = N_\ell \times C_\ell \times J_\ell \times K_\ell \tag{1.12}$$

$$\mathcal{W}_\ell^{\text{fc}} = N_\ell \times C_\ell \times W_\ell \times H_\ell \tag{1.13}$$

For state-of-the-art CNN models, L_c, N_ℓ, and C_ℓ can be quite large. This makes CNNs *computationally and memory intensive*, where for instance, the classification of a single frame using the VGG19 network requires 19.5 billion MAC operations.

It can be observed in the same table that most of the MACs occur on the convolution parts, and consequently, 90% of the execution time of a typical inference is spent on *conv* layers [18]. By contrast, FC layers marginalize most of the weights and thus the size of a given CNN model.

1.2.4.2   Parallelism in CNNs

The high computational workload of CNNs makes their inference a challenging task, especially on low-energy embedded devices. The key solution to this challenge is to leverage on the extensive concurrency they exhibit. These parallelism opportunities can be formalized as follows:

- **Batch Parallelism:** CNN implementations can simultaneously classify multiple frames grouped as a *batch B* in order to reuse the filters in each layer, minimizing the number the memory accesses. However, and as shown in [10], batch parallelism quickly reaches its limits. This is due to the fact that most of the memory transactions result from storing intermediate results and not loading CNN parameters. Consequently, reusing the filters only slightly impacts the overall processing time per image.
- **Inter-layer Pipeline Parallelism:** CNNs have a feed-forward hierarchical structure consisting of a succession of data-dependent layers. These layers can be executed in a pipelined fashion by launching layer (ℓ) before ending the execution of layer ($\ell-1$). This pipelining costs latency but increases throughput.

Moreover, the execution of the most computationally intensive parts (i.e., *conv* layers), exhibits the four following types of concurrency:

- **Inter-FM Parallelism:** Each two-dimensional plane of an FM can be processed separately from the others, meaning that P_N elements of Y^{conv} can be computed in parallel ($0 < P_N < N$).

- **Intra-FM Parallelism:** In a similar way, pixels of a single output FM plane are data-independent and can thus be processed concurrently by evaluating $P_V \times P_U$ values of $Y^{conv}[n]$ $(0 < P_V \times P_U < V \times U)$.
- **Inter-convolution Parallelism:** Depth convolutions occurring in *conv* layers can be expressed as a sum of 2D convolutions, as shown in Equation 1.2. These 2D convolutions can be evaluated simultaneously by computing concurrently P_c elements $(0 < P_c < C)$.
- **Intra-convolution Parallelism:** The 2D convolutions involved in the processing of *conv* layers can be implemented in a pipelined fashion such as in [76]. In this case $P_J \times P_K$ multiplications are implemented concurrently $(0 < P_J \times P_K < J \times K)$.

1.2.4.3 Memory Accesses

As a consequence of the previous discussion, the inference of a CNN shows large vectorization opportunities that can be exploited by allocating multiple computational resources to concurrently process multiple features. However, this parallelization can not accelerate the execution of a CNN if no datacaching strategy is implemented. In fact, memory bandwidth is often the bottleneck when processing CNNs.

In *FC* parts, the execution can be memory-bounded because of the high number of weights that these layers contain, and consequently, the high number of memory reads required.

This is expressed in Equation 1.14, where $\mathcal{M}_\ell^{fc}$ refers to the number of memory accesses occurring in an FC layer ℓ. This number can be written as the sum of memory accesses reading the inputs $\mathbf{X}_\ell^{fc}$, the memory accesses reading the weights (θ_ℓ^{fc}), and the number of memory accesses writing the results $(\mathbf{Y}_\ell^{fc})$.

$$\mathcal{M}_\ell^{fc} = \text{MemRd}(\mathbf{X}_\ell^{fc}) + \text{MemRd}(\theta_\ell^{fc}) + \text{MemWr}(\mathbf{Y}_\ell^{fc}) \qquad (1.14)$$

$$= C_\ell H_\ell W_\ell + N_\ell C_\ell H_\ell W_\ell + N_\ell \qquad (1.15)$$

$$\sim N_\ell C_\ell H_\ell W_\ell \qquad (1.16)$$

Note that the fully connected parts of state-of-the-art models involve large values of N_ℓ and C_ℓ, making the memory reading of weights the most impacting factor, as formulated in Equation 1.16. In this context, batch parallelism can significantly accelerate the execution of CNNs with a large number of FC layers.

In the *conv* parts, the high number of MAC operations results in a high number of memory accesses, as each MAC requires at least 2 memory reads and 1 memory write*. This number of memory accesses accumulates with the high dimensions of data manipulated by *conv* layers, as shown in Equation 1.18. If all these accesses are towards external memory (for instance, DRAM), throughput and energy consumption

---

* This is the best-case scenario of a fully pipelined MAC, where intermediate results do not need to be loaded.

will be highly impacted, because DRAM access engenders high latency and energy consumption, even more than the computation itself [21].

$$\mathcal{M}_\ell^{conv} = \text{MemRd}(\mathbf{X}_\ell^{conv}) + \text{MemRd}(\theta_\ell^{conv}) + \text{MemWr}(\mathbf{Y}_\ell^{conv}) \qquad (1.17)$$

$$= C_\ell H_\ell W_\ell + N_\ell C_\ell J_\ell K_\ell + N_\ell U_\ell V_\ell \qquad (1.18)$$

The number of these DRAM accesses, and thus latency and energy consumption, can be reduced by implementing a memory-caching hierarchy using on-chip memories. As discussed in the next sections, state-of-the-art CNN accelerators employ register files as well as several levels of caches. The former, being the fastest, is implemented at the nearest of the computational capabilities. The latency and energy consumption resulting from these caches is lower by several orders of magnitude than external memory accesses, as pointed out in Sze et al. [22].

1.2.4.4 Hardware, Libraries, and Frameworks

In order to catch the parallelism of CNNs, dedicated hardware accelerators are developed. Most of them are based on GPUs, which are known to perform well on regular parallelism patterns thanks to simd and simt execution models, a dense collection of floating-point computing elements that peak at 12 TFLOPs, and high capacity/bandwidth on/off-chip memories [23]. To support these hardware accelerators, specialized libraries for deep learning are developed to provide the necessary programming abstraction, such as CudNN on Nvidia GPU [24]. Built upon these libraries, dedicated frameworks for deep learning are proposed to improve productivity of conceiving, training, and deploying CNNs, such as Caffe [25] and TensorFlow [26].

Beside GPU implementations, numerous FPGA accelerators for CNNs have been proposed. FPGAs are fine-grained programmable devices that can catch the CNN parallelism patterns with no memory bottleneck, thanks to the following:

1. A high density of hard-wired digital signal processor (DSP) blocks that are able to achieve up to 20 (8 TFLOPs) TMACs [8].
2. A collection of in situ on-chip memories, located next to DSPs, that can be exploited to significantly reduce the number of external memory accesses.

As a consequence, CNNs can benefit from a significant acceleration when running on reconfigurable hardware. This has caused numerous research efforts to study FPGA-based CNN acceleration, targeting both high performance computing (HPC) applications [27] and embedded devices [28].

In the remaining parts of this chapter, we conduct a survey on methods and hardware architectures to accelerate the execution of CNN on FPGA. The next section lists the evaluation metrics used, then Sections 1.4 and 1.5 respectively study the computational transforms and the data-path optimization involved in recent CNN accelerators. Finally, the last section of this chapter details how approximate computing is a key in FPGA-based deep learning, and overviews the main contributions implementing these techniques.

1.3   FPGA-BASED DEEP LEARNING

Accelerating a CNN on an FPGA-powered platform can be seen as an optimization effort that focuses on one or several of the following criteria:

- *Computational Throughput* $(\mathcal{T})$: A large number of the works studied in this chapter focus on reducing the CNN execution times on the FPGA (i.e., the computation latency), by improving the computational throughput of the accelerator. This throughput is usually expressed as the number of MACs an accelerator performs per second. While this metric is relevant in the case of HPC workloads, we prefer to report the throughput as the number of frames an accelerator processes per second (fps), which better suits the embedded vision context. The two metrics can be directly related using Equation 1.19, where $\mathcal{C}$ is defined in Equation 1.8, and refers to the number of computations a CNN involve in order to process a single frame:

$$\mathcal{T}_{(FPS)} = \frac{\mathcal{T}_{(MACS)}}{\mathcal{C}_{(MAC)}} \tag{1.19}$$

- *Classification/Detection Perf.* $(\mathcal{A})$: Another way to reduce CNN execution times is to trade some of their modeling performance in favor of faster execution timings. For this reason, the classification and detection metrics are reported, especially when dealing with *approximate computing* methods. Classification performance is usually reported as top-1 and top-5 accuracies, and detection performance is reported using the mAP50 and mAP75 metrics.
- *Energy and Power Consumption* $(\mathcal{P})$: Numerous FPGA-based acceleration methods can be categorized as either latency-driven or energy-driven. While the former focus on improving the computational throughput, the latter considers the power consumption of the accelerator, reported in watts. Alternatively, numerous latency-driven accelerators can be ported to low-power-range FPGAs and perform well under strict power consumption requirements.
- *Resource Utilization* $(\mathcal{R})$: When it comes to FPGA acceleration, the utilization of the available resources (lut, DSP blocks, sram blocks) is always considered. Note that the resource utilization can be correlated to the power consumption*, but improving the ratio between the two is a technological problem that clearly exceeds the scope of this chapter. For this reason, both power consumption and resources utilization metrics will be reported when available.

An FPGA implementation of a CNN has to satisfy to the former requirements. In this perspective, the literature provides three main approaches to address the problem

---

* At a similar number of memory accesses. These accesses typically play the most dominant role in the power consumption of an accelerator.

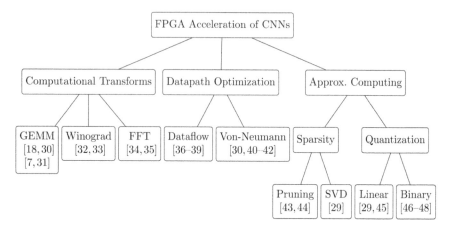

FIGURE 1.2 Main approaches to accelerate CNN inference on FPGAs.

of FPGA-based deep learning. These approaches mainly consists of computational transforms, data-path optimization, and approximate computing techniques, as illustrated in Figure 1.2.

1.4   COMPUTATIONAL TRANSFORMS

In order to accelerate the execution of *conv* and FC layers, numerous implementations rely on computational transforms. These transforms, which operate on the FM and weight arrays, aim at vectorizing the implementations and reducing the number of operations occurring during inference.

Three main transforms can be distinguished. The *im2col* method reshapes the feature and weight arrays in a way to transform depth convolutions into matrix multiplications. The *FFT* method operates on the frequency domain, transforming convolutions into multiplications. Finally, in *Winograd* filtering, convolutions boil down to element-wise matrix multiplications thanks to a tiling and a linear transformation of data.

These computational transforms mainly appear in temporal architectures and are implemented by means of variety of *linear algebra* libraries such OpenBLAS for CPUs* or cuBLAS for GPUs[†]. Besides this, various implementations make use of these transforms to efficiently map CNNs on FPGAs.

This section discusses the three former methods, highlighting their use-cases and computational improvements. For a better understanding, we recall that for each layer ℓ:

- The input feature map is represented as four-dimensional array **X**, in which the dimensions $B \times C \times H \times W$ respectively refer to the batch size, the number of input channels, the height, and the width.

---

* https://www.openblas.net/
[†] https://developer.nvidia.com/cublas

- The weights are represented as four-dimensional array Θ, in which the dimensions $N \times C \times J \times K$ respectively refer to the depth of the output feature map, the depth of the input feature map, the vertical, and the horizontal kernel size.

1.4.1 THE IM2COL TRANSFORMATION

In CPUs and GPUs, a common way to process CNNs is to map *conv* and FC layers as general matrix multiplications (GEMMs). A number of studies generalize this approach to FPGA-based implementations.

For FC layers, in which the processing boils down to a matrix-vector multiplication problem, the GEMM-based implementations find their interest when processing a *batch* of FMs. As mentioned in Section 1.2.4.1, most of the weights of CNNs are employed in the FC parts. Instead of loading these weights multiple times to classify multiple inputs, features extracted from a batch of inputs are concatenated onto a $CHW \times B$ matrix. In this case, the weights are loaded only one time per batch, as depicted in Figure 1.3a. As a consequence, the former Equation 1.16 – which expressed the number of memory accesses occurring on FC layers – becomes the following:

$$\mathcal{M}_\ell^{fc} = \mathrm{MemRd}(\theta_\ell^{fc}) + \mathrm{MemRd}(\mathbf{X}_\ell^{fc}) + \mathrm{MemWr}(\mathbf{Y}_\ell^{fc}) \qquad (1.20)$$

$$= N_\ell C_\ell W_\ell H_\ell + BC_\ell H_\ell W_\ell + BN_\ell \qquad (1.21)$$

$$\sim N_\ell C_\ell H_\ell W_\ell \qquad (1.22)$$

As detailed in Section 1.2.4.2, the vectorization of FC layers is often employed in GPU implementations to increase the computational throughput while maintaining a constant memory bandwidth utilization. The same concept holds true for FPGA implementations [31, 48, 49], which batch the FC layers to map them as GEMMs.

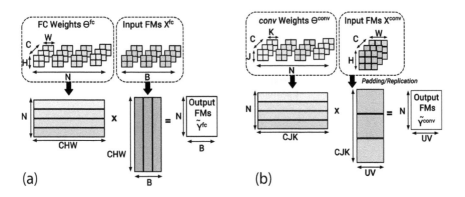

FIGURE 1.3 GEMM-based processing of FC layers (a) and conv layers (b).

3D convolutions can also be mapped as GEMMs using the so-called *im2col* method introduced in [30]. First, this method flattens all the weights of a given *conv* layer onto an $N \times CKJ$ matrix $\tilde{\Theta}$. Second, it rearranges the input feature maps onto a $CKJ \times UV$ matrix $\tilde{X}$, squashing each feature map to a column*. With these reshaped data, the output feature maps $\tilde{Y}$ are computed by multiplying of two former matrices, as illustrated in Figure 1.3b.

$$\tilde{\mathbf{Y}}^{\mathrm{conv}} = \tilde{\Theta}^{\mathrm{conv}} \times \tilde{\mathbf{X}}^{\mathrm{conv}} \tag{1.23}$$

Suda et al. [29] and more recently, Zhang et al. [50] and Guan et al. [51] leverage on *im2col* to derive OpenCL-based FPGA accelerators for CNN. However, this method introduces redundant data in the input FM matrix, which can lead to either inefficiency in storage or complex memory access patterns. As a result, and as pointed out in [22], other strategies to map convolutions have to be considered.

1.4.2 WINOGRAD TRANSFORM

Winograd minimal filtering algorithm, introduced in [52], is a computational transform that can be applied to process convolutions with a stride of 1, which is very common in CNN topologies.

This algorithm is particularly efficient when processing small convolutions (where $K \leq 3$), as advocated in [53]. In this work, authors outperformed the throughput of the conventional *im2col* method by a factor of ×7.2 when executing VGG16 on a TitanX GPU.

In Winograd filtering (Figure 1.4), data is processed by blocks, referred to as *tiles*, as follows:

1. An input FM tile x of size $(u \times u)$ is pre-processed: $\tilde{x} = \mathbf{A}^T x \mathbf{A}$
2. In a similar way, θ, the filter tile of size $(k \times k)$, is transformed into $\tilde{\theta} : \tilde{\theta} = \mathbf{B}^T x \mathbf{B}$

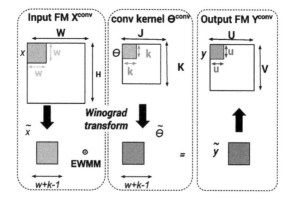

FIGURE 1.4   Winograd filtering $F(u \times u, k \times k)$.

---

* That's what the im2col name refers to: flattening an image to a column.

3. Winograd filtering algorithm, denoted $F(u \times u, k \times k)$, outputs a tile y of size $(u \times u)$ that is computed according to Equation 1.24

$$y = C^T \left[\tilde{\theta} \odot \tilde{x} \right] C \tag{1.24}$$

where A, B, C are transformation matrices defined in the Winograd algorithm [52] and $\odot$ denotes the Hadamard product also known as EWMM.

While a standard filtering requires $u^2 \times k^2$ multiplications, Winograd algorithm, denoted $F(u \times u, k \times k)$, requires $(u+k-1)^2$ multiplications [52]. In the case of tiles of a size $u = 2$ and kernels of size $k = 3$, this corresponds to an arithmetic complexity reduction of $\times 2.25$ [53], and in this case, transform matrices can be written as follows:

$$\mathbf{A}^T = \begin{bmatrix} 1 & 1 & 1 & 0 \\ 0 & 1 & -1 & -1 \end{bmatrix}; \quad \mathbf{B}^T = \begin{bmatrix} 1 & 0 & -1 & 0 \\ 0 & 1 & 1 & 0 \\ 0 & -1 & 1 & 0 \\ 0 & 1 & 0 & -1 \end{bmatrix}$$

$$C = \begin{bmatrix} 1 & 0 & 0 \\ 1/2 & 1/2 & 1/2 \\ 1/2 & -1/2 & 1/2 \\ 0 & 0 & 1 \end{bmatrix} \tag{1.25}$$

Beside this complexity reduction, implementing Winograd filtering in FPGA-based CNN accelerators has two advantages. First, transformation matrices A, B, C can be evaluated offline once u and k are determined. As a result, these transforms become multiplications with the constants that can be implemented by means of lut and shift registers, as proposed in [54].

Second, Winograd filtering can employ the loop optimization techniques discussed in Section 1.5.2 to vectorize the implementation. On one hand, the computational throughput is increased when *unrolling* the computation of the ewmm parts over multiple DSP blocks. On the other hand, memory bandwidth is optimized using loop *tiling* to determine the size of the FM tiles and filter buffers.

First, utilization of Winograd filtering in FPGA-based CNN accelerators is investigated in [32] and delivers a computational throughput of 46 GOPs when executing AlexNet convolutional layers. This performance is significantly improved by a factor of $\times 42$ in [31] when optimizing the data path to support Winograd convolutions (by employing loop unrolling and tiling strategies), and storing the intermediate FM in on-chip buffers (cf Section 1.4).

The same method is employed in [54] to derive a CNN accelerator on a Xilinx ZCU102 device that delivers a throughput of 2.94 TOPs on VGG convolutional layers. The reported throughput corresponds to half of the performance of a TitanX device, with 5.7$\times$ less power consumption [23]*.

---

* Implementation in the TitanX GPU employs Winograd algorithm and 32-bit floating point arithmetic.

1.4.3 Fast Fourier Transform

Fast Fourier Transform (FFT) is a well known algorithm to transform the 2D convolutions into ewmm in the frequency domain, as shown in Equation 1.26:

$$\textbf{conv2D}(X[c], \Theta[n,c]) = \text{IFFT}(\text{FFT}(X[c]) \odot \text{FFT}(\Theta[n,c])) \qquad (1.26)$$

Using FFT to process 2D convolutions reduces the complexity from $O(W^2 \times K^2)$ to $O(W^2 log_2(W))$, which is exploited to derive FPGA-based accelerators and to infer CNN [34]. When compared to standard filtering and Winograd algorithm, FFT finds its interest in convolutions with large kernel size ($K > 5$), as demonstrated in [53, 55]. The computational complexity of FFT convolutions can be further reduced to $O(W log_2(K))$ using the overlap-and-add method [56], which can be applied when the signal size is much larger than the filter size, which is typically the case in *conv* layers ($W >> K$). Works in [33, 57] leverage on the overlap-and-add to implement frequency domain acceleration for *conv* layers on FPGA, which results in a computational throughput of 83 GOPs for AlexNet (Table 1.3).

1.5 DATA-PATH OPTIMIZATIONS

As highlighted in Section 2.4.2, the execution of CNN exhibits numerous sources of parallelism. However, due to the resource limitations of FPGA devices, it might be impossible to fully exploit all the concurrency patterns, especially with the sheer volume of operations involved in deep topologies. In other words, the execution of recent CNN models cannot fully be unrolled sometimes, not even for a single *conv* layer.

To address this problem, the general approach, advocated in state-of-the-art implementations, is to map a limited number of processing elements (PEs) on the FPGA. These PEs are then reused by temporally iterating data through them.

1.5.1 Systolic Arrays

Early FPGA-based accelerators for CNN implemented systolic arrays to accelerate the 2D filtering in convolutions layers [58—61]. As illustrated in Figure 1.5a, systolic arrays employ *a static collection* of PE, typically arranged in a 2-dimensional grid. These PE operate as a co-processor under the control of a central processing unit. The configuration of systolic arrays is *agnostic* to the CNN model, making them inefficient to process large-scale networks for the following three reasons:

First, the static collection of PE can support convolutions only up to a given filter size K_m, where typical values of K_m range from 7 in [59] to 10 in [61]. Therefore, in convolutional layer (ℓ), $K_\ell > K_m$ is not supported by the accelerator. Second, systolic arrays suffer from under-utilization when processing layers in which the kernel size K_ℓ is much smaller then K_m. This is for instance the case in [61], where the processing of 3×3 convolutions uses only 9% of DSP blocks, while the processing of these layers can be further parallelized and thus accelerated. Third and finally, PE in systolic arrays do not usually include memory caches and have to fetch their inputs from

TABLE 1.3

Accelerators Employing Computational Transforms

Method	Entry	Network	Comp (GOP)	Params (M)	Bit-width	Desc.	Device	Freq (MHz)	Through (GOPs)	Power (W)	LUT (K)	DSP	Memory (MB)
Winograd	[33]	AlexNet-C	1.3	2.3	Float 32	OpenCL	Virtex7 VX690T	200	46	–	505	3683	56.3
	[32]	AlexNet-C	1.3	2.3	Float16	OpenCL	Arria10 GX1150	303	1382	44.3	246	1576	49.7
	[55]	VGG16-C	30.7	14.7	Fixed 16	HLS	Zynq ZU9EG	200	3045	23.6	600	2520	32.8
	[55]	AlexNet-C	1.3	2.3	Fixed 16	HLS	Zynq ZU9EG	200	855	23.6	600	2520	32.8
FFT	[34]	AlexNet-C	1.3	2.3	Float 32	–	Stratix5 QPI	200	83	13.2	201	224	4.0
	[34]	VGG19-C	30.6	14.7	Float 32	–	Stratix5 QPI	200	123	13.2	201	224	4.0
GEMM	[30]	AlexNet-C	1.3	2.3	Fixed 16	OpenCL	Stratix5 GXA7	194	66	33.9	228	256	37.9
	[50]	VGG16-F	31.1	138.0	Fixed 16	HLS	Kintex KU060	200	365	25.0	150	1058	14.1
	[50]	VGG16-F	31.1	138.0	Fixed 16	HLS	Virtex7 VX960T	150	354	26.0	351	2833	22.5
	[51]	VGG16-F	31.1	138.0	Fixed 16	OpenCL	Arria10 GX1150	370	866	41.7	437	1320	25.0
	[51]	VGG16-F	31.1	138.0	Float 32	OpenCL	Arria10 GX1150	385	1790	37.5	–	2756	29.0

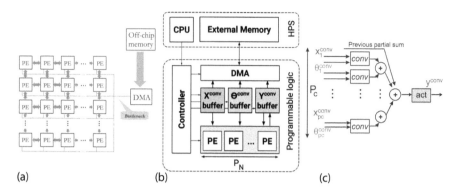

FIGURE 1.5 Generic data paths of FPGA-based CNN accelerators: (a) Static systolic array. (b) Dedicated SIMD accelerator. (c) Dedicated processing element.

an off-chip memory. As a result, the performance of systolic arrays can rapidly be bounded by memory bandwidth of the device.

1.5.2 LOOP OPTIMIZATION IN SPATIAL ARCHITECTURES

Due to the inefficiency of systolic arrays, flexible and dedicated spatial architectures for CNN were mapped on FPGA. The general computation flow in these accelerators is illustrated in Figure 1.5b.

First, FMs and weights are fetched from DRAM to on-chip buffers, and are then *streamed* into the PE. At the end of the PE computation, results are transferred back to on-chip buffers and, if necessary, to the external memory in order to be fetched in their turn to process the next layers. Each PE – as depicted in Figure 1.5c – is configurable and has its own *computational* capabilities by means of DSP blocks, and its own data *caching* capabilities by means of on-chip registers. With this paradigm, the problem of CNN mapping consists of finding the optimal architectural and temporal configuration of PE: in other words, the best number of DSP blocks per PE, the optimal temporal scheduling of data that maximizes the computational throughput.

For convolutional layers, in which the processing is described in Listing 1.1, finding the optimal PE configuration comes down to a loop optimization problem [28, 29, 39, 40, 62–64].

Listing 1.1: Nested Loops

```
// Lb : Batch
for (int b =0;b<B,l++) {
// Ll: Layer
for (int l =0;l<L,l++) {
// Ln: Y Depth
for (int n =0;n<N;n++) {
// Lv: Y Columns
for (int v =0;v<V,v++) {
// Lu: Y Raws
```

```
for (int u =0;u<U,u++) {
// Lc: X Depth
for (int c =0;n<C;c++) {
// Lj: Theta Columns
for (int j =0;j<J,j++) {
// Lk: Theta Raws
for (int k =0;k<K,k++) {
Y[b,l,n,v,u] +=
X[b,l,c,v+j,u+k] *
Theta [l,n,c,j,k]
}}}}}}}
```

Listing 1.2: Loop Tiling in conv layers

```
for (int b =0;b<B,l++){
for (int n =0;n<N;n+= Tn){
for (int v =0;v<V,v+= Tv){
for (int u =0;u<U,u+= Tu){
for (int c =0;n<C;c+= Tc){
// DRAM : Load in on - chip
buffers the tiles :
// X[l,c:c+Tc ,v:v+Tv ,u:u+Tu]
// Theta [l,n:n+Tn ,c:c+Tc ,j,k]
for (int tn =0; tn <Tn;tn ++){
for (int tv =0; tv <Tv ,tv ++){
for (int tu =0; tu <Tu ,tu ++){
for (int tc =0; tn <Tc;tc ++){
for (int j =0;j<J,j++){
for (int k =0;k<K,k++){
Y[l,tn ,tv ,tu] +=
X[l,tc ,tv+j,tu+k] *
Theta [l,tn ,tc ,j,k];
}}}}}} // DRAM : Store output
}}}}
```

This problem is addressed by applying loop optimization techniques such *loop unrolling, loop tiling,* or *loop interchange* to the 7 nested loops of Listing 1.1. In this case, the unroll and tiling factors (respectively P_i and T_i) determine the number of PEs, the computational resources, and the on-chip memory allocated to each PE.

Loop Unrolling

Unrolling a loop L_i with an unrolling factor P_i ($P_i \leq i, i \in \{L,V,U,N,C,J,K\}$) accelerates its execution by allocating multiple computational resources. Each of the parallelism patterns listed in Section 1.2.4.2 can be implemented by unrolling one of the loops of Listing 1.1, as summarized in Table 1.4. For the configuration given in Figure 1.5c, the unrolling factor P_N sets the number of PEs. The remaining factors – P_C, P_K, P_J – determine the number of multipliers, as well as the size of buffer contained in each PE (Figure 1.6).

TABLE 1.4

Loop Optimization Parameters P_i and T_i

Parallelism	Intra layer	Inter FM	Intra FM		Inter conv.	Intra conv.	
Loop	L_L	L_N	L_V	L_U	L_c	L_J	L_K
Unroll Factor	P_L	P_N	P_V	P_U	P_c	P_J	P_K
Tiling Factor	T_L	T_N	T_U	T_U	T_C	T_J	T_K

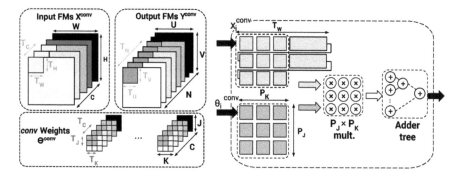

FIGURE 1.6   Loop tiling and unrolling in convolutional layers.

Loop Tiling

In general, the capacity of on-chip memory in current FPGA is not large enough to store the weights and intermediate FM of all CNN layers*. For example, AlexNet's convolutional layers resort to 18.6 Mbits of weights, and generate a total 70.7 Mbits of intermediate feature maps[†]. In contrast, the highest-end Stratix V FPGA provides a maximum of 52 Mbits of on-chip ram.

As a consequence, FPGA-based accelerators resort to external DRAM to store these data. As mentioned in Section 1.2.4.3, DRAM accesses are costly in terms of energy and latency, and data caches must be implemented by means of on-chip buffers and local registers. The challenge is thus to build a data path in a way that every data transferred from DRAM is reused as much as possible.

For *conv* layers, this challenge can be addressed by *tiling* the nested loops of Listing 1.1. *Loop tiling* [66] divides the FM and weights of each layer into multiple groups that can fit into the on-chip buffers. For the configuration given in Figure 1.5c, the size of the buffers containing input FM, weights, and output FM is set according to the tiling factors listed in Table 1.4.

$$\mathcal{B}_X^{conv} = T_C \times T_H \times T_W \qquad (1.27)$$

---

* Exception can be made for [6666], where a large cluster of FPGAs is interconnected and resorts only
  to on-chip memory to store CNN weights and intermediate data.
[†] Estimated by summing the number of outputs for each convolution layer.

$$\mathcal{B}_{\Theta}^{conv} = T_N \times T_C \times T_J \times T_K \qquad (1.28)$$

$$\mathcal{B}_{Y}^{conv} = T_N \times T_V \times T_U \qquad (1.29)$$

With these buffers, the memory accesses occurring in the *conv* layer (cf Equation 1.18) are respectively divided by $\mathcal{B}_{X}^{conv}$, $\mathcal{B}_{\Theta}^{conv}$ and $\mathcal{B}_{Y}^{conv}$, as expressed in Equation 1.30.

$$\mathcal{M}_{\ell}^{conv} = \frac{C_{\ell}H_{\ell}W_{\ell}}{T_C T_H T_W} + \frac{N_{\ell}C_{\ell}J_{\ell}K_{\ell}}{T_N T_C T_J T_K} + \frac{N_{\ell}U_{\ell}V_{\ell}}{T_N T_V T_U} \qquad (1.30)$$

Since the same hardware is reused to accelerate the execution of multiple conv layers with different workloads, the tiling factors are agnostic to the workload of a specific layer, as can be noticed in the denominator of Equation 1.30. As a result, the value of the tiling factors is generally set to optimize the overall performance of a CNN execution.

1.5.3 DESIGN SPACE EXPLORATION

Finding the optimal unrolling and tiling factors for a specific device is a complex problem that is generally solved using brute-force design space exploration [29, 39, 40, 48, 67, 68]. This exploration is driven by an analytical model, in which the inputs are loop factors P_i, T_i and outputs are theoretical predictions of the computational throughput ($\mathcal{T}$), the size of buffers ($\mathcal{B}$), and the number of external memory accesses ($\mathcal{M}$). This model is parametrized by the available resources of a given FPGA platform and the workload of the considered CNN. To select feasible solutions for this optimization problem, most literature approaches rely on the *Roofline* method [69] to accept or reject design solutions that do not match with the maximum computational throughput or the maximum memory bandwidth of a given device (Figures 1.7).

A typical design space exploration driven by the roofline model is illustrated in Figure 1.8. In this graph, each point represents the performance of an explored solution (P_i, T_i). For a given FPGA platform, the attainable bandwidth and computational throughput are respectively reported by the diagonal and horizontal lines. Point A is an invalid solution, as it is above the bandwidth roof, while point A' is feasible but delivers mediocre computational throughput. Acceptable solutions are represented by points C and D, the latter being better than the former since it has lower bandwidth requirements.

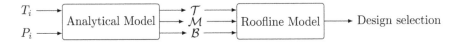

FIGURE 1.7 Design space exploration methodology.

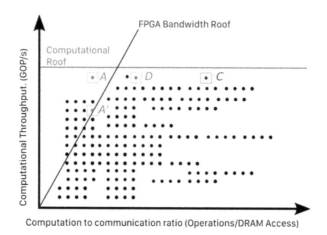

FIGURE 1.8   Example of a design selection driven by the roofline model.

1.5.4   FPGA Implementations

Employing loop optimizations to derive FPGA-based CNN accelerator was first investigated in [39]. In this work, Zhang et al. report a computational throughput of 61.62 GOPs in the execution of AlexNet convolutional layers by unrolling loops L_C and L_N. This accelerator, described with Vivado HLS tools, relies on 32-bit floating-point arithmetic. Works in [68] follow the same unrolling scheme and feature a 16-bit fixed-point arithmetic, resulting in a ×2.2 improvement in terms of computational throughput. Finally, the same unrolling and tiling scheme is employed in recent work [48], where authors report a ×13.4 improvement, thanks to a deeply pipelined FPGA cluster of four Virtex7-XV960t devices.

In all these implementations, loops L_J and L_K are not unrolled because J and K are usually small, especially in recent topologies. Works of Motamedi et al. [40] study the impact of unrolling these loops in AlexNet, where the first convolutional layers use large 11×11 and 5×5 filters. Expanding loop unrolling and tiling to loops L_J and L_K results in a 1.36× improvement in computational throughput vs. [39] on the same VX485T device when using 32-bit floating-point arithmetic. Nevertheless, and as pointed out in [63], unrolling these loops is ineffective for recent CNN models that employ small convolution kernels.

The values of U, V, N can be very large in CNN models. Consequently, unrolling and tiling loops L_U, L_V, L_N can be efficient only for devices with high computational capabilities (i.e., DSP blocks). This is demonstrated in works of Rahman et al. [67] that report an improvement of ×1.22 over [39] when enlarging the design space exploration to loops L_U, L_V, L_N, which comes at the price of very long exploration timing. In order to keep data in on-chip buffer after the execution of a given layer, works of Alwani et al. [62] advocate the use of *fused-layer* accelerators by tiling across layer L_L. As a result, authors are able to remove 95% of DRAM accesses at the cost of 362 KB of extra on-chip memory.

In all these approaches, loops L_N, L_C, L_J, L_K are unrolled in a similar way they are tiled (i.e., $T_i = P_i$). By contrast, the works of Ma et al. [63, 70] fully explore all the design variables searching for optimal loop unroll and tiling factors. More particularly, the authors demonstrate that the input FM and weights are optimally reused when unrolling only computations within a single input FM (i.e., when $P_C = P_J = P_K = 1$). Tiling factors are set in such a way that all the data required to compute an element of Y are fully buffered (i.e., $T_C = C$, $T_K = K$, $T_J = J$). The remaining design parameters are derived after a brute-force design exploration. The same authors leverage on these loop optimizations to build an RTL compiler for CNNs in [71]. To the best of our knowledge, this accelerator outperforms all the previous implementations that are based on loop optimization in terms of computational throughput (Tables 1.5 through 1.7).

1.6 APPROXIMATE COMPUTING OF CNN MODELS

Besides the computational transforms and data-path optimization, the CNN execution can be accelerated when employing approximate computing, which is known to perform efficiently on FPGAs [73].

In the methods detailed in this section, a minimal amount of the CNN accuracy is traded to improve the computational throughput or energy efficiency of the accelerator. Two main strategies are employed. The first implements approximate *arithmetic* to process the CNN layers with a reduced precision. The second aims at reducing the number of operations occurring in CNN models without critically affecting the modeling performance. Note that both approaches can resort to *fine-tuning* in order to compensate the accuracy loss introduced by approximate computing.

1.6.1 APPROXIMATE ARITHMETIC FOR CNNs

Several studies have demonstrated that the precision of both operations and operands in CNN, and more generally in neural networks, can be reduced without critically affecting their predictive performance. This reduction can be achieved by *quantizing* either or both of the CNN inputs, weights, and/or FM using a fixed-point numerical representation.

1.6.1.1 Fixed-Point Arithmetic

In a general way, CNN models are deployed in CPU and GPU using the same numerical precision they were trained with, relying on the *single-precision floating-point* representation. This format employs 32 bits, arranged according to the IEEE754 standard. As current FPGAs support floating operations, various implementations [39, 62, 67] employ such data representation.

Nonetheless, numerous studies such [74–76] demonstrate that the inference of CNNs can be achieved with a reduced precision of operands. More particularly, works in [77, 78] demonstrate the applicability of fixed-point ($F \times P$) arithmetic to *train* and *infer* CNNs. The $F \times P$ representation encodes numbers with a given bit-width b, using i bits for the *integer* part, and f bits for the *fractional* part ($b = i + f$). Note that the value of i is selected according the desired *numerical range*, and the value of f is selected according to the desired numerical *precision*.

TABLE 1.5
Accelerators Employing Loop Optimization

Entry	Network	Comp (GOP)	Params (M)	Bit-width	Desc.	Device	Freq (MHz)	Through (GOPs)	Power (W)	LUT (K)	DSP	Memory (MB)
[40]	AlexNet-C	1.3	2.3	Float 32	HLS	Virtex7 VX485T	100	61.62	18.61	186	2240	18.4
[29]	VGG16SVD-F	30.8	50.2	Fixed 16	RTL	Zynq Z7045	150	136.97	9.63	183	780	17.5
[30]	AlexNet-C	1.3	2.3	Fixed 16	OpenCL	Stratix5 GSD8	120	187.24	33.93	138	635	18.2
[30]	AlexNet-F	1.4	61.0	Fixed 16	OpenCL	Stratix5 GSD8	120	71.64	33.93	272	752	30.1
[30]	VGG16-F	31.1	138.0	Fixed 16	OpenCL	Stratix5 GSD8	120	117.9	33.93	524	1963	51.4
[68]	AlexNet-C	1.3	2.3	Float 32	HLS	Virtex7 VX485T	100	75.16	33.93	28	2695	19.5
[49]	AlexNet-F	1.4	61.0	Fixed 16	HLS	Virtex7 VX690T	150	825.6	126.00	N.R	14400	N.R
[49]	VGG16-F	31.1	138.0	Fixed 16	HLS	Virtex7 VX690T	150	1280.3	160.00	N.R	21600	N.R
[69]	NIN-F	2.2	61.0	Fixed 16	RTL	Stratix5 GXA7	100	114.5	19.50	224	256	46.6
[69]	AlexNet-F	1.5	7.6	Fixed 16	RTL	Stratix5 GXA7	100	134.1	19.10	242	256	31.0
[38]	AlexNet-F	1.4	61.0	Fixed 16	RTL	Virtex7 VX690T	156	565.94	30.20	274	2144	34.8
[63]	AlexNet-C	1.3	2.3	Float 32	HLS	Virtex7 VX690T	100	61.62	30.20	273	2401	20.2
[64]	VGG16-F	31.1	138.0	Fixed 16	RTL	Arria10 GX1150	150	645.25	50.00	322	1518	38.0
[42]	AlexNet-C	1.3	2.3	Fixed 16	RTL	Cyclone5 SEM	100	12.11	N.R	22	28	0.2
[42]	AlexNet-C	1.3	2.3	Fixed 16	RTL	Virtex7 VX485T	100	445	N.R	22	2800	N.R
[72]	NiN	20.2	7.6	Fixed 16	RTL	Stratix5 GXA7	150	282.67	N.R	453	256	30.2
[72]	VGG16-F	31.1	138.0	Fixed 16	RTL	Stratix5 GXA7	150	352.24	N.R	424	256	44.0
[72]	ResNet-50	7.8	25.5	Fixed 16	RTL	Stratix5 GXA7	150	250.75	N.R	347	256	39.3
[72]	NiN	20.2	7.6	Fixed 16	RTL	Arria10 GX1150	200	587.63	N.R	320	1518	30.4
[72]	VGG16-F	31.1	138.0	Fixed 16	RTL	Arria10 GX1150	200	720.15	N.R	263	1518	44.5
[72]	ResNet-50	7.8	25.5	Fixed 16	RTL	Arria10 GX1150	200	619.13	N.R	437	1518	38.5
[73]	AlexNet-F	1.5	7.6	Float 32	N.R	Virtex7 VX690T	100	445.6	24.80	207	2872	37
[73]	VGG16SVD-F	30.8	50.2	Float 32	N.R	Virtex7 VX690T	100	473.4	25.60	224	2950	47

TABLE 1.6

Accelerators Employing Approximate Arithmetic

A×C	Entry	Dataset	Comp (GOP)	Params (M)	Bit-width In/Out	Bit-width FMs	Bit-width θ^{conv}	Bit-width θ^{FC}	Acc (%)	Device	Freq (MHz)	Through. (GOPs)	Power (W)	LUT (K)	DSP	Memory (MB)
FP32	[51]	ImageNet	30.8	138.0	32	32	32	32	90.1	Arria10 GX1150	370	866	41.7	437	1320	25.0
FP16	[32]	ImageNet	30.8	61.0	16	16	16	16	79.2	Arria10 GX1150	303	1382	44.3	246	1576	49.7
	[64]	ImageNet	30.8	138.0	16	16	8	8	88.1	Arria10 GX1150	150	645	N.R	322	1518	38.0
DFP	[72]	ImageNet	30.8	138.0	16	16	16	16	N.R	Arria10 GX1150	200	720	N.R	132	1518	44.5
	[51]	ImageNet	30.8	138.0	16	16	16	16	N.R	Arria10 GX1150	370	1790	N.R	437	2756	29.0
BNN	[91]	Cifar10	1.2	13.4	20	2	1	1	87.7	Zynq Z7020	143	208	4.7	47	3	N.R
	[46]	Cifar10	0.3	5.6	20/16	2	1	1	80.1	Zynq Z7045	200	2465	11.7	83	N.R	7.1
	[93]	MNIST	0.0	9.6	8	2	1	1	98.2	Stratix5 GSD8	150	5905	26.2	364	20	44.2
	[93]	Cifar10	1.2	13.4	8	8	1	1	86.3	Stratix5 GSD8	150	9396	26.2	438	20	44.2
	[93]	ImageNet	2.3	87.1	8	32	1a	1	66.8	Stratix5 GSD8	150	1964	26.2	462	384	44.2
	[94]	Cifar10	1.2	13.4	8	2	2	2	89.4	Xilinx7 VX690T	250	10962	13.6	275	N.R	39.4
TNN	[94]	SVHN	0.3	5.6	8	2	2	2	97.6	Xilinx7 VX690T	250	86124	7.1	155	N.R	12.2
	[94]	GTSRB	0.3	5.6	8	2	2	2	99.0	Xilinx7 VX690T	250	86124	6.6	155	N.R	12.2

TABLE 1.7

Accelerators Employing Pruning and Low Rank Approximation

Reduc.	Entry	Dataset	Comp (GOP)	Params (M)	Removed Param. (%)	Bit-width	Acc (%)	Device	Freq (MHz)	Through. (GOPs)	Power (W)	LUT (K)	DSP	Memory (MB)
SVD	[29]	ImageNet	30.8	50.2	63.6	16 Fixed	88.0	Zynq 7Z045	150	137.0	9.6	183	780	17.50
Pruning	[44]	Cifar10	0.3	13.9	89.3	8 Fixed	91.5	Kintex 7K325T	100	8620.7	7.0	17	145	15.12
	[7]	ImageNet	1.5	9.2	85.0	32 Float	79.7	Stratix 10	500	12000.0	141.2	N.R	N.R	N.R

In the simplest version of fixed-point arithmetic, all the numbers are encoded with the *same* fractional and integer bit-widths. This means that the position of the radix point is similar for all the represented numbers. In this chapter, we refer to this representation as *static* F×P.

When compared to floating point, F×P is known to be more efficient in terms of hardware utilization and power consumption. This is especially true in FPGAs [79], where – for instance – a single DSP block in Intel devices can either implement *one* 32-bit floating-point multiplication or *three* concurrent F×P multiplications of 9 bits [8]. This motivated early FPGA implementations such as [61, 80] to employ fixed-point arithmetic in deriving CNN accelerators. These implementations mainly use a 16-bit Q8.8 format, where 8 bits are allocated to the integer parts, and 8 bits to the fractional part. Note that the same Q8.8 format is used for representing the features and the weights of all the layers.

In order to prevent overflow, the former implementations also *expand* the bit-width when computing weighted sums of convolutions. Equation 1.31 explains how the bit-width is expanded; if b_x bits are used to quantize the input FM and b_Θ bits are used to quantize the weights, an accumulator of b_{acc} bits is required to represent a weighted sum of $C_\ell K_\ell^2$ elements, where:

$$b_{acc} = b_x + b_\Theta + \max_\ell \left[\log_2 \left(C_\ell K_\ell^2 \right) \right] \qquad (1.31)$$

In practice, most FPGA accelerators use 48-bit accumulators, such as in [59, 60] (Figure 1.9).

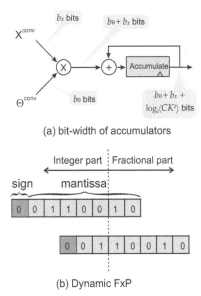

(a) bit-width of accumulators

(b) Dynamic FxP

FIGURE 1.9 Fixed-point arithmetic for CNN accelerators.

1.6.1.2   Dynamic Fixed Point for CNNs

In deep topologies, it can be observed that distinct parts of a network can have significantly different ranges of data. In particular, the features of the deep layers tend to have a much larger numerical range when compared to the features of the first CNN layers.

The histograms of Figure 1.10a depict this phenomenon for AlexNet convolutional layers*. While the CNN inputs (data column) are normalized take their values between 0 and 1, the outputs of the first convolutional layer (conv1 column) have a *wider* numerical range, between 2^{-7} and 2^2. This is even more salient for the fifth convolutional layer, where most of the outputs take their values between 2^{-1} and 2^6. The same problem appears when comparing the numericals of the CNN weights, and CNN activations. In this case, the weights are numerically much *smaller* when compared to the activations, as illustrated in Figure 1.10b[†].

As a consequence, large bit-widths have to be allocated to the integer fractional parts in order to keep a uniform precision across the network while preventing overflow. This expansion badly increases the resource requirements of a given FPGA mapping. As a result, static $F \times P$, with its unique shared fixed exponent, is ill-suited to deep learning, as pointed out in. [81]

To address this problem, works in [77, 81, 82] advocate the use of *dynamic* $F \times P$ [83][‡]. In dynamic $F \times P$, different scaling factors are used to process different parts of the network. In other words, the position of the radix point varies from one layer

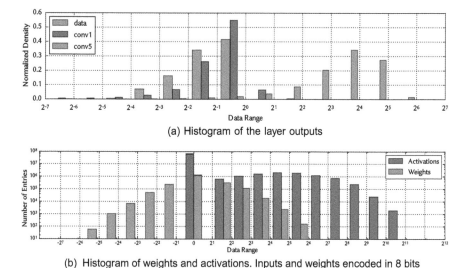

(a) Histogram of the layer outputs

(b) Histogram of weights and activations. Inputs and weights encoded in 8 bits

FIGURE 1.10   Distribution of AlexNet activations and weights.

---

* Code made available at github.com/KamelAbdelouahab/CNN-Data-Distribution.
† This figure deliberately multiplies the weights and activations by a scale factor of $2^7 - 1$ to emulate an 8-bit quantization.
‡ Another approach to address this problem is to use custom floating point representations, as detailed in [31].

to another. More particularly, weights, weighted sums, and outputs of each layer are assigned distinct integer and fractional bit-widths.

The optimal values of these bit-widths (i.e., the ones that deliver the best trade-off between accuracy loss and computational load) for each layer can be derived after a profiling process performed by dedicated frameworks that support F×P. Among these frameworks, Ristretto [81] and FixCaffe [84] are compatible with Caffe, while TensorFlow natively supports 8-bit computations. Most of these tools can *fine-tune* a given CNN model to improve the accuracy of the quantized network.

In particular, the works in [85] demonstrate the efficiency dynamic of F×P, pointing out how the inference of AlexNet is possible using 6 bits in dynamic F×P instead of 16 bits with a conventional fixed-point format.

1.6.1.3 FPGA Implementations

The FPGA-based CNN accelerator proposed in [29] is built upon this quantization scheme and employs different precisions to represent the FM, convolution kernels, and FC weights with 16, 8, and 10 bits, respectively. Without fine-tuning, the authors report a drop of 1% in the classification accuracy of AlexNet. In a similar way, Qiu et al. employ F×P to quantize the VGG network with respectively 8 bits for the weights, 8 bits for activations, and 4 bits for FC layers, resulting in an accuracy drop of 2%. In all these accelerators, dynamic quantization is supported by means of data shift modules [28, 82]. Finally, the accelerator in [41] relies on the Ristretto framework [81] to derive an AlexNet model wherein the data is quantized in 16 bits with distinct integer bit-widths per layer*.

1.6.1.4 Extreme Quantization and Binary Networks

Training and inferring CNNs with *extremely compact data representations* is an area that has recently gained a lot of research interest. Early works of Courbariaux et al. in BinaryConnect [86] demonstrate the feasibility of training neural networks using *binary* weights, i.e., weights with either a value of $-\theta$ or θ encoded in 1 bit. BinaryConnect lowers the bandwidth requirements of a network by a factor of ×32 at the price of an accuracy loss, evaluated at 19.2% on ImageNet[†]. The same authors go further in their investigations in [15] and propose BNNs that represent both feature maps and weights with only 1 bit. In these networks, negative values are represented as 0, while positive values are represented as 1. BNNs greatly simplify the processing of convolutions, boiling down the computations of MACs into bitwise XNOR operations followed by a pop-count (see Figure 1.11b). Moreover, the authors use the *sign* function as activation and apply batch normalization before applying the activation, which reduces the information lost during binarization (see Figure 1.11a). In turn, a higher drop in classification accuracy occurs when using BNNs, evaluated at 29.8% for ImageNet. This accuracy drop is then lowered to 11% by Rastegari et al., using different scale factors for binary weights (i.e., $-_{\theta 1}$ or $+_{\theta 2}$).

---

* Since the same PEs are reused to process different layers, the same bit width is used with a variable radix point for each layer.

[†] When compared to an exact 32-bit implementation of AlexNet.

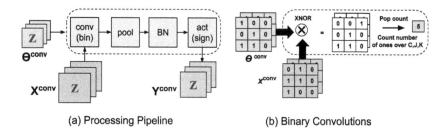

(a) Processing Pipeline   (b) Binary Convolutions

FIGURE 1.11   Binary neural networks.

Beside BNNs, *pseudo-binary networks*, such as DoReFa-Net [87] and quantized neural networks (QNNs) [88], reduce the accuracy drop to 6.5% when employing a slightly expanded bit-width (2 bits) to represent the intermediate FM. Similarly, in TTQ [89], weights are constrained to three values (2 bits) $-_{\theta1}$, 0, $-_{\theta2}$, but FMs are represented in a 32-bit float scheme. As a consequence, the efficiency gain of TTQ is not as high as in BNNs. In turn, TTQ achieves comparable accuracy on ImageNet, within 0.7% of full-precision.

In FPGAs, BNNs benefit from a significant acceleration, as the processing of "binary" convolutions can be mapped on XNOR gates followed by a pop-count operation, as depicted in Figure 1.11b. Furthermore, and as suggested in [7], a pop-count operation can be implemented using lookup tables in a way that convolutions are processed only with logical elements. Thus, the DSP blocks can be used to process the batch norm calculation (Equation 1.7, which can be formulated as a linear transform in order to reduce the number of operations). This approach is followed in the implementation of [90] to derive an FPGA-based accelerator for BNNs that achieves 207.8 GOPs while only consuming 4.7 W and 3 DSP blocks to classify the Cifar10 dataset.

For the same task, works in [45, 91] use a smaller network configuration* and reach a throughput of 2.4 TOPs when using a larger Zynq 7Z045 device with 11W power consumption.

For ImageNet classification, binary net implementation of [92] delivers an overall throughput of 1.9 TOPs on a Stratix V GSD device. In all these works, the first layer is not binarized to achieve better classification accuracy. As pointed out in [92], the performance in this layer can be improved when using a higher number of DSP blocks. Finally, an accelerator for TTQ proposed in [93] achieves a peak performance of 8.36 TMACs when classifying the Cifar10 dataset with a 2-bit precision.

1.6.2 REDUCED COMPUTATIONS

In addition to approximate arithmetic, several studies attempt to reduce the number of operations involved in CNNs. For FPGA-based implementations, two main strategies are investigated: *weight pruning*, which increases the *sparsity* of the model, and *low-rank approximation* of filters, which reduces the number of multiplications occurring in the inference.

---

* The network topology used in this work involves 90% fewer computations and achieves 7% less classification accuracy on Cifar10.

1.6.2.1 Weight Pruning

As highlighted in [94], CNNs as overparametrized networks and a large amount of the weights can be removed – or *pruned* – without critically affecting the classification accuracy. In its simplest form, pruning is performed according to the magnitude, such that the lowest values of the weights are truncated to zero [95]. In a more recent approach, weight removal is driven by energy consumption of a given node of the graph, which is 1.74× more efficient than magnitude-based approaches [96]. In both cases, pruning is followed by a fine-tuning of the remaining weights in order to improve the classification accuracy. This is for instance the case in [97], where pruning removes respectively 53% and 85% of the weights in AlexNet *conv* and FC layers for less then 0.5% accuracy loss (Figure 1.12).

1.6.2.2 Low Rank Approximation

Another way to reduce the computations occurring in CNNs is to maximize the number of *separable filters*. A 2D-separable filter, denoted θ^{sep}, has a unitary rank*, and can be expressed as two successive 1D filters ($\theta^{\text{sep}}_{J\times1}$ then $\theta^{\text{sep}}_{1\times K}$). Filter decomposition reduces the number of multiplications from $J \times K$ to $J + K$. This is illustrated in Figure 1.13, where the 3×3 averaging filter is separable, and can thus be decomposed into two successive one-dimensional convolutions.

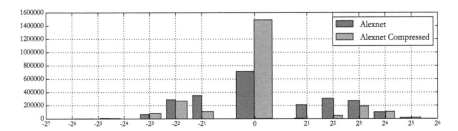

FIGURE 1.12 Histogram of conv weights in a compressed AlexNet model[†].

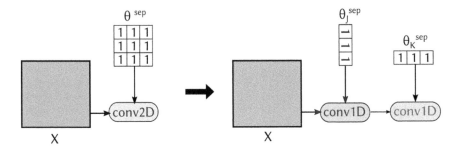

FIGURE 1.13 Example of a separable filter.

---

* Meaning that rank(θ^{sep}) = 1.
† Pruned filters treated as zero-valued weights.

The same concept expands to depth convolutions, where a separable filter requires $C+J+K$ multiplications instead of $C \times J \times K$ multiplications.

Nonetheless, only a small proportion of CNN filters are separable. To increase this proportion, a first approach is to force the convolution kernels to be separable by penalizing *high-rank filters* when training the network [98]. Alternatively, and after the training, the weights Θ of a given layer can be approximated into a small set of r *low-rank filters*. In this case, $r \times (C+J+K)$ multiplications are required to process a single depth convolution.

Finally, CNN computations can be reduced further by decomposing the weight matrix $\tilde{\Theta}$ through single-value decomposition (SVD). As demonstrated in the early works of [99], SVD greatly reduces the resource utilization of a given 2D-filter implantation. Moreover, SVD also finds its interest when processing FC layers and convolutions that employ the *im2col* method (cf Section 1.4.1). In a similar way to pruning, low rank approximation or SVD is followed by a fine-tuning in order to counterbalance the drop in classification accuracy.

1.6.2.3 FPGA Implementations

In FPGA implementations, SVD is applied on FC layer to significantly reduce the number of weights, such as in [28], where the authors derive a VGG16-SVD model that achieves 87.96% accuracy on ImageNet with 63% fewer parameters.

Alternatively, one can take advantage of the numerous research efforts given to accelerate Sparse GEMM on FPGA [100]. In this case, the challenge is to determine the optimal format of matrices that maximizes the chance to detect and skip zero computations, such compressed sparse column (CSC) or compressed sparse row (CSR) formats*. Based on this, Sze et al. [22] advocate the use of the CRC to process CNN. Indeed, this format requires lower memory bandwidths when the output matrix is smaller then the input, which is typically the case in CNNs where $N < CJK$, as in Figure 1.3b.

However, this efficiency of CRC format is valid only for extremely sparse matrices (typically with $\leq 1\%$ of non-zeros), while in practice, pruned CNN matrices are not that sparse (typically, $\leq 4\% - 80\%$ of non-zeros). Therefore, works in [7] propose a *zero skip scheduler* that identifies zero elements and skips them in the scheduling of the MAC processing. As a consequence, the number of cycles required to compute the sparse GEMM is reduced. For AlexNet layers, the zero skip scheduler results in a 4× speedup. The same authors project a throughput of 12 TOPs for pruned CNN in the next Intel Stratix10 FPGAs, which outperforms the computational throughput of state-of-the-art GPU implementations by 10%.

1.7   CONCLUSIONS

In this chapter, a number of methods and tools have been compared that aim at porting convolutional neural networks onto FPGAs. At the network level, approximate computing and data-path optimization methods have been covered, while at

---

* This format represents a matrix by three one-dimensional arrays, that respectively contain nonzero values, row indices, and column indices.

the neuron level, the optimizations of convolutional and fully connected layers have been detailed and compared. All the different degrees of freedom offered by FPGAs (custom data types, local data streams, dedicated processors, etc.) are exploited by the presented methods. Moreover, algorithmic and data-path optimizations can and should be jointly implemented, resulting in additive hardware performance gains.

CNNs are by nature overparameterized and support particularly well approximate computing techniques such as weight pruning and fixed-point computation. Approximate computing already constitutes a key to CNN acceleration over hardware and will certainly continue driving the performance gains in the years to come.

BIBLIOGRAPHY

1. Y. LeCun, Y. Bengio, and G. Hinton, "Deep learning," *Nature*, vol. 521, no. 7553, pp. 436–444, 2015.
2. O. Russakovsky, J. Deng, H. Su, J. Krause, S. Satheesh, S. Ma, Z. Huang, A. Karpathy, A. Khosla, M. Bernstein, A. C. Berg, and L. Fei-Fei, "ImageNet large scale visual recognition challenge," *International Journal of Computer Vision*, vol. 115, no. 3, pp. 211–252, September 2014.
3. R. Girshick, "Fast R-CNN," in *Proceedings of the IEEE International Conference on Computer Vision - ICCV '15*, 2015, pp. 1440–1448.
4. J. Long, E. Shelhamer, and T. Darrell, "Fully convolutional networks for semantic segmentation," in *Proceedings of the IEEE Conference on Computer Vision and Pattern Recognition - CVPR '15*, 2015, pp. 3431–3440.
5. Y. Zhang, M. Pezeshki, P. Brakel, S. Zhang, C. L. Y. Bengio, and A. Courville, "Towards end-to-end speech recognition with deep convolutional neural networks," arXiv preprint, vol. arXiv:1701, 2017.
6. K. Simonyan and A. Zisserman, "Very deep convolutional networks for large-scale image recognition," arXiv preprint, vol. arXiv:1409, pp. 1–14, 2014.
7. E. Nurvitadhi, S. Subhaschandra, G. Boudoukh, G. Venkatesh, J. Sim, D. Marr, R. Huang, J. OngGeeHock, Y. T. Liew, K. Srivatsan, and D. Moss, "Can FPGAs beat GPUs in accelerating next-generation deep neural networks?" in *Proceedings of the ACM/SIGDA International Symposium on Field-Programmable Gate Arrays - FPGA '17*, 2017, pp. 5–14.
8. Intel FPGA, "Intel Stratix 10 variable precision DSP blocks user guide," pp. 4–5, 2017.
9. G. Lacey, G. Taylor, and S. Areibi, "Deep learning on FPGAs: Past, present, and future," arXiv e-print, 2016.
10. S. I. Venieris, A. Kouris, and C.-S. Bouganis, "Toolflows for mapping convolutional neural networks on FPGAs," *ACM Computing Surveys*, vol. 51, no. 3, pp. 1–39, June 2018.
11. Y. LeCun, L. Bottou, Y. Bengio, and P. Haffner, "Gradient based learning applied to document recognition," in *Proceedings of the IEEE*, vol. 86, no. 11, pp. 2278–2324, 1998.
12. A. Krizhevsky, I. Sutskever, and G. E. Hinton, "ImageNet classification with deep convolutional neural networks," in *Advances in Neural Information Processing Systems - NIPS'12*, 2012, p. 19.
13. D. H. Hubel and T. N. Wiesel, "Receptive fields, binocular interaction and functional architecture in the cat's visual cortex," *The Journal of Physiology*, vol. 160, no. 1, pp. 106–154, 1962.
14. S. Ioffe and C. Szegedy, "Batch normalization: Accelerating deep network training by reducing internal covariate shift," in *Proceedings of the International Conference on Machine Learning - ICML '15*, F. Bach and D. Blei, Eds., vol. 37, 2015, pp. 448–456.

15. I. Hubara, M. Courbariaux, D. Soudry, R. El-Yaniv, and Y. Bengio, "Binarized neural networks," in *Advances in Neural Information Processing Systems – NIPS '16*, February 2016, pp. 4107–4115.

16. C. Szegedy, W. Liu, Y. Jia, P. Sermanet, S. Reed, D. Anguelov, D. Erhan, V. Vanhoucke, and A. Rabinovich, "Going deeper with convolutions," in *Proceedings of the IEEE Conference on Computer Vision and Pattern Recognition - CVPR '15*, pp. 1–9, 2015.

17. K. He, X. Zhang, S. Ren, and J. Sun, "Deep residual learning for image recognition," in *Proceedings of the IEEE Conference on Computer Vision and Pattern Recognition - CVPR '16*, 2016, pp. 770–778.

18. J. Cong and B. Xiao, "Minimizing computation in convolutional neural networks," in *Proceedings of the International Conference on Artificial Neural Networks - ICANN '14*. Springer, 2014, pp. 281–290.

19. A. Canziani, A. Paszke, and E. Culurciello, "An analysis of deep neural network models for practical applications," arXiv e-print, May 2016.

20. R. G. Shoup, "Parameterized convolution filtering in a field programmable gate array," in *Proceedings of the International Workshop on Field Programmable Logic and Applications on More FPGAs*, 1994, pp. 274–280.

21. M. Horowitz, "Computing's energy problem (and what we can do about it)," in *Proceedings of the IEEE International Solid-State Circuits - ISSCC '14*. IEEE, February 2014, pp. 10–14.

22. V. Sze, Y.-H. Chen, T.-J. Yang, and J. Emer, "Efficient processing of deep neural networks: A tutorial and survey," *Proceedings of the IEEE*, vol. 105, no. 12, pp. 2295–2329, December 2017.

23. Nvidia, "GPU-based deep learning inference: A performance and power analysis," White Paper, 2015. https://www.nvidia.com/

24. S. Chetlur, C. Woolley, P. Vandermersch, J. Cohen, J. Tran, B. Catanzaro, and E. Shelhamer, "cuDNN: Efficient primitives for deep learning," arXiv e-print, 2014.

25. Y. Jia, E. Shelhamer, J. Donahue, S. Karayev, J. Long, R. Girshick, S. Guadarrama, and T. Darrell, "Caffe: Convolutional architecture for fast feature embedding," in *Proceedings of the ACM International Conference on Multimedia – MM '14*, 2014, pp. 675–678.

26. M. Abadi, P. Barham, J. Chen, Z. Chen, A. Davis, J. Dean, M. Devin, S. Ghemawat, G. Irving, M. Isard, M. Kudlur, J. Levenberg, R. Monga, S. Moore, D. G. Murray, B. Steiner, P. Tucker, V. Vasudevan, P. Warden, M. Wicke, Y. Yu, X. Zheng, and G. Brain, "TensorFlow: A system for large-scale machine learning," in *Proceedings of the USENIX Symposium on Operating Systems Design and Implementation - OSDI '16*, 2016, pp. 265–284.

27. K. Ovtcharov, O. Ruwase, J.-y. Kim, J. Fowers, K. Strauss, and E. Chung, "Accelerating deep convolutional neural networks using specialized hardware," White Paper, pp. 3–6, February 2015.

28. J. Qiu, J. Wang, S. Yao, K. Guo, B. Li, E. Zhou, J. Yu, T. Tang, N. Xu, S. Song, Y. Wang, and H. Yang, "Going deeper with embedded FPGA platform for convolutional neural network," in *Proceedings of the ACM/SIGDA International Symposium on Field-Programmable Gate Arrays - FPGA '16*. ACM, 2016, pp. 26–35.

29. N. Suda, V. Chandra, G. Dasika, A. Mohanty, Y. Ma, S. Vrudhula, J.-s. Seo, and Y. Cao, "Throughput-optimized openCL-based FPGA accelerator for large-scale convolutional neural networks," in *Proceedings of the ACM/SIGDA International Symposium on Field-Programmable Gate Arrays - FPGA '16*, 2016, pp. 16–25.

30. K. Chellapilla, S. Puri, and P. Simard, "High performance convolutional neural networks for document processing," in *Proceedings of the International Workshop on Frontiers in Handwriting Recognition – FHR '06*. Suvisoft, October 2006.

31. U. Aydonat, S. O'Connell, D. Capalija, A. C. Ling, and G. R. Chiu, "An openCL(TM) deep learning accelerator on arria 10," in *Proceedings of the ACM/SIGDA International Symposium on Field-Programmable Gate Arrays - FPGA '17*, ACM, Ed. ACM, 2017, pp. 55–64.

32. R. DiCecco, G. Lacey, J. Vasiljevic, P. Chow, G. Taylor, and S. Areibi, "Caffeinated FPGAs: FPGA framework for convolutional neural networks," in *Proceedings of the International Conference on Field- Programmable Technology - FPT '16*, pp. 265–268, 2016.

33. C. Zhang and V. Prasanna, "Frequency domain acceleration of convolutional neural networks on CPU-FPGA shared memory system," in *Proceedings of the ACM/SIGDA International Symposium on Field- Programmable Gate Arrays - FPGA '17*, 2017, pp. 35–44.

34. J. H. Ko, B. A. Mudassar, T. Na, and S. Mukhopadhyay, "Design of an energy-efficient accelerator for training of convolutional neural networks using frequency-domain computation," in *Proceedings of the Annual Conference on Design Automation - DAC '17*, 2017.

35. S. Venieris and C. Bouganis, "FpgaConvNet: A framework for mapping convolutional neural networks on FPGAs," in *Proceedings of the IEEE Annual International Symposium on Field-Programmable Custom Computing Machines - FCCM '16*, 2016, pp. 40–47.

36. H. Sharma, J. Park, D. Mahajan, E. Amaro, J. K. Kim, C. Shao, A. Mishra, and H. Esmaeilzadeh, "From high-level deep neural models to FPGAs," in *Proceedings of the International Symposium on Microarchitecture - MICRO '16*, 2016, pp. 1–12.

37. H. Li, X. Fan, L. Jiao, W. Cao, X. Zhou, and L. Wang, "A high performance FPGA-based accelerator for large-scale convolutional neural networks," in *Proceedings of the International Conference on Field Programmable Logic and Applications - FPL '16*. IEEE, August 2016, pp. 1–9.

38. K. Abdelouahab, M. Pelcat, J. Serot, C. Bourrasset, and F. Berry, "Tactics to directly map CNN graphs on embedded FPGAs," *IEEE Embedded Systems Letters*, vol. 9, no. 4, pp. 113–116, December 2017.

39. C. Zhang, P. Li, G. Sun, Y. Guan, B. Xiao, and J. Cong, "Optimizing FPGA-based accelerator design for deep convolutional neural networks," in *Proceedings of the ACM/SIGDA International Symposium on Field-Programmable Gate Arrays - FPGA '15*, ser. FPGA, 2015, pp. 161–170.

40. M. Motamedi, P. Gysel, V. Akella, and S. Ghiasi, "Design space exploration of FPGA-based deep convolutional neural networks," in *Proceedings of the Asia and South Pacific Design Automation Conference – ASPDAC '16*, January 2016, pp. 575–580.

41. M. Motamedi, P. Gysel, and S. Ghiasi, "PLACID: A platform for FPGA-based accelerator creation for DCNNs," *ACM Transactions on Multimedia Computing, Communications, and Applications*, vol. 13, no. 4, pp. 62:1–62:21, September 2017.

42. P. Molchanov, S. Tyree, T. Karras, T. Aila, and J. Kautz, "Pruning convolutional neural networks for resource efficient learning," arXiv preprint, 2017.

43. T. Fujii, S. Sato, H. Nakahara, and M. Motomura, "An FPGA realization of a deep convolutional neural network using a threshold neuron pruning," in *Proceedings of the International Symposium on Applied Reconfigurable Computing – ARC '16*, vol. 9625, 2017, pp. 268–280.

44. S. Zhou, Y. Wu, Z. Ni, X. Zhou, H. Wen, and Y. Zou, "DoReFaNet: Training low bit-width convolutional neural networks with low bitwidth gradients," arXiv e-print, 2016.

45. Y. Umuroglu, N. J. Fraser, G. Gambardella, M. Blott, P. Leong, M. Jahre, and K. Vissers, "FINN: A framework for fast, scalable binarized neural network inference," in *Proceedings of the ACM/SIGDA International Symposium on Field-Programmable Gate Arrays - FPGA '17*, 2017, pp. 65–74.

46. R. Andri, L. Cavigelli, D. Rossi, and L. Benini, "YodaNN: An ultralow power convolutional neural network accelerator based on binary weights," in *2016 IEEE Computer Society Annual Symposium on VLSI (ISVLSI)*, pp. 236–241, July 2016.

47. R. Zhao, W. Ouyang, H. Li, and X. Wang, "Saliency detection by multicontext deep learning," in *Proceedings of the IEEE Conference on Computer Vision and Pattern Recognition - CVPR '15*, 2015, pp. 1265–1274.

48. C. Zhang, D. Wu, J. Sun, G. Sun, G. Luo, and J. Cong, "Energy-efficient CNN implementation on a deeply pipelined FPGA cluster," in *Proceedings of the International Symposium on Low Power Electronics and Design - ISLPED '16*, 2016, pp. 326–331.

49. C. Zhang, Z. Fang, P. Zhou, P. Pan, and J. Cong, "Caffeine: Towards uniformed representation and acceleration for deep convolutional neural networks," in *Proceedings of the International Conference on Computer-Aided Design - ICCAD '16*. ACM, 2016, pp. 1–8.

50. J. Zhang and J. Li, "Improving the performance of openCL-based FPGA accelerator for convolutional neural network," in *Proceedings of the ACM/SIGDA International Symposium on Field-Programmable Gate Arrays - FPGA '17*, 2017, pp. 25–34.

51. Y. Guan, H. Liang, N. Xu, W. Wang, S. Shi, X. Chen, G. Sun, W. Zhang, and J. Cong, "FP-DNN: An automated framework for mapping deep neural networks onto FPGAs with RTL-HLS hybrid templates," in *Proceedings of the IEEE Annual International Symposium on Field- Programmable Custom Computing Machines - FCCM '17*. IEEE, 2017, pp. 152–159.

52. S. Winograd, *Arithmetic Complexity of Computations*. SIAM, 1980, vol. 33.

53. A. Lavin and S. Gray, "Fast algorithms for convolutional neural networks," arXiv e-print, vol. arXiv: 150, September 2015.

54. L. Lu, Y. Liang, Q. Xiao, and S. Yan, "Evaluating fast algorithms for convolutional neural networks on FPGAs," in *Proceedings of the IEEE Annual International Symposium on Field-Programmable Custom Computing Machines - FCCM '17*, 2017, pp. 101–108.

55. J. Bottleson, S. Kim, J. Andrews, P. Bindu, D. N. Murthy, and J. Jin, "ClCaffe: OpenCL accelerated caffe for convolutional neural networks," in *Proceedings of the IEEE International Parallel and Distributed Processing Symposium – IPDPS '16*, 2016, pp. 50–57.

56. T. Highlander and A. Rodriguez, "Very efficient training of convolutional neural networks using fast fourier transform and overlap-and- add," arXiv preprint, pp. 1–9, 2016.

57. H. Zeng, R. Chen, C. Zhang, and V. Prasanna, "A framework for generating high throughput CNN implementations on FPGAs," in *Proceedings of the ACM/SIGDA International Symposium on Field- Programmable Gate Arrays - FPGA '18*. ACM Press, 2018, pp. 117–126.

58. M. Sankaradas, V. Jakkula, S. Cadambi, S. Chakradhar, I. Durdanovic, E. Cosatto, and H. P. Graf, "A massively parallel coprocessor for convolutional neural networks," in *Proceedings of the IEEE International Conference on Acoustics, Speech and Signal Processing - ICASSP '17*. IEEE, July 2009, pp. 53–60.

59. C. Farabet, C. Poulet, J. Y. Han, Y. LeCun, D. R. Tobergte, and S. Curtis, "CNP: An FPGA-based processor for convolutional networks," in *Proceedings of the International Conference on Field Programmable Logic and Applications - FPL '09*, pp. 32–37, 2009.

60. S. Chakradhar, M. Sankaradas, V. Jakkula, and S. Cadambi, "A dynamically configurable coprocessor for convolutional neural networks," *ACM SIGARCH Computer Architecture News*, vol. 38, no. 3, pp. 247–257, June 2010.

61. V. Gokhale, J. Jin, A. Dundar, B. Martini, and E. Culurciello, "A 240 G-ops/s mobile coprocessor for deep neural networks," in *Proceedings of the IEEE Conference on Computer Vision and Pattern Recognition - CVPR '14*, June 2014, pp. 696–701.

62. M. Alwani, H. Chen, M. Ferdman, and P. Milder, "Fused-layer CNN accelerators," in *Proceedings of the Annual International Symposium on Microarchitecture - MICRO '16*, vol. 2016, December 2016.

63. Y. Ma, Y. Cao, S. Vrudhula, and J.-s. Seo, "Optimizing loop operation and dataflow in FPGA acceleration of deep convolutional neural networks," in *Proceedings of the ACM/SIGDA International Symposium on Field-Programmable Gate Arrays - FPGA '17*, 2017, pp. 45–54.

64. V. Gokhale, A. Zaidy, A. Chang, and E. Culurciello, "Snowflake: An efficient hardware accelerator for convolutional neural networks," in *Proceedings of the IEEE International Symposium on Circuits and Systems - ISCAS '17*. IEEE, May 2017, pp. 1–4.

65. Microsoft, "Microsoft unveils Project Brainwave for real-time AI," 2017. https://www.microsoft.com/en-us/research/blog/microsoft-unveils-project-brainwave/

66. S. Derrien and S. Rajopadhye, "Loop tiling for reconfigurable accelerators," in *Proceedings of the International Conference on Field Programmable Logic and Applications - FPL '01*. Springer, 2001, pp. 398–408.

67. R. Atul, L. Jongeun, and C. Kiyoung, "Efficient FPGA acceleration of convolutional neural networks using logical-3D compute array," in *Proceedings of the Design, Automation & Test in Europe Conference & Exhibition - DATE '16*. IEEE, 2016, pp. 1393–1398.

68. Y. Ma, N. Suda, J. S. Seo, and S. Vrudhula, "Scalable and modularized RTL compilation of Convolutional Neural Networks onto FPGA," in *Proceedings of the 26th International Conference on Field Programmable Logic and Applications (FPL)*, pp. 1–8. IEEE, 2016.

69. S. Williams, A. Waterman, and D. Patterson, "Roofline: An insightful visual performance model for multicore architectures," *Communications of the ACM*, vol. 52, no. 4, p. 65, April 2009.

70. Y. Ma, M. Kim, Y. Cao, S. Vrudhula, and J.-s. Seo, "End-to-end scalable FPGA accelerator for deep residual networks," in *Proceedings of the IEEE International Symposium on Circuits and Systems - ISCAS '17*. IEEE, May 2017, pp. 1–4.

71. Y. Ma, Y. Cao, S. Vrudhula, and J.-s. Seo, "An automatic RTL compiler for high-throughput FPGA implementation of diverse deep convolutional neural networks," in *Proceedings of the International Conference on Field Programmable Logic and Applications - FPL '17*. IEEE, September 2017, pp. 1–8.

72. Z. Liu, Y. Dou, J. Jiang, J. Xu, S. Li, Y. Zhou, and Y. Xu, "Throughput-optimized FPGA accelerator for deep convolutional neural networks," *ACM Transactions on Reconfigurable Technology and Systems*, vol. 10, no. 3, pp. 1–23, 2017.

73. S. Mittal, "A survey of techniques for approximate computing," *ACM Computing Surveys*, vol. 48, no. 4, pp. 1–33, March 2016.

74. S. Anwar, K. Hwang, and W. Sung, "Fixed point optimization of deep convolutional neural networks for object recognition," in *2015 IEEE International Conference on Acoustics, Speech and Signal Processing (ICASSP)*. IEEE, April 2015.

75. S. Gupta, A. Agrawal, P. Narayanan, K. Gopalakrishnan, and P. Narayanan, "Deep learning with limited numerical precision," in *Proceedings of the International Conference on Machine Learning - ICML '15*, 2015, pp. 1737–1746.

76. D. Lin, S. Talathi, and V. Annapureddy, "Fixed point quantization of deep convolutional networks," in *Proceedings of the International Conference on Machine Learning - ICML '16*, 2016, pp. 2849–2858.

77. M. Courbariaux, Y. Bengio, and J.-P. David, "Training deep neural networks with low precision multiplications," arXiv e-print, December 2014.

78. S. Zhou, Y. Wang, H. Wen, Q. He, and Y. Zou, "Balanced quantization: An effective and efficient approach to quantized neural networks," *Journal of Computer Science and Technology*, vol. 32, pp. 667–682, 2017.

79. J.-P. David, K. Kalach, and N. Tittley, "Hardware complexity of modular multiplication and exponentiation," *IEEE Transactions on Computers*, vol. 56, no. 10, pp. 1308–1319, October 2007.

80. C. Farabet, B. Martini, B. Corda, P. Akselrod, E. Culurciello, and Y. LeCun, "NeuFlow: A runtime reconfigurable dataflow processor for vision," in *Proceedings of the IEEE Conference on Computer Vision and Pattern Recognition - CVPR '11*. IEEE, June 2011, pp. 109–116.

81. P. Gysel, M. Motamedi, and S. Ghiasi, "Hardware-oriented approximation of convolutional neural networks," arXiv preprint, 2016, p. 8.

82. A. Kouris, S. I. Venieris, and C.-S. Bouganis, "CascadeCNN: Pushing the performance limits of quantisation in convolutional neural networks," in *Proceedings of the International Conference on Field Programmable Logic and Applications - FPL '18*, pp. 155–1557, July 2018.

83. D. Williamson, "Dynamically scaled fixed point arithmetic," in *Proceedings of the IEEE Pacific Rim Conference on Communications, Computers and Signal Processing Conference*. IEEE, 1991, pp. 315–318.

84. S. Guo, L. Wang, B. Chen, Q. Dou, Y. Tang, and Z. Li, "FixCaffe: Training CNN with low precision arithmetic operations by fixed point caffe," in *Proceedings of the International Workshop on Advanced Parallel Processing Technologies - APPT '17*. Springe, August 2017, pp. 38–50.

85. P. Gysel, J. Pimentel, M. Motamedi, and S. Ghiasi, "Ristretto: A framework for empirical study of resource-efficient inference in convolutional neural networks," *IEEE Transactions on Neural Networks and Learning Systems*, vol. 29, pp. 1–6, 2018.

86. M. Courbariaux, Y. Bengio, and J.-P. David, "BinaryConnect: Training deep neural networks with binary weights during propagations," in *Advances in Neural Information Processing Systems – NIPS '15*, 2015, pp. 3123–3131.

87. H. Nakahara, T. Fujii, and S. Sato, "A fully connected layer elimination for a binarizec convolutional neural network on an FPGA," in *Proceedings of the International Conference on Field Programmable Logic and Applications - FPL '17*. IEEE, September 2017, pp. 1–4.

88. I. Hubara, M. Courbariaux, D. Soudry, R. El-Yaniv, and Y. Bengio, "Quantized neural networks: Training neural networks with low precision weights and activations," *Journal of Machine Learning Research*, vol. 18, pp. 187:1–187:30, September 2018.

89. C. Zhu, S. Han, H. Mao, and W. J. Dally, "Trained ternary quantization," in *Proceedings of the International Conference on Learning Representations – ICLR '17*, December 2017.

90. R. Zhao, W. Song, W. Zhang, T. Xing, J.-H. Lin, M. Srivastava, R. Gupta, and Z. Zhang, "Accelerating binarized convolutional neural networks with software-programmable FPGAs," in *Proceedings of the ACM/SIGDA International Symposium on Field-Programmable Gate Arrays - FPGA '17*, 2017.

91. N. J. Fraser, Y. Umuroglu, G. Gambardella, M. Blott, P. Leong, M. Jahre, and K. Vissers, "Scaling binarized neural networks on reconfigurable logic," in *Proceedings of the Workshop on Parallel Programming and Run-Time Management Techniques for Many-core Architectures and Design Tools and Architectures for Multicore Embedded Computing Platforms - PARMA-DITAM '17*. ACM, 2017, pp. 25–30.

92. S. Liang, S. Yin, L. Liu, W. Luk, and S. Wei, "FP-BNN: Binarized neural network on FPGA," *Neurocomputing*, vol. 275, pp. 1072–1086, January 2018.

93. A. ProstBoucle, A. Bourge, F. Ptrot, H. Alemdar, N. Caldwell, and V. Leroy, "Scalable high-performance architecture for convolutional ternary neural networks on FPGA," in *Proceedings of the International Conference on Field Programmable Logic and Applications - FPL '17*, pp. 1–7, July 2017.

94. B. Liu, M. Wang, H. Foroosh, M. Tappen, and M. Pensky, "Sparse convolutional neural networks," in *Proceedings of the IEEE Conference on Computer Vision and Pattern Recognition - CVPR '15*, 2015, pp. 806–814.

95. S. Han, J. Pool, J. Tran, and W. J. Dally, "Learning both weights and connections for efficient neural network," in *Advances in Neural Information Processing Systems – NIPS '15*, 2015, pp. 1135–1143.

96. T.-J. Yang, Y.-H. Chen, and V. Sze, "Designing energy-efficient convolutional neural networks using energy-aware pruning," in *Proceedings of the IEEE Conference on Computer Vision and Pattern Recognition - CVPR '17*, pp. 5687–5695, 2017.

97. S. Han, H. Mao, and W. J. Dally, "Deep compression - compressing deep neural networks with pruning, trained quantization and huffman coding," in *Proceedings of the International Conference on Learning Representations – ICLR '16*, 2016, pp. 1–13.

98. A. Sironi, B. Tekin, R. Rigamonti, V. Lepetit, and P. Fua, "Learning separable filters," *IEEE Transactions on Pattern Analysis and Machine Intelligence*, vol. 37, no. 1, pp. 94–106, 2015.

99. C. Bouganis, G. Constantinides, and P. Cheung, "A novel 2D filter design methodology for heterogeneous devices," in *Proceedings of the Annual IEEE Symposium on Field-Programmable Custom Computing Machines - FCCM '05*. IEEE, 2005, pp. 13–22.

100. R. Dorrance, F. Ren, and D. Markovi, "A scalable sparse matrix-vector multiplication kernel for energy-efficient sparse-blas on FPGAs," in *Proceedings of the ACM/SIGDA International Symposium on Field- Programmable Gate Arrays - FPGA '14*. ACM, 2014, pp. 161–170.

2 Object Detection with Convolutional Neural Networks

Kaidong Li, Wenchi Ma, Usman Sajid,
Yuanwei Wu, and Guanghui Wang

CONTENTS

2.1 INTRODUCTION

Deep learning was first proposed in 2006 [1]; however, it did not attract much attention until 2012. With the development of computational power and a large number of labeled datasets [2, 3], deep learning has proven to be very effective in extracting intrinsic structural and high-level features. Significant progress has been made in solving various problems in the artificial intelligence community [4], especially in areas where the data are multidimensional and the features are difficult to hand-engineer, including speech recognition [5, 6, 7], natural language processing [8, 9, 10], and computer vision [11, 12, 13, 14]. Deep learning also shows dominance on areas like business analysis [15], medical diagnostic [16, 17], art creation [18], and image translation [19, 20]. Since little human input is required, deep learning will continue to make more and more impact in the future with the increase of computing power and exploding data growth.

Among all these areas, computer vision has witnessed remarkably successful applications of deep learning [21, 22, 23]. Vision, as the most important sense in terms of navigation and recognition for humans, provides the most information about the surroundings. Thanks to the rapid development of digital cameras in the last 10 years, image sensors have become one of the most accessible types of hardware. Researchers tend to explore the possibility of using computer vision in applications like autonomous driving, visual surveillance, facial recognition [24], etc. The demand in these areas, in return, draws tremendous attention and provides huge resources to the computer vision field. As a result, it is developing at a speed that has been rarely seen in other research fields. Every month, sometimes even within a week, we have seen record performance achieved on datasets like ImageNet [11], PASCAL VOC [25], and COCO [26].

Object detection is a fundamental step for many computer vision applications [23, 27, 28, 29]. The architecture of convolutional neural networks (CNN) designed for object detection is usually used as the first step to extract objects' spatial and classification information. Then more modules are added for specific applications. For example, tracking shares a very similar task to object detection with the addition of a temporal module. Therefore, the performance of object detection will affect almost all other computer vision research. A huge amount of effort has been put into its development [30, 31, 32].

In this chapter, we present a brief overview of the recent developments in object detection using convolutional neural networks (CNNs). Several classical CNN-based detectors are presented. Some developments are based on the detector architectures [33, 34], while others are focused on solving certain problems, like model degradation and small-scale object detection [35, 36, 37]. The chapter also presents some performance comparison results of different models on several benchmark datasets. Through a discussion of these models, we hope to give readers a general idea about the developments of CNN-based object detection.

2.1.1 MAJOR EVOLUTION

To talk about object detection, we have to mention image classification. Image classification is the task of assigning a label to an input image from a fixed set of categories. The assigned label usually corresponds to the most salient object in the image. It works best when the object is centered and dominating in the image frame. However, most images contain multiple objects, scattering in the frame with different scales. One label is far from enough to describe the contextual meaning of images. Therefore, object detection is introduced not only to output multiple labels corresponding to an image but also to generate the spatial region associated with each label.

Two-stage models. Inspired by image classification, it is natural for researchers to detect objects by exploring a two-stage approach. In most two-stage models, the first stage is for region proposal, followed by image classification on the proposed regions. With this modification, additional work is done on localization, with few changes to previous classification models. For the region proposal part, some systems employ a sliding window technique, like Deformable Parts Models (DPM) [38] and OverFeat [39]. With certain strategies, these methods usually apply classifiers on windows at different locations with different scales. Another region proposal is selective search [40], which is adopted by regions with CNN features (R-CNN) [41]. It selectively extracts around 2,000 bottom-up region proposals [41], which greatly reduces the regions needed in sliding window methods.

The two-stage models, generally speaking, yield higher accuracy. With each stage doing its specific task, these models perform better on objects with various sizes. However, with the demands for real-time detection, two-stage models show its weakness in processing speed.

One-stage models. With more and more powerful computing ability available, CNN layers are becoming deeper and deeper. Researchers are able to utilize one-stage methods, like YOLO (You Only Look Once) [34] and SSD (single-shot multi-box detector) [42], with faster detection speed and similar or sometimes even higher accuracy. One-stage models usually divide the image into $N \times N$ grids. Each grid cell is responsible for the object whose center falls into that grid. Thus, the output is an $N \times N \times S$ tensor. Each of the S feature maps is an $N \times N$ matrix, with each element describing the feature on its corresponding grid cell. A common design of these feature maps for one object consists of the following $S = (5 + C)$ values:

- Four values for the bounding-box dimensions (x coordinate, y coordinate, height, and width);
- One value for the possibility of this grid cell containing the object;
- C values indicating which class this object belongs to.

One-stage models, thanks to their simpler architecture, usually require less computational resource. Some recent networks could achieve more than 150 frames per second (fps) [43]. However, the trade-off is accuracy, especially for small-scale objects as shown in Table 2.2 in Section 2.5.1. Another drawback is the class imbalance during training. In order to detect B objects in a grid, the output tensor we mentioned above includes information for $N \times N \times B$ number of anchor boxes. But among them, only a few contain objects. The ratio most of the time is about 1,000:1 [36]. This results in low efficiency during training.

2.1.2 OTHER DEVELOPMENT

With the general architecture established, new detector development mainly focuses on certain aspects to improve the performance.

Studies [44, 12] show strong evidence that network depth is crucial in CNN model performance. However, when we expect the model to converge for a loss function during training, the problem of vanishing/exploding gradients prevents it from behaving this way. The study from Bradley [65] back in 2010 found that the magnitude of gradient update decreases exponentially from output layer to input layer. Therefore, training becomes inefficient towards the input layers. This shows the difficulty of training very deep models. On the other hand, research shows that simply adding additional layers will not necessarily result in better detection performance. Monitoring the training process shows that added layers are trained to become identity maps. Therefore, it can only generate models whose performance is equal to a shallower network at most after a certain amount of layers. To address this issue, skip connections are introduced into the networks [35, 46, 47] to pass information between nonadjacent layers.

Feature pyramid networks (FPN) [48], which output feature maps at different scales, can detect objects at very different scales. This idea could also be found when the features were hand-engineered [49]. Another problem is the trailing accuracy in one-stage detector. According to Lin et al. [36], the lower accuracy is caused by extreme foreground-background class imbalance during training. To address this issue, Lin et al. introduced RetinaNet [36] by defining focal loss to reduce the weight of background loss.

2.2 TWO-STAGE MODEL

A two-stage model usually consists of regional proposals' extraction and classification. It is an intuitive idea to do this after the success of image classification. This type of model could use the proven image classification network after the region proposal stage. In addition, to some extent, the two steps resemble how humans receive visual information and arrange attentions on regions of interest (RoI). In this section, we will introduce the R-CNN, fast R-CNN, and Faster R-CNN models [41, 50, 33], and discuss the improvements of each model.

2.2.1 REGIONS WITH CNN FEATURES (R-CNN)

R-CNN was developed using a multi-stage process, which is shown in Figure 2.1. It generally can be divided into two stages.

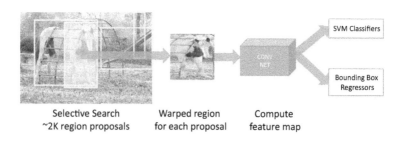

Selective Search   Warped region   Compute
~2K region proposals   for each proposal   feature map

FIGURE 2.1 An overview of the R-CNN detection system. The system generates $2k$ proposals. Each warped proposal produces feature maps. Then the system classifies each proposal and refines the bounding-box prediction.

The first stage is the region proposal extraction. The R-CNN model utilizes selective search [40], which takes the entire image as input and generates around 2,000 class-independent region proposals. In theory, R-CNN is able to work with any region proposal methods. Selective search is chosen because it performs well and has been employed by other detection models. The second stage starts with a CNN that takes a fixed-dimension image as input and generates a fixed-length feature vector as output. The input comes from the regional proposals. Each proposal is wrapped into the required size regardless of its original size and aspect ratio. Then using the CNN feature vector, pre-trained class-specific linear support vector machines (SVMs) are applied to calculate the class scores. Girshick et al. [41] conducted an error analysis, and based on the analysis result, a bounding-box regression is added to reduce the localization errors using the CNN feature vector. The regressor, which is class specific, can refine bounding-box prediction.

2.2.2   FAST R-CNN

R-CNN [50] has achieved great improvement compared to the previous best methods. Let us take the results on PASCAL VOC 2012 as an example. R-CNN has a 30% relative performance increase over the previous best algorithm. Apart from all the progress, R-CNN suffers from two major limitations. The first one is the space and time cost in training. In the second stage, the features for the 2,000 region proposals are all extracted separately using CNN and stored to the disk. With VGG16, VOC07 trainval dataset of 5,000 images requires 2.5 GPU-days*, and the memory expands to hundreds of GB. The second limitation is the slow detection speed. For each test image, it takes a GPU 47 seconds with VGG16. This weakness prevents the R-CNN model from real-time applications.

From the analysis above, both of the problems are caused by individually calculating a huge number of feature vectors for region proposals. Fast R-CNN is designed to overcome this inefficiency. The architecture of fast R-CNN is illustrated in Figure 2.2. The first stage remains unchanged, and generates region proposals

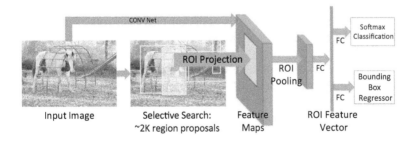

FIGURE 2.2   An overview of Fast R-CNN structure. Feature maps of the entire image are extracted. RoI is projected to the feature maps and pooled into a fixed size. Then, RoI feature vector is computed using fully connected layers. Using the vector, softmax classification probability and bounding-box regression are calculated as outputs for each proposal.

* A unit of computational complexity equivalent to what a single GPU can complete in a day.

using selective search. For the second stage, instead of generating feature vectors for each region proposal separately, fast R-CNN calculates one feature map from the entire image and uses a RoI pooling layer to extract a fixed-length feature vector for each region proposal. Each RoI feature vector is processed by a sequence of fully connected layers and forked into two branches. The first branch calculates class scores with a softmax layer, and the other branch is again a bounding-box regressor that refines the bounding-box estimation.

In fast R-CNN, the RoI pooling layer is the key part where the region proposals are translated into fixed-dimension feature vectors. This layer takes the feature map of the entire image and RoI as input. A four-element tuple (r,c,h,w) represents a RoI, with (r,c) and (h,w) specifying its top-left corner and its height and width respectively [50]. Then, the (h,w) dimension of RoI is divided into an $H \times W$ grid, where H and W are layer hyper-parameters. Standard max pooling is conducted to each grid cell in each feature map channel.

With similar mAP (mean average precision) levels, fast R-CNN is very successful in improving the training/testing speed. For large-scale objects defined in He et al. [35], fast R-CNN is 8.8 times faster in training and 146 times faster in detection [50]. It achieves 0.32s/image performance, which is a significant leap towards real-time detection.

2.2.3 FASTER R-CNN

To further increase the detection speed, researchers discover that the region proposal computation is the bottleneck of performance improvement. Before fast R-CNN, the time taken at the second stage is significantly more than that at the region proposal stage. Considering that selective search [40] works robustly and is a very popular method, improving region proposal seems unnecessary. However, with the improvement in fast R-CNN, selective search now is an order slower compared to classification stage. Even with some proposal methods that are balanced between speed and quality, region proposal stage still costs as much as the second stage. To solve this issue, Ren et al. [33] introduced a Region proposal network (RPN), which shares the feature map of the entire image with stage two.

The first step in Faster R-CNN is a shared CNN network. The RPN then takes the shared feature maps as input and generates region proposals with scores indicating how confident the network is that there are objects in them. In the RPN, a small network will slide over the input feature map as indicated in Figure 2.3. Each sliding window will generate a lower-dimensional vector. This lower-dimensional vector will be the input of two sibling fully connected layers, a bounding-box regression layer (reg layer), and a bounding-box classification layer (cls layer). To increase the nonlinearity, Rectified Linear Units (ReLUs) are applied to the output of the CNN layer. For each sliding window, k region proposals will be generated. Correspondingly, the reg layer will have $4k$ outputs for bounding-box coordinates, and the cls layer will have $2k$ probabilities of object/not-object for each bounding-box proposals. The k proposals are parameterized relative to k reference boxes, called anchors. The parameters include the scales and aspect ratio.

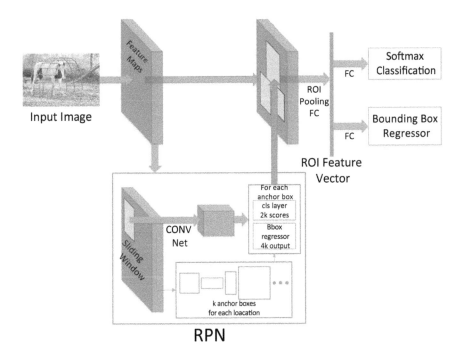

FIGURE 2.3   An overview of the Faster R-CNN framework. Region proposal network (RPN) shares feature maps with a fast R-CNN network; a small convolutional network slides over the shared feature maps to produce region proposal; and each location has k anchor boxes.

To train RPN, binary class labels are assigned to each anchor. Positive is assigned to two situations.

- Anchors with highest Intersection-over-Union (IoU) with ground-truth boxes;
- Anchors with IoU over 0.7 with ground-truth boxes.

Non-positive is given to anchors with IoU lower than 0.3 for all ground-truth boxes. During training, only positive or non-positive anchors contribute to loss functions. With the introduction of RPN, the total time to complete object detection on GPU is 198 ms using VGG as the CNN layer. Compared to the selective search, it is almost 10 times faster. The proposal stage is improved from 1,510 ms to only 10 ms. Combined, the new Faster R-CNN achieves 5 fps.

Two-stage detectors usually have higher accuracy in comparison to one-stage detectors. Most of the models at the top detection dataset leader board are two-stage models. Region proposals with regressor to refine localization means that they inherently produce much better results for bounding-box predictions. In terms of speed, these detectors offered far from real-time performance when they were first introduced. With recent development, they are getting closer by simplifying the architecture. It evolves from models with multiple computationally heavy steps to models with a single shared CNN. After the Faster R-CNN [33] was published, the networks are mainly based on shared feature maps. For example, R-FCN (region-based fully convolutional networks) [51] and Cascade R-CNN [52] focus on localization accuracy, and at the same time, they can both achieve around 10 fps. Region-based two-stage detectors are becoming not only more accurate but also faster. Therefore, it becomes one of the main branches for object detection.

2.3   ONE-STAGE MODEL

From the development of R-CNN, we can observe the success of sharing CNN in almost all the steps of a detection network. Region proposal, classification, and bounding-box refinement all take the output of the last shared CNN. This indicates that, if designed properly, the feature maps contain information about all of the above tasks. It is natural to explore the possibility of implementing detection as a single regression problem.

2.3.1   You Only Look Once (YOLO)

YOLO [34] is a simple model that takes the entire image as input and simultaneously generates multiple bounding boxes and class probabilities. Since YOLO takes input from the entire image, it reasons globally for detection, which means each detection takes the background and other objects into consideration.

YOLO divides input images into $S \times S$ grids. If an object center falls into a grid cell, this grid cell is responsible for that object's detection. Each grid cell will predict B bounding boxes, and the bounding boxes are described by five parameters: x, y, w, h, and the *confidence*, where (x,y) represents the center of the box relative to the bounds of the grid cell, and (w,h) represents the width and height relative to the dimensions of the whole image, respectively. The confidence is defined as:

$$Pr(\text{Object}) * \text{IOU}_{\text{pred}}^{\text{truth}}$$

which reflects how confident the prediction is that an object exists in this bounding box and how accurate the bounding box is relative to the ground-truth box. The final parameter is C, conditional class probabilities. It represents the conditional probabilities of the grid cell containing an object, regardless of how many bounding boxes a grid cell has. Therefore, the final predictions are an $S \times S \times (B \times 5 + C)$ tensor.

The network design is shown in Table 2.1. It is inspired by GoogLeNet [12], followed by 2 fully connected layers: FC1 of 4,096 dimensional vector, and the final output FC2 of $7 \times 7 \times 30$ tensor. In the paper, the authors also introduced a fast version with only 79 convolutional layers.

During training, the loss function is shown in Equation 2.1, where 1_{ij}^{obj} denotes whether the jth prediction in cell i should make a prediction, and 1_{i}^{obj} denotes whether there's an object in cell i [34]. From the function, we can see that it penalizes the

TABLE 2.1

The Architecture of the Convolutional Layers of YOLO [34]

Stage	Image	conv1	conv2	conv3	conv4		conv5		conv6
Filter/		7×7	3×3	1×1	1×1		1×1		3×3
Size		64/2	192	128	256	×4	512	×2	1024
Stride				3×3	3×3		3×3		3×3
				256	512		1024		1024
				1×1	1×1		3×3		
				256	512		1024		
				3×3	3×3		3×3		
				512	1024		1024/2		
Maxpool		2×2	2×2	2×2	2×2				
Stride		2	2	2	2				
Output	448×448	112×112	56×56	28×28	14×14		7×7		7×7
	3	192	256	512	1024		1024		1024

error only when an object exists and when a prediction is actually responsible for the ground truth.

$$\text{Loss} = \text{Loss}_{\text{bounding box}} + \text{Loss}_{\text{confidence}} + \text{Loss}_{\text{classification}} \tag{2.1}$$

where

$$\text{Loss}_{\text{bounding box}} = \lambda_{\text{coord}} \sum_{i=0}^{s^2} \sum_{j=0}^{B} 1_{ij}^{\text{obj}} \left[(x_i - \hat{x}_i)^2 + (y_i - \hat{y}_i)^2 \right]$$

$$+ \lambda_{\text{coord}} \sum_{i=0}^{s^2} \sum_{j=0}^{B} 1_{ij}^{\text{obj}} \left[\left(\sqrt{w_i} - \sqrt{\hat{w}_i} \right)^2 + \left(\sqrt{y_i} - \sqrt{\hat{y}_i} \right)^2 \right],$$

$$\text{Loss}_{\text{confidence}} = \sum_{i=0}^{s^2} \sum_{j=0}^{B} 1_{ij}^{\text{obj}} \left(C_i - \hat{C}_i \right)^2$$

$$+ \lambda_{\text{noobj}} \sum_{i=0}^{s^2} \sum_{j=0}^{B} 1_{ij}^{\text{obj}} \left(C_i - \hat{C}_i \right)^2,$$

$$\text{Loss}_{\text{classification}} = \sum_{i=0}^{s^2} 1_i^{\text{obj}} \sum_{c \in \text{classes}} \left(p_i(c) - \hat{p}_i(c) \right)^2.$$

2.3.2 YOLOv2 AND YOLO9000

YOLOv2 [53] is an improved model based on YOLO. The original YOLO backbone network is replaced by a simpler Darknet-19, and the fully connected layers at the end are removed. Redmon et al. also tested other different design changes and applied modifications only with accuracy increases. YOLO9000 [53], as the name suggests, can detect 9,000 object categories. Based on a slightly modified version of YOLOv2, this is achieved with a joint training strategy on classification and detection datasets. A WorldTree hierarchy [34] is used to merge the ground-truth classes from different datasets.

In this section, we will discuss some of the most effective modifications. Batch normalization can help to reduce the effect of internal covariate shifts [54], thus accelerating convergence during training. By adding batch normalization to all CNN layers, the accuracy is improved by 2.4%.

In YOLO, the classifier is trained on the resolution of 224×224. At the stage of detection, the resolution is increased to 448×448. In YOLOv2, during the last 10 epochs of classifier training, the image is changed to a full 448 resolution, so the detection training can focus on object detection rather than adapting to the new resolution. This gives a 4% mAP increase. While trying anchor boxes with YOLO, the issue of instability is exposed. YOLO predicts the box by generating offsets to each anchor. The most efficient training is when all objects are predicted by the closest anchor with a minimal number of offsets. However, without offset constrains, an original anchor could predict an object at any location. Therefore in YOLOv2 [53], a logistic activation constraint on offset is introduced to limit the predicted bounding box near the original anchor. This makes the network more stable and increases mAP by 4.8%.

2.3.3   YOLOv3

YOLOv3 [43] is another improved detector based on YOLOv2. The major improvement is on the convolutional network. Based on the idea of Feature Pyramid Networks [48], YOLOv3 predicts the boxes at 3 different scales, which helps to detect small objects. The idea of skip connection, as discussed in the following section, is also added into the design. As a result, the network becomes larger, with 53 layers compared to its original 19 layers. Thus, it is called Darknet-53 [43]. Another achievement is that Darknet-53 runs at the highest measured floating point operation speed, which is an indication that the network is better at utilizing GPU resources.

2.3.4   SINGLE-SHOT DETECTOR (SSD)

SSD [42] was introduced to make one-stage detector run in real time with comparative accuracy to the region proposal detectors. It is faster than YOLO [34] and generates competitive accuracy compared to the latest two-stage models like Faster R-CNN [33]. The architecture of SSD, which is shown in Figure 2.4, has three distinctive features. First, it uses multi-scale feature maps. From Figure 2.4, we can observe that the output layers decrease in size progressively after a truncated base network. The layers are chosen to be output and perform detection at different scales. It allows prediction to be made at different scales. The shallower layers with more details yield better results for smaller objects, while deeper layers with information about the background are suited for larger objects. Second, the network in SSD is fully convolutional, unlike YOLO [34], which employs fully connected layers at the end. The third is the default bounding boxes and aspect ratios. In SSD, the image

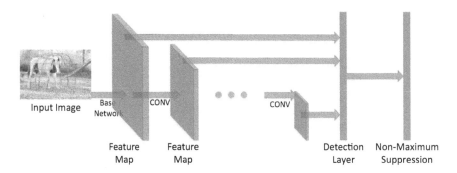

FIGURE 2.4   SSD architecture. The first CONV layer, which is called the base network, is truncated from a standard network. The detection layer computes confidence scores for each class and offsets to default boxes.

is divided into grid cells. Each cell in the feature maps associates a set of default bounding boxes and aspect ratios. SSD then computes category confidence scores and bounding-box offsets to those default bounding boxes for each set. During prediction, SSD performs detection for objects with different sizes on the feature maps with various scales (Figure 2.4).

One-stage models inherently have the advantage of speed. In recent years, models with more than 150 fps have been published, while a fast two-stage model [33] only achieves around 20 fps. To make the prediction more accurate, networks combine ideas from both one-stage and two-stage models to find the balance between speed and accuracy. For example, Faster R-CNN [33] resembles the one-stage model in sharing one major CNN network. One-stage models normally produce prediction from the entire feature map. So they are good at taking context information into consideration. However, this also means they are not very sensitive to smaller objects. Methods like deconvolutional single-shot detector (DSSD) [55] and RetinaNet [36] have been proposed to fix this problem from different viewpoints. DSSD learned from FPN [48] modifies its CNN to predict objects at different scales. RetinaNet develops a unique loss function to focus on hard objects. A detailed comparison of results from different models on the PASCAL VOC dataset is given in Table 2.2.

2.4   OTHER DETECTOR ARCHITECTURES

As we discussed at the beginning of this chapter, emerging detectors published after 2016 mainly focus on the performance improvement of detection. Some focus on optimizing the feature extraction ability of backbone networks [35, 56, 46, 48]. Some methods improve performance by modifying certain metrics, such as the loss function in RetinaNet [36] and a new IoU definition given in GIoU [57]. With new measuring metrics introduced in Lin et al. [26] and a comprehensive analysis performed after 2017, detection of objects with different scales has become a major focus [48, 58, 59, 37].

In this section, we are going to discuss two examples, ResNet [35] and RetinaNet [36]. ResNet [35] addresses some problems caused by increasing network layers, while RetinaNet [36] tries to optimize the training efficiency.

TABLE 2.2
Detection Results on PASCAL VOC 2007 Test Set

Detector	mAP (%)	fps
R-CNN [41]	58.5	0.02
Fast R-CNN [50]	70.0	0.5
Faster R-CNN (with VGG16) [33]	73.2	7
Faster R-CNN (with ZF) [33]	62.1	18
YOLO [34]	63.4	45
Fast YOLO [34]	52.7	155
YOLO (with VGG16) [34]	66.4	21
YOLOv2 288×288 [53]	69.0	91
YOLOv2 352×352 [53]	73.1	81
YOLOv2 416×416 [53]	76.8	67
YOLOv2 480×480 [53]	77.8	59
YOLOv2 544×544 [53]	78.6	40

2.4.1 DEEP RESIDUAL LEARNING (RESNET)

Degradation is the detection accuracy drop when the network depth increases. Theoretically, this should not happen. If we assume the optimal network for a problem domain consists of a certain number of layers, an even deeper network should be as good as the optimal one at least. The reason is that the added layers can be trained as identity mappings, which simply take the input and pass it to the next layer. The fact that degradation exists in practice implies some systems are more difficult to learn [35]. ResNets was published in December 2015. The effort evolves around assisting the network to find the desired mappings. The key block diagram for ResNets is shown in Figure 2.5.

The desired mapping without a skip connection is denoted as $\mathcal{H}(x) = \mathcal{F}(x) + x$. By introducing the skip connection, which is an identity connection, the layers are now forced to fit the new mapping, $\mathcal{F}(x) = \mathcal{H}(x) - x$. $\mathcal{F}(x)$ is defined as the residual mapping. In addition, the paper hypothesizes that optimizing the residual mapping is easier than the original mapping. For example, when a layer's original optimal

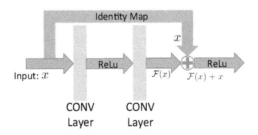

FIGURE 2.5 ResNet building block. Convert the underlying mapping from $\mathcal{H}(x)$ to $\mathcal{F}(x)$, where $\mathcal{H}(x) = \mathcal{F}(x) + x$.

solution is an identity mapping, the modified layer is going to fit zero by adding the skip connection. The hypothesis is proven in the test results, as discussed in the next session.

Another attractive property of ResNets is that the design does not introduce any extra parameters or computational complexity. This is desirable in two ways. First, it helps the network to achieve better performance without extra cost. Second, it facilitates fair comparison between plain networks and residual networks. During the test, we can run same networks with and without the skip connection. Therefore, the performance difference is caused just by the skip connection.

Based on a 34-layer plain network, the skip connection is added as indicated in Figure 2.5. When the input and output dimensions are the same, the skip connection can be added directly. And when the skip connection goes across different dimensions, the paper proposes two options: (i) identity mapping with extra zeros paddings; and (ii) using a projection shortcut, which is the projection in the plain network block. With either option, a stride of 2 will be performed.

The significance of this network design is to reduce the difficulties in finding the optimal mappings of each layer. Without this design, researchers have to design architectures with different depths, train on the dataset, and then compare the detection accuracy to narrow down the optimal depth. The process is very time-consuming, and different problem domains may have various optimal depths. The common training strategy, which is introduced in R-CNN [41], is supervised pre-training on a large auxiliary dataset, followed by domain-specific fine-tuning on a small dataset. On the domain-specific dataset, there is a high chance that the network architecture is not optimal. Plus, it is practically impossible to find the optimal depth for each specific problem. However, with ResNets, we can safely add more layers and expect better performance. We can rely on the networks and training to find the best performance model as long as the optimal model is shallower. Another advantage of ResNets is its efficiency, which achieves performance improvement without adding any computational complexity.

2.4.2 RETINANET

With all the advancements made in recent years, both two-stage methods and one-stage methods have their own pros and cons. One-stage methods are usually faster in detection, but trailing in detection accuracy. According to Lin et al. [36], the lower accuracy is mainly caused by the extreme foreground-background class imbalance during training. For the one-stage detector, it must sample much more candidate locations in an image. In practice, the number of candidate locations normally goes up to around 100k covering different positions, scales, and aspect ratios. Candidate locations are dominated by background examples, which makes the training very inefficient. In two-stage detectors, however, most background regions are filtered out by region proposals.

To address this issue, Lin et al. proposed RetinaNet [36]. The problem of a normal one-stage network is the overwhelming number of easy background candidates. During training, it contributes little to the improvement of accuracy. Candidates can be easily classified as background, which means little information will be added to the model. Since the number of them is dominating, it will overpower other hard foreground candidates, resulting in degenerating models. A regular cross-entropy (CE) loss for binary classification is shown in Equation 2.2.

$$CE(p, y) = -\log(p_t), \text{ where } p_t = \begin{cases} p, & \text{if } y = 1 \\ 1 - p, & \text{otherwise} \end{cases} \qquad (2.2)$$

To reduce the imbalance between the easy background and hard foreground samples, Lin et al. [36] introduced the following focal loss.

$$FL(p_t) = -(1 - p_t)^\gamma \log(p_t) \qquad (2.3)$$

When a sample is misclassified with a high score, p_t will have small value. Therefore, the loss is not affected that much. But when a sample is correctly classified with a high confidence, p_t is close to 1, which will significantly down-weigh its effect. In addition, a weighting factor α is employed to balance the importance of positive/negative examples.

$$FL(p_t) = -\alpha_t(1 - p_t)^\gamma \log(p_t) \qquad (2.4)$$

Both models in this section address the problems encountered during experiments. ResNet [35] has become one of the most widely used backbone networks. Its model is provided in most of the popular deep learning frameworks, such as Caffe, PyTorch, Keras, MXNet, etc. RetinaNet [36] achieves remarkable performance improvement with minimal addition in computational complexity by designing a new loss function. More importantly, it shows a new direction in optimizing the detector. Modifying evaluation metrics could result in significant accuracy increase. Rezatofighi et al. [57] in 2019 proposed Generalized Intersection-over-Union (GIoU) to replace the current IoU and used GIoU as the loss function. Before the study of Rezatofighi et al. [57], all CNN detectors calculate IoU during the test to evaluate the result, while employing other metrics as loss function during training to optimize. Other recent detectors proposed after 2017 mainly focus on detecting objects at various scales, especially at small scale [52, 58, 60]. Among them, the trend in design is refinement based on previous successful detectors [59, 37].

2.5   PERFORMANCE COMPARISON ON DIFFERENT DATASETS

In this section, we will present some comparative evaluation results for different models on three benchmark datasets: PASCAL VOC [25], MS COCO [26], and VisDrone-DET2018 [61].

2.5.1   PASCAL VOC

The PASCAL VOC [25] dataset is one of the pioneering works in generic object detection, which is designed to provide a standardized testbed for object detection, image classification, object segmentation, person layout, and action classification [62]. The latest version is PASCAL VOC 2012. It has 20 classes and 11,530 images in the trainval dataset with 27,450 RoI annotated objects. The initial 2007 version was the first dataset with 20 classes of objects.

To compare the performance of different networks on this dataset, we choose the PASCAL VOC 2007 testing results from models trained on PASCAL VOC 2007 + 2012. The comparison results of different models are shown in Table 2.2.

Link between one-stage and two-stage models. From Table 2.2, we can tell that the speed advantage of the one-stage detector is obvious. The first version of YOLO is almost three times faster than the fastest R-CNN detector. Looking back at the evolution of these detectors, we can see that every modification from R-CNN leading to one-stage detectors actually fits in a trend. R-CNN has two stages, with each stage having its own separate computations. The improvements are made by eliminating duplicate calculations and combining shared features. The final version of R-CNN could be viewed as one main CNN network doing most of the work and some small regressors and CNN layers generating predictions. This architecture has become very similar to the one-stage model.

Speed and accuracy. For the same model, if we want to achieve higher accuracy, speed usually needs to be compromised. So it is important to find the balance for specific applications. The test results show two ways to modify. A more complex classifier or a higher-resolution image usually yields higher accuracy. At the same time, more time will be needed to detect the objects.

2.5.2 MS COCO

MS COCO [26] is a large-scale object detection, segmentation, and captioning dataset with 330,000 images. It aims at addressing three core problems at scene understanding: detection of non-iconic views (or non-canonical perspectives) of objects, contextual reasoning between objects, and the precise 2D localization of objects [26]. COCO defines 12 metrics to evaluate the performance of a detector, which gives a more detailed and insightful look. Since MS COCO is a relatively new dataset, not all detectors have official results. We only compare the results of YOLOv3 and RetinaNet in Table 2.3. Although the two models were tested under different GPUs, M40 and Titan X, their performances are almost identical [43].

Performance difference with higher IoU threshold. From Table 2.3, we can see that YOLOv3 performs better than RetinaNet when the IoU threshold is 50%, while at 75% IoU, RetinaNet has better performance, which indicates that RetinaNet has higher localization accuracy. This aligns with the effort by Lin et al. [36]. The focal loss is designed to put more weight on learning hard examples.

2.5.3 VisDrone-DET2018

VisDrone-DET2018 dataset [61] is a special dataset that consists of images from different drone platforms. It has 8,599 images in total, including 6,471 for training, 548 for validation, and 1,580 for testing. The images feature diverse real-world scenes, collected in various scenarios (across 14 different cities in different

TABLE 2.3

Detection Results on MS COCO*

Detector	Backbone	Scale (pixels)	AP	AP$_{50}$	AP$_{75}$	fps
RetinaNet [36]	ResNet-101-FPN	600	36.0	55.2	38.7	8.2
YOLOv3 [43]	Darknet-53	608	33.0	57.9	34.4	19.6

* AP$_{50}$ and AP$_{75}$ Are Average Precision with IOU Thresholds of 0.5 and 0.75, Respectively. AP (Average Precision) Is the Average over 10 Thresholds ([0.5:0.05:0.95]). RetinaNet Is Measured on an Nvidia M40 GPU [36], while YOLOv3 Is Tested on an Nvidia Titan X [43].

regions), and under various environmental and lighting conditions [62]. This dataset is challenging since most of the objects are small and densely populated as shown in Figure 2.6.

To check the performance of different models on this dataset, we implemented the latest version of each detector and show the result in Tables 2.4 and 2.5. The results are calculated on the VisDrone validation set since the test set is not publicly available.

From Table 2.4, it is evident that Faster R-CNN [33] performs significantly better than YOLOv3 [43]. At IoU=0.5, Faster R-CNN has 10% higher accuracy. YOLO detectors inherently would struggle in datasets like this. The fully connected layers at the end take the entire feature map as input. It enables YOLO to have enough contextual information. However, it lacks local details. In addition, each grid cell can predict only a certain number of objects, which is determined before training starts. So they have a hard time in the small and densely populated VisDrone-DET2018

FIGURE 2.6   Six sample images from the VisDrone-DET2018 dataset. They are pictures taken from drones. The rectangular boxes on each image are ground truth. It has a large number of small objects at distance.

TABLE 2.4
Performance Metrics on VisDrone-DET2018*

Detector	Iterations	AP	Score
*YOLOv3 832×832	40 *k*	AP @ [IoU=0.50:0.05:0.95	maxDets=500]	18.74
		AP @ [IoU=0.50	maxDets=500]	29.77
		AP @ [IoU=0.75	maxDets=500]	19.53
		AR @ [IoU=0.50:0.05:0.95	maxDets=1]	0.91
		AR @ [IoU=0.50:0.05:0.95	maxDets=10]	5.63
		AR @ [IoU=0.50:0.05:0.95	maxDets=100]	27.16
		AR @ [IoU=0.50:0.05:0.95	maxDets=500]	27.42
*Faster R-CNN	80 *k*	AP @ [IoU=0.50:0.05:0.95	maxDets=500]	24.33
		AP @ [IoU=0.50	maxDets=500]	39.74
		AP @ [IoU=0.75	maxDets=500]	24.78
		AR @ [IoU=0.50:0.05:0.95	maxDets=1]	0.73
		AR @ [IoU=0.50:0.05:0.95	maxDets=10]	6.11
		AR @ [IoU=0.50:0.05:0.95	maxDets=100]	34.26
		AR @ [IoU=0.50:0.05:0.95	maxDets=500]	44.62

* AR @ [maxDets=1] and [maxDets=10] mean the maximum recall given 1 detection per image and 10 detections per image, respectively.

TABLE 2.5
RetinaNet on VisDrone-DET2018 without Tuning

Pedestrian	Person	Bicycle	Car	Van
12.6	3.6	4.6	50.2	21.0
Truck	Tricycle	Awning-tricycle	Bus	Motor
17.8	10.7	4.9	32.1	11.7

dataset. It can be noted that the gap is reduced to 5% at IoU=0.75, which means YOLOv3 is catching up in terms of localization precision. We suggest this is the result of adding prediction at 3 different scales in YOLOv3 and detecting on high-resolution (832×832) images.

Table 2.5 shows the mAPs for each class from RetinaNet [36]. It shows that recent detectors still need to be improved for small and morphologically similar objects. The VisDrone dataset is a unique dataset, which has many potential real-life applications. Actually, most practical applications have to face different challenging situations, like bad exposure, lack of lighting, and a saturated number of objects. More investigation has to be done to develop more effective models to handle these complex real-life applications.

The evolution of object detection is partially linked to the availability of labeled large-scale datasets. The existence of various datasets helps train the neural network to extract more effective features. The different characteristics of datasets motivate researchers to focus on different problems. In addition, new metrics proposed by the dataset help us to better evaluate the detector performance. For example, MS COCO style mAP [26] helps us to understand the performance related to localization accuracy.

2.6   CONCLUSION

In this chapter, we have made a brief review of CNN-based object detection by presenting some of the most typical detectors and network architectures. The detector starts with a two-stage multi-step architecture and evolves to a simpler one-stage model. In the latest models, even two-stage ones employ an architecture that shares a single CNN feature map so as to reduce the computational load [52]. Some models gain accuracy increase by fusing different ideas into one detector [47, 59]. In addition to the direct effort on detectors, training strategy is also an important factor to produce high-quality results [41, 63]. With recent developments, most detectors have decent performance in both accuracy and efficiency. In practical applications, we need to make a trade-off between the accuracy and the speed by choosing a proper set of parameters and network structures. Although great progress has been made in object detection in past years, some challenges still need to be addressed, like occlusion [64] and truncation. In addition, better-designed datasets, like VisDrone, need to be developed for specific practical applications.

BIBLIOGRAPHY

1. G. E. Hinton, S. Osindero, and Y.-W. Teh, "A fast learning algorithm for deep belief nets," *Neural Computation*, vol. 18, no. 7, pp. 1527–1554, 2006.
2. Y. Bengio, A. Courville, and P. Vincent, "Representation learning: A review and new perspectives," *IEEE Transactions on Pattern Analysis and Machine Intelligence*, vol. 35, no. 8, pp. 1798–1828, 2013.
3. I. Arel, D. C. Rose, T. P. Karnowski, et al., "Deep machine learning-a new frontier in artificial intelligence research," *IEEE Computational Intelligence Magazine*, vol. 5, no. 4, pp. 13–18, 2010.
4. Y. LeCun, Y. Bengio, and G. Hinton, "Deep learning," *Nature*, vol. 521, no. 7553, p. 436, 2015.
5. G. Hinton, L. Deng, D. Yu, G. E. Dahl, A.-r. Mohamed, N. Jaitly, A. Senior, V. Vanhoucke, P. Nguyen, T. N. Sainath, et al., "Deep neural networks for acoustic modeling in speech recognition: The shared views of four research groups," *IEEE Signal Processing Magazine*, vol. 29, no. 6, pp. 82–97, 2012.
6. T. N. Sainath, A.-r. Mohamed, B. Kingsbury, and B. Ramabhadran, "Deep convolutional neural networks for lvcsr," in *IEEE International Conference on Acoustics, Speech and Signal Processing (ICASSP)*, pp. 8614–8618, IEEE, 2013.
7. G. E. Dahl, D. Yu, L. Deng, and A. Acero, "Context-dependent pre-trained deep neural networks for large-vocabulary speech recognition," *IEEE Transactions on Audio, Speech, and Language Processing*, vol. 20, no. 1, pp. 30–42, 2012.
8. R. Collobert, J. Weston, L. Bottou, M. Karlen, K. Kavukcuoglu, and P. Kuksa, "Natural language processing (almost) from scratch," *Journal of Machine Learning Research*, vol. 12, pp. 2493–2537, 2011.
9. I. Sutskever, O. Vinyals, and Q. V. Le, "Sequence to sequence learning with neural networks," in *Advances in Neural Information Processing Systems*, pp. 3104–3112, 2014.

10. R. Socher, C. C. Lin, C. Manning, and A. Y. Ng, "Parsing natural scenes and natural language with recursive neural networks," in *Proceedings of the 28th International Conference on Machine Learning (ICML-11)*, pp. 129–136, 2011.

11. A. Krizhevsky, I. Sutskever, and G. E. Hinton, "Imagenet classification with deep convolutional neural networks," in *Advances in Neural Information Processing Systems*, pp. 1097–1105, 2012.

12. C. Szegedy, W. Liu, Y. Jia, P. Sermanet, S. Reed, D. Anguelov, D. Erhan, V. Vanhoucke, and A. Rabinovich, "Going deeper with convolutions," in *Proceedings of the IEEE Conference on Computer Vision and Pattern Recognition*, pp. 1–9, 2015.

13. L. He, G. Wang, and Z. Hu, "Learning depth from single images with deep neural network embedding focal length," *IEEE Transactions on Image Processing*, vol. 27, no. 9, pp. 4676–4689, 2018.

14. J. Gao, J. Yang, G. Wang, and M. Li, "A novel feature extraction method for scene recognition based on centered convolutional restricted boltzmann machines," *Neurocomputing*, vol. 214, pp. 708–717, 2016.

15. M. M. Najafabadi, F. Villanustre, T. M. Khoshgoftaar, N. Seliya, R. Wald, and E. Muharemagic, "Deep learning applications and challenges in big data analytics," *Journal of Big Data*, vol. 2, no. 1, p. 1, 2015.

16. G. Litjens, T. Kooi, B. E. Bejnordi, A. A. A. Setio, F. Ciompi, M. Ghafoorian, J. A. Van Der Laak, B. Van Ginneken, and C. I. Sánchez, "A survey on deep learning in medical image analysis," *Medical Image Analysis*, vol. 42, pp. 60–88, 2017.

17. X. Mo, K. Tao, Q. Wang, and G. Wang, "An efficient approach for polyps detection in endoscopic videos based on faster R-CNN," in *2018 24th International Conference on Pattern Recognition (ICPR)*, pp. 3929–3934, IEEE, 2018.

18. A. Elgammal, B. Liu, M. Elhoseiny, and M. Mazzone, "CAN: Creative adversarial networks, generating "art" by learning about styles and deviating from style norms," arXiv preprint arXiv:1706.07068, 2017.

19. W. Xu, S. Keshmiri, and G. Wang, "Adversarially approximated autoencoder for image generation and manipulation," *IEEE Transactions on Multimedia*, doi:10.1109/TMM.2019.2898777, 2019.

20. W. Xu, S. Keshmiri, and G. Wang, "Toward learning a unified many-to-many mapping for diverse image translation," *Pattern Recognition*, doi:10.1016/j.patcog.2019.05.017, 2019.

21. W. Ma, Y. Wu, Z. Wang, and G. Wang, "MDCN: Multi-scale, deep inception convolutional neural networks for efficient object detection," in *2018 24th International Conference on Pattern Recognition (ICPR)*, pp. 2510–2515, IEEE, 2018.

22. Z. Zhang, Y. Wu, and G. Wang, "Bpgrad: Towards global optimality in deep learning via branch and pruning," in *Proceedings of the IEEE Conference on Computer Vision and Pattern Recognition*, pp. 3301–3309, 2018.

23. L. Liu, W. Ouyang, X. Wang, P. Fieguth, J. Chen, X. Liu, and M. Pietikäinen, "Deep learning for generic object detection: A survey," arXiv preprint arXiv:1809.02165, 2018.

24. F. Cen and G. Wang, "Dictionary representation of deep features for occlusion-robust face recognition," *IEEE Access*, vol. 7, pp. 26595–26605, 2019.

25. M. Everingham, L. Van Gool, C. K. Williams, J. Winn, and A. Zisserman, "The pascal visual object classes (voc) challenge," *International Journal of Computer Vision*, vol. 88, no. 2, pp. 303–338, 2010.

26. T.-Y. Lin, M. Maire, S. Belongie, J. Hays, P. Perona, D. Ramanan, P. Dollár, and C. L. Zitnick, "Microsoft coco: Common objects in context," in *European Conference on Computer Vision*, pp. 740–755, Springer, 2014.

27. S. P. Bharati, S. Nandi, Y. Wu, Y. Sui, and G. Wang, "Fast and robust object tracking with adaptive detection," in *2016 IEEE 28th International Conference on Tools with Artificial Intelligence (ICTAI)*, pp. 706–713, IEEE, 2016.

28. Y. Wu, Y. Sui, and G. Wang, "Vision-based real-time aerial object localization and tracking for UAV sensing system," *IEEE Access*, vol. 5, pp. 23969–23978, 2017.

29. S. P. Bharati, Y. Wu, Y. Sui, C. Padgett, and G. Wang, "Real-time obstacle detection and tracking for sense-and-avoid mechanism in UAVs," *IEEE Transactions on Intelligent Vehicles*, vol. 3, no. 2, pp. 185–197, 2018.

30. Y. Wei, X. Pan, H. Qin, W. Ouyang, and J. Yan, "Quantization mimic: Towards very tiny CNN for object detection," in *Proceedings of the European Conference on Computer Vision (ECCV)*, pp. 267–283, 2018.

31. K. Kang, H. Li, J. Yan, X. Zeng, B. Yang, T. Xiao, C. Zhang, Z. Wang, R. Wang, X. Wang, et al., "T-CNN: Tubelets with convolutional neural networks for object detection from videos," *IEEE Transactions on Circuits and Systems for Video Technology*, vol. 28, no. 10, pp. 2896–2907, 2018.

32. W. Chu and D. Cai, "Deep feature based contextual model for object detection," *Neurocomputing*, vol. 275, pp. 1035–1042, 2018.

33. S. Ren, K. He, R. Girshick, and J. Sun, "Faster R-CNN: Towards real-time object detection with region proposal networks," in *Advances in Neural Information Processing Systems*, pp. 91–99, 2015.

34. J. Redmon, S. Divvala, R. Girshick, and A. Farhadi, "You only look once: Unified, real-time object detection," in *Proceedings of the IEEE Conference on Computer Vision and Pattern Recognition*, pp. 779–788, 2016.

35. K. He, X. Zhang, S. Ren, and J. Sun, "Deep residual learning for image recognition," in *Proceedings of the IEEE Conference on Computer Vision and Pattern Recognition*, pp. 770–778, 2016.

36. T.-Y. Lin, P. Goyal, R. Girshick, K. He, and P. Dollár, "Focal loss for dense object detection," in *IEEE Transactions on Pattern Analysis and Machine Intelligence*, p. 1, 2018.

37. Q. Zhao, T. Sheng, Y. Wang, Z. Tang, Y. Chen, L. Cai, and H. Ling, "M2det: A single-shot object detector based on multi-level feature pyramid network," *CoRR*, vol. abs/1811.04533, 2019.

38. P. Felzenszwalb, D. McAllester, and D. Ramanan, "A discriminatively trained, multi-scale, deformable part model," in *IEEE Conference on Computer Vision and Pattern Recognition*, pp. 1–8, IEEE, 2008.

39. P. Sermanet, D. Eigen, X. Zhang, M. Mathieu, R. Fergus, and Y. LeCun, "Overfeat: Integrated recognition, localization and detection using convolutional networks," arXiv preprint arXiv:1312.6229, 2013.

40. J. R. Uijlings, K. E. Van De Sande, T. Gevers, and A. W. Smeulders, "Selective search for object recognition," *International Journal of Computer Vision*, vol. 104, no. 2, pp. 154–171, 2013.

41. R. Girshick, J. Donahue, T. Darrell, and J. Malik, "Rich feature hierarchies for accurate object detection and semantic segmentation," in *Proceedings of the IEEE Conference on Computer Vision and Pattern Recognition*, pp. 580–587, 2014.

42. W. Liu, D. Anguelov, D. Erhan, C. Szegedy, S. Reed, C.-Y. Fu, and A. C. Berg, "Ssd: Single shot multibox detector," in *European Conference on Computer Vision*, pp. 21–37, Springer, 2016.

43. J. Redmon and A. Farhadi, "Yolov3: An incremental improvement," arXiv preprint arXiv:1804.02767, 2018.

44. K. Simonyan and A. Zisserman, "Very deep convolutional networks for large-scale image recognition," arXiv preprint arXiv:1409.1556, 2014.

45. X. Glorot and Y. Bengio, "Understanding the difficulty of training deep feedforward neural networks," in *Proceedings of the Thirteenth International Conference on Artificial Intelligence and Statistics*, pp. 249–256, 2010.

46. G. Huang, Z. Liu, L. Van Der Maaten, and K. Q. Weinberger, "Densely connected convolutional networks," in *Proceedings of the IEEE Conference on Computer Vision and Pattern Recognition*, pp. 4700–4708, 2017.

47. Y. Chen, J. Li, H. Xiao, X. Jin, S. Yan, and J. Feng, "Dual path networks," in *Advances in Neural Information Processing Systems*, pp. 4467–4475, 2017.
48. T.-Y. Lin, P. Dollár, R. Girshick, K. He, B. Hariharan, and S. Belongie, "Feature pyramid networks for object detection," in *Proceedings of the IEEE Conference on Computer Vision and Pattern Recognition*, pp. 2117–2125, 2017.
49. D. G. Lowe, "Object recognition from local scale-invariant features," in *The Proceedings of the Seventh IEEE International Conference on Computer Vision*, vol. 99, no. 2, pp. 1150–1157, IEEE, 1999.
50. R. Girshick, "Fast R-CNN," in *Proceedings of the IEEE International Conference on Computer Vision*, pp. 1440–1448, 2015.
51. J. Dai, Y. Li, K. He, and J. Sun, "R-FCN: Object detection via region-based fully convolutional networks," in *Advances in Neural Information Processing Systems*, pp. 379–387, 2016.
52. Z. Cai and N. Vasconcelos, "Cascade R-CNN: Delving into high quality object detection," in *Proceedings of the IEEE Conference on Computer Vision and Pattern Recognition*, pp. 6154–6162, 2018.
53. J. Redmon and A. Farhadi, "Yolo9000: Better, faster, stronger," in *Proceedings of the IEEE Conference on Computer Vision and Pattern Recognition*, pp. 7263–7271, 2017.
54. S. Ioffe and C. Szegedy, "Batch normalization: Accelerating deep network training by reducing internal covariate shift," in *Proceedings of the 32nd International Conference on Machine Learning*, vol. 37, pp. 448–456, 2015.
55. C.-Y. Fu, W. Liu, A. Ranga, A. Tyagi, and A. C. Berg, "DSSD: Deconvolutional single shot detector," arXiv preprint arXiv:1701.06659, 2017.
56. C. Szegedy, V. Vanhoucke, S. Ioffe, J. Shlens, and Z. Wojna, "Rethinking the inception architecture for computer vision," in *Proceedings of the IEEE Conference on Computer Vision and Pattern Recognition*, pp. 2818–2826, 2016.
57. H. Rezatofighi, N. Tsoi, J. Gwak, A. Sadeghian, I. Reid, and S. Savarese, "Generalized intersection over union: A metric and a loss for bounding box regression," arXiv preprint arXiv:1902.09630, 2019.
58. P. Zhou, B. Ni, C. Geng, J. Hu, and Y. Xu, "Scale-transferrable object detection," in *Proceedings of the IEEE Conference on Computer Vision and Pattern Recognition*, pp. 528–537, 2018.
59. S. Zhang, L. Wen, X. Bian, Z. Lei, and S. Z. Li, "Single-shot refinement neural network for object detection," in *Proceedings of the IEEE Conference on Computer Vision and Pattern Recognition*, pp. 4203–4212, 2018.
60. B. Singh and L. S. Davis, "An analysis of scale invariance in object detection snip," in *Proceedings of the IEEE Conference on Computer Vision and Pattern Recognition*, pp. 3578–3587, 2018.
61. P. Zhu, L. Wen, X. Bian, L. Haibin, and Q. Hu, "Vision meets drones: A challenge," arXiv preprint arXiv:1804.07437, 2018.
62. P. Zhu, L. Wen, D. Du, X. Bian, H. Ling, Q. Hu, Q. Nie, H. Cheng, C. Liu, X. Liu, et al., "VisDrone-DET2018: The vision meets drone object detection in image challenge results," In *Proceedings of the European Conference on Computer Vision (ECCV)*, pp. 437–468, 2018.
63. J. S. Bergstra, R. Bardenet, Y. Bengio, and B. Kégl, "Algorithms for hyper-parameter optimization," in *Advances in Neural Information Processing Systems*, pp. 2546–2554, 2011.
64. B. Wu and R. Nevatia, "Detection of multiple, partially occluded humans in a single image by bayesian combination of edgelet part detectors," in *Tenth IEEE International Conference on Computer Vision*, vol. 1, pp. 90–97, 2005.
65. Bradley, David M. "Learning in modular systems," No. CMU-RI-TR-09-26. CARNEGIE-MELLON UNIV PITTSBURGH PA ROBOTICS INST, 2010.

3 Efficient Convolutional Neural Networks for Fire Detection in Surveillance Applications

Khan Muhammad, Salman Khan,
and Sung Wook Baik

CONTENTS

3.1 INTRODUCTION

Recently, a variety of sensors have been introduced for different applications such as setting off a fire alarm [1], detecting vehicle obstacles [2], visualizing the interior of the human body for diagnosis [3–5], monitoring animals and ships, and surveillance. Of these applications, surveillance has primarily attracted the attention of researchers due to the enhanced embedded processing capabilities of cameras. Using smart surveillance systems, various abnormal events such as road accidents, fires, medical emergencies, etc. can be detected at early stages, and the appropriate authority can be autonomously informed [6, 7]. A fire is an abnormal event that can cause significant damage to lives and property within a very short time. Such disasters, the causes of which include human error and system failures, result in severe loss of human life and other damage [8]. In June 2013, fire disasters killed 19 firefighters and ruined 100 houses in Arizona. Similarly, another forest fire in August 2013 in California ruined an area of land the size of 1,042 km^2, causing a loss of US$127.35 million. According to an annual disaster report, fire disasters alone affected 494,000 people and resulted in a loss of US$3.1 billion in 2015. According to [9], at least 37 people were killed and about 130 injured in a fire at Sejong Hospital, Miryang, South Korea, in December 2017, and a fire in Greece killed around 80 people, injured hundreds, and destroyed about 1,000 homes in July 2018. It is important to detect fires at early stages utilizing smart surveillance cameras, to avoid such disasters.

Two broad categories of approach can be identified for fire detection: traditional fire alarms and vision sensor-assisted fire detection. Traditional fire alarm systems are based on sensors that require close proximity for activation, such as infrared and optical sensors. To overcome the limitations of existing approaches, numerous vision sensor-based methods have been explored by researchers in this field; these have many other advantages, e.g., less human interference, faster response, affordable cost, and larger surveillance coverage. Despite these advantages, there are still some issues with these systems, e.g., the complexity of the scenes under observation, irregular lighting, and low-quality frames; researchers have made several efforts to address these aspects, taking into consideration both color and motion features. For instance, [8, 10–17] used color features for fire detection by exploring different color models including HIS [8], YUV [13], YCbCr [14], RGB [15], and YUC [10]. The major issue with these methods is their high rate of false alarms. Several attempts have been made to solve this issue by combing the color information with motion and analyses of fire's shape and other characteristics. However, maintaining a well-balanced trade-off between accuracy, false alarms, and computational efficiency still have remained a challenge. In addition, the above-mentioned methods fail to detect small fires or fires at larger distances.

In this chapter, two major issues related to fire detection have been investigated: image representation with feature vectors and feature map selection for fire localization and contextual information extraction. The proposed solutions for image representation used highly discriminative deep convolutional features to construct a robust representation. The extensive fine-tuning process enabled the feature extraction pipeline to focus on fire regions in images, thereby effectively capturing the essential features needed for fire and ignoring irrelevant and trivial or often misleading features. Further details about feature extraction, feature map selection, and contextual information extraction are provided in the respective sections of the chapter.

Section 1.2 presents an overview of image representation techniques for fire detection based on traditional hand-crafted features and recent deep learning approaches. Section 1.3 introduces the main contributions of this chapter to fire detection literature for both indoor and outdoor surveillance. Section 1.4 contains experimental results on benchmark datasets and evaluations from different perspectives of the proposed methods included in this chapter. Finally, Section 1.5 concludes this chapter by focusing on the key findings, strengths, and weaknesses of the proposed methods along with some future research directions.

3.2 RELATED WORK

The available literature dictates that fire can be detected either by traditional sensors or by using visible light camera. A broader classification of different fire detection methods is shown in Figure 3.1. Details about these methods are described in subsequent sections.

3.2.1 TRADITIONAL SENSOR-BASED FIRE DETECTION

The recent advancements in technology have resulted in a variety of sensors for different applications such as wireless capsule sensors for visualization of the interior of a human body [3], vehicle sensors for obstacle detection [18], and fire alarm sensors [1]. The current fire alarm sensors such as infrared, ion, and optical sensors need close proximity of the heat, fire, radiation, or smoke for activation; hence, such sensors are not considered as good candidates for environments of a critical nature. Among fire deaths, 78% occurred only due to home fires [19]. One of the main reasons is the delayed escape for disabled people, as traditional fire alarm systems need strong fires or close proximity, failing to generate an alarm on time for such people.

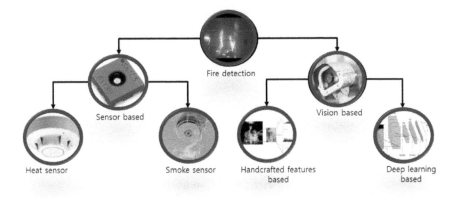

FIGURE 3.1   General classification of fire detection methods.

This necessitates the existence of effective fire alarm systems for surveillance. The majority of the research is conducted for fire detection using cameras and visual sensors, whose details are given in the next section.

3.2.2 Vision-Based Fire Detection

The conventional fire alerting systems are based on optical sensors, needing close proximity to the fire, thus failing to facilitate in providing additional information, e.g., burning degree, fire location, and size. In addition to this, such systems involve much human intervention such as visiting the fire location for confirming the fire in case of any fire alarm. To cope with these limitations, many visual sensor-based fire detection systems have been presented. Visual sensor-based fire detection systems are motivated by several encouraging advantages, including the following: 1) low cost due to existing setup of installed cameras for surveillance, 2) monitoring of larger regions, 3) comparatively fast response time due to elimination of waiting time for heat diffusion, 4) fire confirmation without visiting the fire location, 5) flexibility for detection of smoke and flames by adjustment of certain parameters, and 6) availability of fire details such as size, location, and burning degree. Due to these characteristics, such systems have attracted the attention of many researchers and as a result, many fire detection methods have been investigated based on numerous visual features, achieving reasonable performance with specific constraints.

3.2.2.1 Color- and Motion-Based Fire Detection Methods

Color is one of the important clues for determining fire or flame in a video frame. Researchers have presented several methods for detection of fire by exploiting color features in numerous color models. Chen et al. [8] investigated the dynamic behavior and irregularity of flames in both RGB and HSI color spaces for fire detection. Since their method considers the frame difference during prediction, it fails to differentiate real fire from fire-like moving outliers and objects. Another method is presented by Celik et al. [14], who used YCbCr with specific rules for separating the chrominance component from luminance. The method has potential to detect flames with good accuracy but only at short distances and larger fire sizes. Summarizing the color-based methods, it can be noted that such methods are sensitive to brightness and shadows. As a result, the number of false warnings produced by these methods is high. Because of this, such methods may interpret red vehicles or people wearing red clothes as fire due to the predominance of red. Later on, possible solutions were introduced based on the fact that fire changes its shape continuously, which can differentiate it from moving rigid objects. Kolesov et al. [18] used optimal mass transport optical flow as a low-dimensional descriptor along with RGB color channels and neural networks for differentiating between smoke and a similarly colored white wall, as well as fire from a similarly colored background.

3.2.2.2 Deep Learning-Based Fire Detection Methods

Deep learning has shown encouraging performance for solving different problems in the areas of computer vision [20–24], image classification [25–28], action recognition [29, 30], object tracking and detection, medical image analysis [31], and image indexing and retrieval [32, 33]. Recently, some preliminary results for fire detection using CNNs have also been reported. It is therefore important to highlight and analyze their performance in comparison with our proposed works. Frizzi et al. [34] presented a CNN-based method for fire and smoke detection. This work is based on a limited number of images, having no comparison with existing methods that could prove its performance. Sharma et al. [35] explored VGG16 and Resnet50 for fire detection. The dataset is very small (651 images only) and the reported testing accuracy is less than 93%. This work is compared with that of Frizzi et al. [34] with testing accuracy of 50%. Another recent work has been presented in Zhong et al. [36], where a flame detection algorithm based on CNN processed the video data generated by an ordinary camera monitoring a scene. Firstly, a candidate target area extraction algorithm was used for dealing with the suspected flame area and for improving the efficiency of recognition. Secondly, the extracted feature maps of candidate areas were classified by the designed deep neural network model based on CNN. Finally, the corresponding alarm signal was obtained by the classification results. Experiments conducted on a homemade database show that this method can identify fire and achieve reasonable accuracy but with a higher false alarm rate. The method can process 30 fps on a system with the following specifications: standard desktop PC equipped with OctaCore, CPU 3.6 GHz, and 8 GB RAM.

It can be observed that some of the methods are too naïve: their execution time is fast, but they compromise on accuracy, producing a large number of false alarms. Conversely, some methods have achieved good fire detection accuracies but their execution time is too high; hence, they cannot be applied in real-world environments, especially in critical areas where minor delay can lead to huge disasters. Therefore, for more accurate and early detection of fire, we need a robust mechanism that can detect fire during varying conditions and can send the important keyframes and alerts immediately to disaster management systems.

3.3 CNN-BASED FIRE DETECTION METHODS

The maturity of sensors has resulted in several useful applications for e-health, smarter surveillance, law enforcement, and disaster management. Among these applications, disaster management is of a critical nature and has attracted much attention from both academia and industry. Disaster management is mainly based on smoke/fire detection systems [37–39], for which researchers have presented both traditional and learned representation-based methods as described in Section 1.2. In order to avoid fire disasters, it is important to detect fires at early stages utilizing smart surveillance cameras and analyze the fire scene for collecting further necessary information. For this purpose, three methods [7, 40, 41] are described in this section.

3.3.1 EFD: EARLY FIRE DETECTION FOR EFFICIENT DISASTER MANAGEMENT

Early fire detection in the context of disaster management systems during surveillance of public areas, forests, and nuclear power plants can result in the prevention of ecological, economic, and social damage. However, early detection is a challenging problem due to varying lighting conditions, shadows, and movement of fire-colored

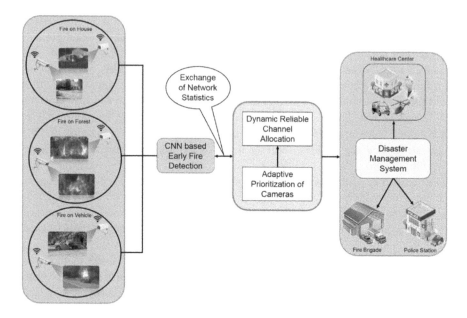

FIGURE 3.2 Early fire detection using CNN with reliable communication for effective disaster management [7].

objects. Thus, there is a need for an algorithm that can achieve better accuracy in the aforementioned scenarios while minimizing the number of false alarms. To achieve this goal, we explored deep CNNs and devised a fine-tuned architecture for early fire detection during surveillance for effective disaster management. Our system is overviewed in Figure 3.2.

3.3.1.1 EFD Architecture

The successful use of CNNs for various tasks motivated us to investigate them for fire detection and fire scene analysis. Thus, a model with similar architecture to the AlexNet [42] with modifications according to the problem of fire detection is used as shown in Figure 3.3. The output of the last fully connected layer is fed into the Softmax classifier, which computes probabilities for the two classes.

3.3.1.2 EFD Training and Fine-Tuning

As previously discussed, a typical CNN consists of three different neural layers, i.e., convolutional, pooling, and fully connected layers. All of these layers play different roles in the overall modeling process. The loss cost between the predicted output computed bypassing input image and labels via network layers is calculated; this is designed to simplify the final output. To select the best model, the data are trained over various models and the performance of the transfer learning approach was evaluated; a 4–5% improvement was predicted. In the context of fire detection, a fine-tuned AlexNet model was used on our fire dataset by modifying the last fully connected layer and keeping a slower learning rate. Model fine-tuning was performed for 10 epochs, achieving an improvement of about 5% in classification accuracy, compared to the freshly trained model. After training and the fine-tuning

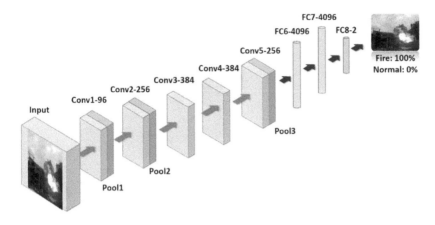

FIGURE 3.3 Architecture of the proposed CNN [7].

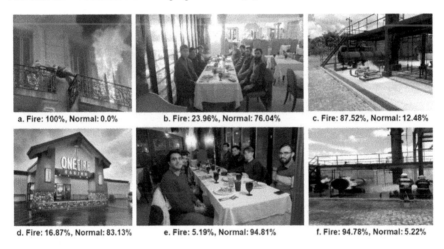

a. Fire: 100%, Normal: 0.0%   b. Fire: 23.96%, Normal: 76.04%   c. Fire: 87.52%, Normal: 12.48%

d. Fire: 16.87%, Normal: 83.13%   e. Fire: 5.19%, Normal: 94.81%   f. Fire: 94.78%, Normal: 5.22%

FIGURE 3.4 Sample query images along with their probabilities for CNN-based fire detection [7].

process, a target model that can be used for prediction of fire at early stages was achieved. For testing, the query image is passed through the proposed model, which results in probabilities for both classes: fire and non-fire (normal). Based on the higher probability, the image is assigned to its appropriate class. Examples of query images along with their probabilities are shown in Figure 3.4.

3.3.2 BEA: BALANCING EFFICIENCY AND ACCURACY

The research since the last decade has been focused on traditional feature extraction methods for flame detection. The issues with such methods are their time-consuming process of feature engineering and their low performance for flame detection. Such methods also generate high numbers of false alarms, especially in surveillance with shadows, varying lighting, and fire-colored objects. To cope with such issues, deep learning architectures have been extensively studied and explored for early flame detection. Motivated by the

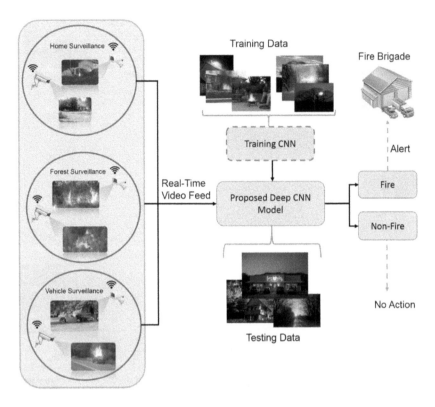

FIGURE 3.5   Early flame detection in surveillance videos using deep CNN [40].

recent improvements in embedded processing capabilities and the potential of deep features, numerous CNNs have been investigated to improve flame detection accuracy and minimize the false warning rate. In this section, we propose a cost-effective fire detection CNN architecture for surveillance videos. The model is inspired from GoogleNet architecture, considering its reasonable computational complexity and suitability for the intended problem compared to other computationally expensive networks such as "AlexNet." To balance the efficiency and accuracy, the model is fine-tuned considering the nature of the target problem and fire data. An overview of our framework for flame detection in CCTV surveillance networks is given in Figure 3.5.

3.3.2.1   BEA Framework

In this section, the architectural details of this method have been described. For the intended classification of images into fire and non-fire, a model similar to GoogleNet [43] with amendments as per our problem is used. The inspirational reasons for using GoogleNet rather than other models such as AlexNet include its better classification accuracy, small model size, and suitability of implementation on Field Programmable Gate Arrays (FPGAs) and other low-memory hardware architectures. The intended architecture consists of 100 layers with 2 main convolutions, 4 max pooling, one average pooling, and 7 inception modules as given in Figure 3.6.

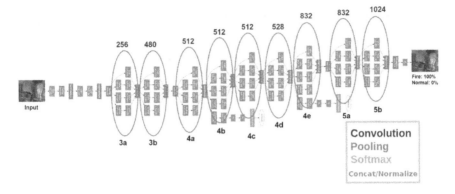

FIGURE 3.6 Architectural overview of the proposed deep CNN [40].

The size of the input image is 224×224×3 pixels passed through the proposed CNN architecture visualized in Figure 3.6. The motivation for using the proposed architecture is to avoid an uncontrollable increase in the computational complexity and network flexibility and significant increases in the number of units at each stage. To achieve this, a dimensionality reduction mechanism is applied before computation-hungry convolutions of patches with larger sizes. Further, the architecture is modified according to the problem of fire classification by keeping the number of output classes to 2, i.e., fire and non-fire. The better results for using this architecture over benchmark datasets with prediction scores for fire and non-fire classes are shown in Figure 3.7.

a. Fire: 86.04%, Normal: 13.96%   b. Fire: 99.82%, Normal: 0.18%   c. Fire: 61.17%, Normal: 38.83%

d. Fire: 2.07%, Normal: 97.93%   e. Fire: 37.99%, Normal: 62.01%   f. Fire: 0.27%, Normal: 99.73%

FIGURE 3.7 Probability scores and predicted labels produced by the proposed deep CNN framework for different images from benchmark datasets [40].

3.3.3   FLCIE: Fire Localization and Contextual Information Extraction

In this section, an energy-friendly and computationally efficient CNN architecture is proposed, inspired by the SqueezeNet [44] architecture for fire detection, localization, and semantic understanding of the scene of the fire. It uses smaller convolutional kernels and contains no fully connected layers, which helps keep the computational requirements to a minimum. Despite its low computational needs, the proposed solution achieves accuracies that are comparable to other, more complex models, mainly due to its increased depth. Moreover, this work shows how a trade-off can be reached between fire detection accuracy and efficiency, by considering the specific characteristics of the problem of interest and the variety of fire data.

3.3.3.1   FLCIE Architecture

Details of the proposed architecture are described in this section. A model with an architecture similar to that of SqueezeNet is modified in accordance with our target problem of fire detection. The original model was trained on the ImageNet dataset and is capable of classifying 1,000 different objects. In the fire detection case, however, this architecture is used to detect fire and non-fire images. This was achieved by reducing the number of neurons in the final layer from 1,000 to 2. Keeping the rest of the architecture similar to the original was done in an attempt to reuse the parameters to solve the fire detection problem more effectively. During the experiments, a transfer learning strategy was also explored in an attempt to further improve the accuracy. Interestingly, an improvement in classification accuracy of approximately 5% for the test data after fine-tuning was achieved. A set of sample images with their predicted class labels and probability scores is given in Figure 3.8.

3.3.3.2   Feature Map Selection for Localization

In this section, the process of localizing fire regions in video frames is described. To localize a fire in a sample image, the framework is employed as given in Figure 3.10.

a. Fire: 99.53%, Normal: 0.47%   b. Fire: 99.98%, Normal: 0.02%   c. Fire: 99.46%, Normal: 0.54%

d. Fire: 14.46%, Normal: 85.54%   e. Fire: 40.91%, Normal: 59.09%   f. Fire: 13.56%, Normal: 86.44%

FIGURE 3.8   Prediction scores for a set of query images using the proposed deep CNN [45].

FIGURE 3.9 Sample images and the corresponding localized fire regions using our approach. The first row shows the original images, while the second row shows the localized fire regions [45].

First, a prediction is obtained from our deep CNN. In non-fire cases, no further action is performed; in the case of fire, further processing of its localization is performed, as given in Algorithms 1 and 2.

After analyzing all the feature maps of the different layers of our proposed CNN using Algorithm 1, feature maps 8, 26, and 32 of the "Fire2/Concat" layer were found to be sensitive to fire regions and to be appropriate for fire localization. Therefore, these three feature maps are fused and binarization is applied to segment the fire [46]. A set of sample fire images with their segmented regions is given in Figure 3.9.

Algorithm 1. Feature Map Selection Algorithm for Localization

Input: Training samples (TS), ground truth (GT), and the proposed deep CNN model (CNN-M)

1. Forward propagate TS through CNN-M
2. Select the feature maps F_N from layer L of CNN-M
3. Resize *GT* and F_N to 256×256 pixels
4. Compute mean activations map $F_{MA(i)}$ for F_N
5. Binarize each feature map F_i as follows:

$$F(x,y)_{bin(i)} = \begin{cases} 1, & F(x,y)_i > F_{MA(i)} \\ 0, & \text{Otherwise} \end{cases}$$

6. Calculate the hamming distance HD_i between *GT* and each feature map $F_{bin(i)}$ as follows:

$$HD_i = \left| F_{bin(i)} - GT \right|$$

This results in $TS \times F_N$ hamming distances
7. Calculate the sum of all resultant hamming distances, and shortlist the minimum hamming distances using threshold T
8. Select appropriate feature maps according to the shortlisted hamming distances

Output: Feature maps sensitive to fire

3.3.3.3 Contextual Information Extraction Mechanism

The fire detection methods presented so far in the literature can either detect fire or localize fire regions with reasonable accuracy. To the best of our knowledge, there is no published work that can extract contextual information from the fire scenes. "Contextual information" refers to anything that is of interest to fire brigades and disaster management systems. Whenever a person reports to a fire brigade, the following information is collected by them:

1. What is on fire?
2. Are there any people in the building?
3. What type of building is on fire?
4. Are there any hazards in or around the building?
5. Is evacuation in progress?
6. What is the address of the premises?
7. What is your phone number?
8. Other information as relevant.

Algorithm 2. Fire Localization Algorithm

Input: Image I of the video sequence and the proposed deep CNN model (CNN-M)

1. Select a frame from the video sequence and forward propagate it through CNN-M
2. **IF** predicted label = non-fire **THEN**
   No action
   ELSE
   a) Extract feature maps 8, 26, and 32 (F_8, F_{26}, F_{32}) from the "Fire2/Concat" layer of CNN-M
   b) Calculate mean activations map (F_{MA}) for F_8, F_{26}, and F_{32}
   c) Apply binarization on F_{MA} through threshold T as follows:

$$F_{Localize} = \begin{cases} 1, & F_{MA} > T \\ 0, & \text{Otherwise} \end{cases}$$

   d) Segment fire regions from F_{MA}
   END

Output: Binary image with segmented fire $I_{Localize}$

The proposed method can extract similar contextual information as well as more useful information (such as size and expansion of fire) automatically by processing the video stream of surveillance cameras. To achieve such information, the segmented fire regions of Algorithm 2 are further processed to determine the severity level/burning degree of the scene under observation and find the zone of influence (ZOI) from the input fire image. The burning degree can be determined from the number of pixels in the segmented fire.

The zone of influence can be calculated by subtracting the segmented fire regions from the original input image. The resultant zone of influence image is then passed from the original SqueezeNet model [44], which predicts its label from 1,000 objects. The object information can be used to determine the situation in the scene, such as a fire in a house, a forest, or a vehicle. This information along with the severity of the fire can be reported to the fire brigade to take appropriate action.

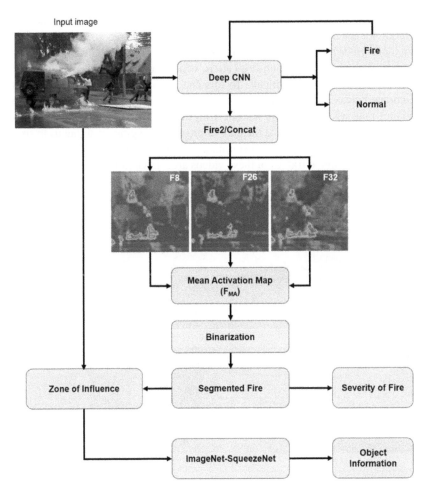

FIGURE 3.10   Fire localization and contextual information extraction using the proposed deep CNN [45].

3.4   EXPERIMENTAL RESULTS AND DISCUSSION

In this section, the proposed methods (EFD, BEA, and FLCIE) are evaluated both qualitatively and quantitatively on challenging benchmark datasets. Two sets of evaluation metrics are used for the evaluation of each method. Performance and various other aspects of the presented methods are highlighted in the relevant sections.

3.4.1   DATASETS

This section presents the benchmark datasets used for experimental results and evaluation. The main focus during experiments is on two famous benchmark dataset: Foggia's video dataset [10] and Chino's dataset [47]. The first dataset contains a total of 31 videos consisting of indoor as well outdoor scenarios, where 14 videos belong to the fire class and the remaining 17 videos belong to the non-fire class. The main motivation for selecting this dataset is that it contains videos recorded in both indoor and outdoor environments. Furthermore, this dataset is very challenging because of the fire-like objects in the non-fire videos, which can be predicated as fire, making the classification more difficult. To this end, color-based methods may fail to differentiate between real fire and scenes with red objects. Similarly, motion-based techniques may wrongly classify a scene with mountains containing smoke, cloud, or fog. These compositions made the dataset more challenging, enabling us to stress our framework and investigate its performance in various situations of the real environment. Representative frames of these videos are shown in Figure 3.11.

The second dataset (DS2) [47] is comparatively small but very challenging. There are 226 images in this dataset, out of which 119 images contain fire while

FIGURE 3.11   Sample images from DS1 videos, containing fire and without fires [7].

(i) Fire004   (ii) Fire048

(iii) Fire114   (iv) not_Fire078

(v) not_fire035   (vi) not_fire069

FIGURE 3.12 Sample images from dataset DS2.

the remaining 107 are fire-like images containing sunsets, fire-like lights, sunlight coming through windows, etc. A set of selected images from this dataset is shown in Figure 3.12.

3.4.2 EXPERIMENTAL SETUP

The models used in the proposed methods were trained using a system with specifications as follows: Intel Core i5 CPU equipped with 64 GB RAM with Ubuntu OS, NVidia GeForce GTX TITAN X (Pascal) having 12 GB onboard memory, and Caffe deep learning framework [48]. The rest of the experiments were conducted using MATLAB R2015a with a Core i5 system containing 8 GB RAM.

Furthermore, two different sets of evaluation metrics were employed to evaluate the performance of each method from all perspectives. The first set of metrics contains true positives (accuracy), false positives, and false negatives as used in Muhammad et al. [49]. Dividing the number of images identified as fire by the system by the number of actual fire images provides the "true positive" (TP) rate. "False negatives" (FN) are calculated by dividing the number of fire images identified as non-fire by the system by the number of actual fire images. The "false positives" (FP) or false alarm rate is achieved by dividing the number of non-fire images predicted as fire by the system by the total number of non-fire images. For better evaluation of the performance, another set of metrics including precision [4], recall, and F-measure [50] is also employed. These metrics are computed as follows:

$$\text{Precision} = \frac{\text{TP}}{(\text{TP} + \text{FP})} \tag{3.1}$$

$$\text{Recall} = \frac{\text{TP}}{(\text{TP} + \text{FN})} \tag{3.2}$$

$$F - \text{Measure} = 2 * \frac{(\text{Precision} * \text{Recall})}{(\text{Precision} + \text{Recall})} \tag{3.3}$$

3.4.3 EXPERIMENTAL RESULTS

This section explains in detail the experiments conducted for the performance evaluation of our proposed methods, i.e., EFD, BEA, and FLCIE. First, experiments performed on various datasets of literature are provided and comparisons are made of all the proposed works with state-of-the-art methods. Next, the results of fire localization are compared with the existing methods. Finally, the experiments for contextual information extraction from the object that is on fire are presented.

3.4.3.1 Results and Comparison

For the training and testing phases of the experiments, the experimental strategy of Foggia et al. [10] are followed, where 20% and 80% of the data are used for training and testing, respectively. Using this strategy, we trained our proposed three methods, i.e., EFD, BEA, and FLCIE, as discussed in Section 1.3. The collected experimental results using Dataset1 are given in Table 3.1.

For the selection of existing literature for assessment, a criterion was employed based on the features used in the related works, years of publication, and the datasets under consideration. Then, the proposed work was compared with the selected fire detection algorithms, as shown in Table 3.1. The results show that Celik et al. [14] and Foggia et al. [10] are the best algorithms in terms of false negatives. However, the results of these methods are not impressive in terms of the other metrics of false positives and accuracy. From the perspective of false positives, the algorithm of

TABLE 3.1
Comparison of Various Fire Detection Methods for Dataset1

Technique	False Positives (%)	False Negatives (%)	Accuracy (%)
FLCIE after FT	8.87	2.12	**94.50**
FLCIE before FT	9.99	10.39	89.8
BEA after FT	**0.054**	1.5	94.43
BEA before FT	0.11	5.5	88.41
EFD after FT	9.07	2.13	94.39
EFD before FT	9.22	10.65	90.06
Foggia et al. [10]	11.67	**0**	93.55
De Lascio et al. [51]	13.33	0	92.86
Habibuglu et al. [16]	5.88	14.29	90.32
Rafiee et al. (RGB) [15]	41.18	7.14	74.20
Rafiee et al. (YUV) [15]	17.65	7.14	87.10
Celik et al. [14]	29.41	**0**	83.87
Chen et al. [8]	11.76	14.29	87.10

Habibuglu et al. [16] performs best and dominates the other methods. However, its false-negative rate is 14.29%, the worst result of all the methods examined. The accuracy of the other four methods is also better than this method, with the most recent method [10] being the best. However, the false positive score of 11.67% is still high, and the accuracy could be further improved. To achieve a high accuracy and a low false positive rate, the use of deep features for fire detection is explored. For this reason, three different methods, i.e., EFD, BEA, and FLCIE, are proposed. In EFD, first we used the AlexNet architecture without fine-tuning, which resulted in an accuracy of 90.06% and reduced false positives from 11.67% to 9.22%. In the baseline AlexNet architecture, the weights of kernels are initialized randomly and are modified during the training process considering the error rate and accuracy. The strategy of transfer learning [52] is also applied, whereby the weights from a pre-trained AlexNet model are initialized with a low learning rate of 0.001 and modified the last fully connected layer according to the problem of fire detection. Interestingly, an improvement in accuracy of 4.33% and reductions in false negatives and false positives of up to 8.52% and 0.15%, respectively, are obtained.

The false alarm score is still high and needs further improvement. Therefore, deep learning architecture GoogleNet is explored for this purpose in BEA. Initially, a GoogleNet model is trained with its default kernel weights, which resulted in an accuracy of 88.41% and a false positive score of 0.11%. The baseline GoogleNet architecture randomly initializes the kernel weights, which are tuned according to the accuracy and error rate during the training process. In an attempt to improve the accuracy, transfer learning [52] by initializing the weights from pre-trained GoogleNet model and keeping the learning rate threshold to 0.001 is explored. Further, the last fully connected layer as per the nature of the intended problem is changed. With this fine-tuning process, the false alarm rate is reduced from 0.11% to 0.054% and the false-negative score from 5.5% to 1.5%.

TABLE 3.2

Comparison of Different Fire Detection Methods for Dataset2

Technique		Precision	Recall	F-Measure
FLCIE	After FT	**0.86**	0.97	**0.91**
	Before FT	0.84	0.87	0.85
BEA	After FT	0.80	0.93	0.86
	Before FT	**0.86**	0.89	0.88
EFD	After FT	0.82	**0.98**	0.89
	Before FT	0.85	0.92	0.88
Chino et al. (BoWFire) [47]		0.51	0.65	0.57
Rudz et al. [53]		0.63	0.45	0.52
Rossi et al. [54]		0.39	0.22	0.28
Celik et al. [14]		0.55	0.54	0.54
Chen et al. [8]		0.75	0.15	0.25

Although the results of EFD and BEA are good compared to other existing methods, there are still certain limitations. Firstly, the size of this model is comparatively large, thereby restricting its implementation in CCTV networks. Secondly, the accuracy is still low and would be problematic for fire brigades and disaster management teams.

With these strong motivations, SqueezeNet, a lightweight architecture, was explored for this problem in FLCIE. The experiments for this new architecture were repeated, and an improvement of 0.11% in accuracy was achieved. Furthermore, the rate of false alarms was reduced from 9.07% to 8.87%. The rate of false negatives remained almost the same. Finally, the major achievement of the proposed framework was the reduction of the model size from 238 MB to 3 MB, thus saving an extra 235 MB, which can greatly minimize the cost of CCTV surveillance systems.

For further experimentation, Dataset2 was considered and compared with the results from four other fire detection algorithms in terms of their relevancy, datasets, and years of publication as shown in Table 3.2.

To ensure a fair evaluation and a full overview of the performance of our approach, we considered another set of metrics (precision, recall, and F-measure) as used by Chino et al. [47]. In a similar way to the experiments on Dataset1, we tested Dataset2 using EFD, BEA, and FLCIE. For the fine-tuned models in EFD and BEA, F-measure scores of 0.89 and 0.86, respectively, were achieved. Further improvement was achieved using FLCIE, increasing the F-measure score to 0.91 and the precision to 0.86. It is evident from Table 3.2 that FLCIE achieved better results than the state-of-the-art methods, confirming the effectiveness of the proposed deep CNN framework.

3.4.3.2 FLCIE Results

In this section, the proposed approach (FLCIE) is assessed with respect to fire localization. The true positive and false positive rates are computed in order to have a better and unbiased evaluation of our proposed fire localization module. The feature

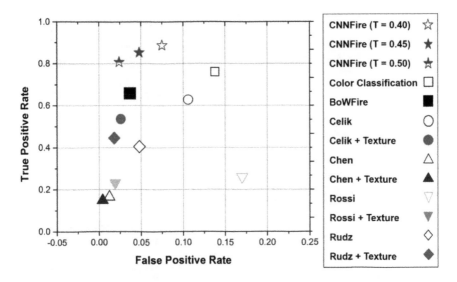

FIGURE 3.13 Comparison of the CNNFire approach with other methods [45].

maps from different convolutional and pooling layers of our proposed CNN architecture are used to localize fire in an image precisely. Next, the number of overlapping fire pixels in the detection maps and ground truth images are used as true positives. Similarly, the number of non-overlapping fire pixels in the detection maps are also determined and then interpreted as false positives. The localization results are compared with those of several state-of-the-art methods such as Chen et al. [8], Celik et al. [14], Rossi et al. [54], Rudz et al. [53], and Chino et al. (BoWFire) [47], as shown in Figure 3.13. Figures 3.14 and 3.15 show the results of all methods for a sample image from Dataset2. The results of BoWFire, color classification, Celik et al., and Rudz et al. are almost the same. Rossi et al. gives the worst results in this case, and Chen et al. is better than Rossi et al. The results from CNNFire are similar to the ground truth.

3.4.3.3   Experiments for Contextual Information Extraction

Along with fire detection and localization, the proposed approach is able to determine the intensity of the fire detected in an input image accurately. To achieve this target, the zone of influence (ZOI) inside the input image along with the segmented fire regions is extracted. The ZOI image is then advanced to the pre-trained SqueezeNet model on the ImageNet dataset comprising 1,000 classes. The label assigned by the SqueezeNet model to the ZOI image is then combined with the severity of the fire for reporting to the fire brigade. A set of sample cases from this experiment is given in Figure 3.16.

In Figure 3.16, the first column shows input images with labels predicted by the CNN model and their probabilities, with the highest probability taken as the final class label; the second column shows three feature maps (F8, F26, and F32) selected by EFD; the third column highlights the results for each image using BEA; the fourth column shows the severity of the fire and ZOI images with a label assigned

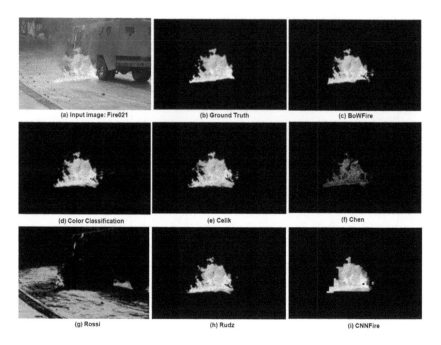

FIGURE 3.14 Visual fire localization results of our CNNFire approach and other fire localization methods [45].

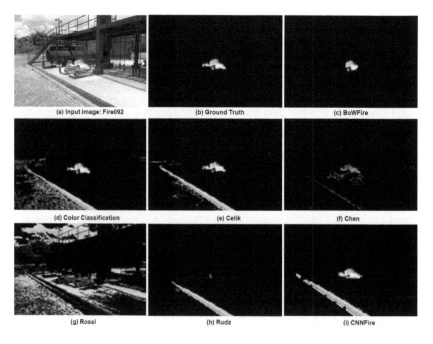

FIGURE 3.15 Fire localization results from our CNNFire and other schemes with false positives [45].

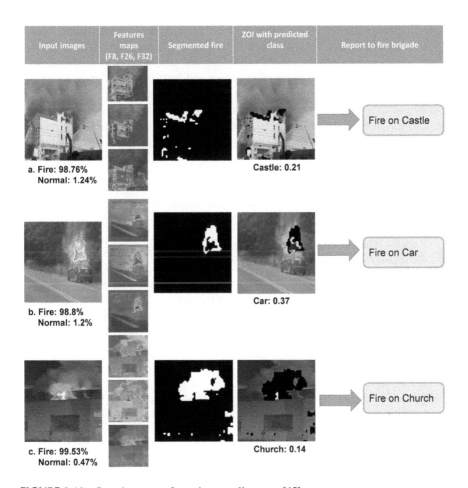

FIGURE 3.16 Sample outputs from the overall system [45].

by the SqueezeNet model; and the final column shows the alert that should be sent to emergency services, such as the fire brigade.

3.5 CONCLUSIONS

Recently, convolutional neural networks (CNNs) have shown substantial progress in surveillance applications; however, their key roles in early fire detection, localization, and fire scene analysis have not been investigated. With this motivation, CNNs were investigated for addressing the aforementioned problems in this chapter for fire detection and localization, thereby achieving better detection accuracy with a minimum false alarm rate compared to existing methods. Several highly efficient and intelligent methods that selected discriminative deep features for fire scene analysis in both indoor and outdoor surveillance have been presented.

In this chapter, a total of three CNN-based frameworks have been proposed. The first framework was proposed for early fire detection using CNNs in surveillance applications. A cost-effective fire detection CNN architecture for surveillance videos was described in the second framework. Finally, an efficient deep CNN-based fire detection system was proposed for fire detection and localization. In the future, we plan to study the feature maps of other recent and lightweight architectures (e.g., ShuffleNet) for localizing fire regions in video frames, helping determining fire severity and burning degree. Further, we will also explore its possibility for extracting semantic information about the fire scene as well as its integration with smoke detection systems for detailed surveillance of sensitive areas.

BIBLIOGRAPHY

1. B. C. Ko, K.-H. Cheong, and J.-Y. Nam, "Fire detection based on vision sensor and support vector machines," *Fire Safety Journal*, vol. 44, pp. 322–329, 2009.
2. V. D. Nguyen, H. Van Nguyen, D. T. Tran, S. J. Lee, and J. W. Jeon, "Learning framework for robust obstacle detection, recognition, and tracking," *IEEE Transactions on Intelligent Transportation Systems*, vol. 18, pp. 1633–1646, 2017.
3. I. Mehmood, M. Sajjad, and S. W. Baik, "Mobile-cloud assisted video summarization framework for efficient management of remote sensing data generated by wireless capsule sensors," *Sensors*, vol. 14, pp. 17112–17145, 2014.
4. K. Muhammad, M. Sajjad, M. Y. Lee, and S. W. Baik, "Efficient visual attention driven framework for key frames extraction from hysteroscopy videos," *Biomedical Signal Processing and Control*, vol. 33, pp. 161–168, 2017.
5. R. Hamza, K. Muhammad, Z. Lv, and F. Titouna, "Secure video summarization framework for personalized wireless capsule endoscopy," *Pervasive and Mobile Computing*, vol. 41, pp. 436–450, 2017/10/01/ 2017.
6. K. Muhammad, R. Hamza, J. Ahmad, J. Lloret, H. H. G. Wang, and S. W. Baik, "Secure surveillance framework for IoT systems using probabilistic image encryption," *IEEE Transactions on Industrial Informatics*, vol. 14(8), pp. 3679–3689, 2018.
7. K. Muhammad, J. Ahmad, and S. W. Baik, "Early fire detection using convolutional neural networks during surveillance for effective disaster management," *Neurocomputing*, vol. 288, pp. 30–42, 2018.
8. C. Thou-Ho, W. Ping-Hsueh, and C. Yung-Chuen, "An early fire-detection method based on image processing," in *2004 International Conference on Image Processing, 2004. ICIP '04*, vol. 3, 2004, pp. 1707–1710.
9. http://www.bbc.com/news/world-asia-42828023 (Visited 31 January, 2018, 9 AM).
10. P. Foggia, A. Saggese, and M. Vento, "Real-time fire detection for video-surveillance applications using a combination of experts based on color, shape, and motion," *IEEE Transactions on Circuits and Systems for Video Technology*, vol. 25(9), pp. 1545–1556, 2015.
11. B. U. Töreyin, Y. Dedeoğlu, U. Güdükbay, and A. E. Cetin, "Computer vision based method for real-time fire and flame detection," *Pattern Recognition Letters*, vol. 27(1), pp. 49–58, 2006.
12. D. Han, and B. Lee, "Development of early tunnel fire detection algorithm using the image processing," in *International Symposium on Visual Computing*, 2006, pp. 39–48.
13. G. Marbach, M. Loepfe, and T. Brupbacher, "An image processing technique for fire detection in video images," *Fire Safety Journal*, vol. 41(4), pp. 285–289, 2006.
14. T. Celik, and H. Demirel, "Fire detection in video sequences using a generic color model," *Fire Safety Journal*, vol. 44(2), pp. 147–158, 2009.
15. A. Rafiee, R. Dianat, M. Jamshidi, R. Tavakoli, and S. Abbaspour, "Fire and smoke detection using wavelet analysis and disorder characteristics," in *2011 3rd International Conference on Computer Research and Development*, 2011, pp. 262–265.

16. Y. H. Habiboğlu, O. Günay, and A. E. Çetin, "Covariance matrix-based fire and flame detection method in video," *Machine Vision and Applications*, vol. 23(6), pp. 1103–1113, 2012.

17. A. Sorbara, E. Zereik, M. Bibuli, G. Bruzzone, and M. Caccia, "Low cost optronic obstacle detection sensor for unmanned surface vehicles," in *2015 IEEE Sensors Applications Symposium (SAS)*, 2015, pp. 1–6.

18. I. Kolesov, P. Karasev, A. Tannenbaum, and E. Haber, "Fire and smoke detection in video with optimal mass transport based optical flow and neural networks," in *2010 IEEE International Conference on Image Processing*, 2010, pp. 761–764.

19. H. J. G. Haynes, "Fire loss in the United States during 2015," http://www.nfpa.org/, 2016.

20. K. Muhammad, T. Hussain, and S. W. Baik, "Efficient CNN based summarization of surveillance videos for resource-constrained devices," *Pattern Recognition Letters*, 2018.

21. M. Sajjad, S. Khan, T. Hussain, K. Muhammad, A. K. Sangaiah, A. Castiglione, et al., "CNN-based anti-spoofing two-tier multi-factor authentication system," *Pattern Recognition Letters*, 126(2019): 123–131.

22. M. Hassaballah, A. A. Abdelmgeid, and H. A. Alshazly, "Image features detection, description and matching," in *Image Feature Detectors and Descriptors*, Awad, A. I., Hassaballah, M., (Eds.): Springer, 2016, pp. 11–45.

23. A. I. Awad, and M. Hassaballah, *Image Feature Detectors and Descriptors*. Studies in Computational Intelligence. Springer International Publishing, Cham, 2016.

24. F. U. M. Ullah, A. Ullah, K. Muhammad, I. U. Haq, and S. W. Baik, "Violence detection using spatiotemporal features with 3D Convolutional Neural Network," *Sensors*, vol. 19(11), p. 2472, 2019.

25. M. Sajjad, S. Khan, Z. Jan, K. Muhammad, H. Moon, J. T. Kwak, et al., "Leukocytes classification and segmentation in microscopic blood smear: A resource-aware healthcare service in smart cities," *IEEE Access*, vol. 5, pp. 3475–3489, 2017.

26. I. U. Haq, K. Muhammad, A. Ullah, and S. W. Baik, "DeepStar: Detecting starring characters in movies," *IEEE Access*, 7(2019): 9265–9272.

27. M. Hassaballah, H. A. Alshazly, and A. A. Ali, "Ear recognition using local binary patterns: A comparative experimental study," *Expert Systems with Applications*, vol. 118, pp. 182–200, 2019.

28. A. I. Awad, and K. Baba, "Singular point detection for efficient fingerprint classification," *International Journal on New Computer Architectures and Their Applications (IJNCAA)*, vol. 2, pp. 1–7, 2012.

29. A. Ullah, K. Muhammad, J. D. Ser, S. W. Baik, and V. Albuquerque, "Activity recognition using temporal optical flow convolutional features and multi-layer LSTM," *IEEE Transactions on Industrial Electronics*, vol. 66(12), pp. 9692–9702, 2019.

30. A. Ullah, J. Ahmad, K. Muhammad, M. Sajjad, and S. W. Baik, "Action recognition in video sequences using deep Bi-directional LSTM with CNN features," *IEEE Access*, vol. 6, pp. 1155–1166, 2018.

31. M. Sajjad, S. Khan, K. Muhammad, W. Wu, A. Ullah, and S. W. Baik, "Multi-grade brain tumor classification using deep CNN with extensive data augmentation," *Journal of Computational Science*, vol. 30, pp. 174–182, 2019.

32. J. Ahmad, K. Muhammad, J. Lloret, and S. W. Baik, "Efficient conversion of deep features to compact binary codes using Fourier decomposition for multimedia big data," *IEEE Transactions on Industrial Informatics*, vol. 14(7), pp. 3205–3215, 2018.

33. J. Ahmad, K. Muhammad, S. Bakshi, and S. W. Baik, "Object-oriented convolutional features for fine-grained image retrieval in large surveillance datasets," *Future Generation Computer Systems*, vol. 81, pp. 314–330, 2018.

34. S. Frizzi, R. Kaabi, M. Bouchouicha, J. Ginoux, E. Moreau, and F. Fnaiech, "Convolutional neural network for video fire and smoke detection," in *IECON 2016 - 42nd Annual Conference of the IEEE Industrial Electronics Society*, 2016, pp. 877–882.

35. J. Sharma, O.-C. Granmo, M. Goodwin, and J. T. Fidje, "Deep convolutional neural networks for fire detection in images," in *International Conference on Engineering Applications of Neural Networks*, 2017, pp. 183–193.

36. Z. Zhong, M. Wang, Y. Shi, and W. Gao, "A convolutional neural network-based flame detection method in video sequence," *Signal, Image and Video Processing*, vol. 12(8), pp. 1619–1627, 2018.

37. K. Muhammad, S. Khan, M. Elhoseny, S. H. Ahmed, and S. W. Baik, "Efficient fire detection for uncertain surveillance environment," *IEEE Transactions on Industrial Informatics*, vol. 15(5), pp. 3113–3122, 2019.

38. S. Khan, K. Muhammad, S. Mumtaz, S. W. Baik, and V. H. C. d. Albuquerque, "Energy-efficient deep CNN for smoke detection in foggy IoT environment," *IEEE Internet of Things Journal*, vol. 6(6), pp. 9237–9245, 2019.

39. K. Muhammad, S. Khan, V. Palade, I. Mehmood, and V. H. C. D. Albuquerque, "Edge intelligence-assisted smoke detection in foggy surveillance environments," *IEEE Transactions on Industrial Informatics*, pp. 1–1, 2019.

40. K. Muhammad, J. Ahmad, I. Mehmood, S. Rho, and S. W. Baik, "Convolutional neural networks based fire detection in surveillance videos," *IEEE Access*, vol. 6, pp. 18174–18183, 2018.

41. K. Muhammad, J. Ahmad, Z. Lv, P. Bellavista, P. Yang, and S. W. Baik, "Efficient deep CNN-based fire detection and localization in video surveillance applications," *IEEE Transactions on Systems, Man, and Cybernetics: Systems*, vol. 49(7), pp. 1419–1434, 2018.

42. A. Krizhevsky, I. Sutskever, and G. E. Hinton, "Imagenet classification with deep convolutional neural networks," *Advances in Neural Information Processing Systems*, pp. 1097–1105, 2012.

43. C. Szegedy, W. Liu, Y. Jia, P. Sermanet, S. Reed, D. Anguelov, et al., "Going deeper with convolutions," in *Proceedings of the IEEE Conference on Computer Vision and Pattern Recognition*, 2015, pp. 1–9.

44. F. N. Iandola, M. W. Moskewicz, K. Ashraf, S. Han, W. J. Dally, and K. Keutzer, "SqueezeNet: AlexNet-level accuracy with 50x fewer parameters and< 1MB model size," arXiv preprint arXiv:1602.07360, 2016.

45. K. Muhammad, J. Ahmad, Z. Lv, P. Bellavista, P. Yang, and S. W. Baik, "Efficient deep CNN-based fire detection and localization in video surveillance applications," *IEEE Transactions on Systems, Man, and Cybernetics: Systems*, vol. 49(7), pp. 1419–1434, 2018.

46. K. Wattanachote, and T. K. Shih, "Automatic dynamic texture transformation based on a new motion coherence metric," *IEEE Transactions on Circuits and Systems for Video Technology*, vol. 26(10), pp. 1805–1820, 2015.

47. D. Y. Chino, L. P. Avalhais, J. F. Rodrigues, and A. J. Traina, "BoWFire: Detection of fire in still images by integrating pixel color and texture analysis," in *2015 28th SIBGRAPI Conference on Graphics, Patterns and Images*, 2015, pp. 95–102.

48. Y. Jia, E. Shelhamer, J. Donahue, S. Karayev, J. Long, R. Girshick, et al., "Caffe: Convolutional architecture for fast feature embedding," in *Proceedings of the 22nd ACM International Conference on Multimedia*, 2014, pp. 675–678.

49. K. Muhammad, J. Ahmad, Z. Lv, P. Bellavista, P. Yang, and S. W. Baik, "Efficient deep CNN-based fire detection and localization in video surveillance applications," *IEEE Transactions on Systems, Man, and Cybernetics: Systems*, vol. 49(7), pp. 1419–1434, 2018.

50. K. Muhammad, J. Ahmad, M. Sajjad, and S. W. Baik, "Visual saliency models for summarization of diagnostic hysteroscopy videos in healthcare systems," *SpringerPlus*, vol. 5(1), p. 1495, 2016.

51. R. Di Lascio, A. Greco, A. Saggese, and M. Vento, "Improving fire detection reliability by a combination of videoanalytics," in *International Conference Image Analysis and Recognition*, 2014, pp. 477–484.

52. S. J. Pan, and Q. Yang, "A survey on transfer learning," *IEEE Transactions on Knowledge and Data Engineering*, vol. 22(10), pp. 1345–1359, 2010.

53. S. Rudz, K. Chetehouna, A. Hafiane, H. Laurent, and O. Séro-Guillaume, "Investigation of a novel image segmentation method dedicated to forest fire applicationsThis paper is dedicated to the memory of Dr Olivier Séro-Guillaume (1950–2013), CNRS Research Director," *Measurement Science and Technology*, vol. 24(7), p. 075403, 2013.

54. L. Rossi, M. Akhloufi, and Y. Tison, "On the use of stereovision to develop a novel instrumentation system to extract geometric fire fronts characteristics," *Fire Safety Journal*, vol. 46(1–2), pp. 9–20, 2011.

4 A Multi-biometric Face Recognition System Based on Multimodal Deep Learning Representations

Alaa S. Al-Waisy, Shumoos Al-Fahdawi, and Rami Qahwaji

CONTENTS

4.1   INTRODUCTION

A reliable and efficient identity administration system is a vital component in many governmental and civilian high-security applications that provide services to only legitimately registered users [1]. Many pattern recognition systems have been used for establishing the identity of an individual based on different biometric traits (e.g., physiological, behavioral, and/or soft biometric traits) [2–3]. In a biometric system, these traits establish a substantial and strong connection between the users and their identities [4]. Face recognition has received considerably more attention in the research community than other biometric traits, due to its wide range of commercial and governmental applications, its low cost, and the ease of capturing the face image in a non-intrusive manner. This is unlike other biometric systems (e.g., fingerprint recognition), which can cause some health risks by transferring viruses and/or some infectious diseases from one user to the other by using the same sensor device to capture the biometric trait from multiple users [5]. Despite these advantages, designing and implementing a face recognition system is considered a challenging task in the image processing, computer vision, and pattern recognition fields. Therefore, different factors and problems need to be addressed when developing any biometric system based on the face image, especially when face images are taken in unconstrained environments. Such environments are challenging due to the large intrapersonal variations, such as changes in facial expression, illumination, multiple views, aging, occlusion from wearing glasses and hats, and the small interpersonal differences [6]. Basically, the performance of the face recognition system depends on two fundamental stages: feature extraction and classification. The correct identification of a person's identity in the latter stage is dependent on the discriminative power of the extracted facial features in the former stage. Thus, extracting and learning powerful and highly discriminating facial features to minimize intrapersonal variations and interpersonal similarities remains a challenging task [7].

Recently, considerable attention has been paid to employing multi-biometric systems in many governmental and private sectors due to their ability to significantly improve the recognition performance of biometric systems. Moreover, multimodal systems have the following advantages [8–11]: (i) improving population coverage; (ii) improving the biometric system's throughput; (iii) deterring spoofing attacks; (iv) maximizing the interpersonal similarities and minimizing the intrapersonal variations; and (v) providing a high degree of flexibility allowing people to choose to provide either a subset or all of their biometric traits depending on the nature of the implemented application and the user's convenience. As illustrated in Figure 4.1, to satisfy the multi-biometric system concept, the face biometric trait can be engaged in four out of five different types of multi-biometric systems.

In this chapter, a novel multi-biometric system for identifying a person's identity using two discriminative deep learning approaches is proposed based on the combination of a convolutional neural network (CNN) and deep belief network (DBN) to address the problem of unconstrained face recognition. CNN is one of the most powerful supervised deep neural networks (DNNs), which is widely used to resolve many tasks in image processing, computer vision, and pattern recognition with high ability to automatically extract discriminative features from input images.

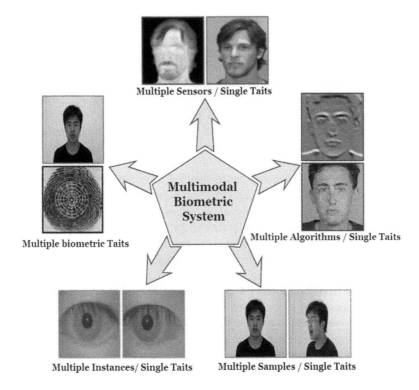

FIGURE 4.1 The five different types of a multimodal biometric system, adapted from Ross and Jain [10].

Furthermore, CNN is considered to be invariant to image deformations, such as translation, rotation, and scaling [12]. On the other hand, DBN, which is one of the most popular unsupervised DNNs, has been effectively applied to learn hierarchical feature representations from unlabeled data in a wide range of fields, including face recognition [13], speech recognition [14], audio classification [15], and natural language understanding [16]. Finally, both CNNs and DBNs have a number of advantages compared to other DNNs, for instance simpler architecture, adaptability, fast convergence, and fewer free parameters. The main contributions of the work presented here can be summarized as follows:

1. A novel local facial feature-based approach based on a combination of the fractal dimension transformation (FDT) and the discriminative restricted Boltzmann machines (DRBM) as a nonlinear classifier to efficiently model the joint distribution of inputs and target classes is proposed and termed the FDT-DRBM approach. The proposed FDT-DRBM approach has managed to produce an illumination-invariant representation of the face image. It also managed to efficiently extract the local facial features (e.g., mouth, nose, eyes, and eyebrows) along with the face texture roughness and fluctuations in the surface efficiently by exploiting the fractal dimension properties, such as self-similarity.

2. An efficient and real-time multi-biometric face recognition system based on building deep learning representations and fusing the results obtained from two discriminative deep learning approaches (i.e., CNN and DBN) is proposed and termed the *FaceDeepNet* system. To the best of our knowledge, this is the first work that investigates the potential use of the CNN and the DBN together for fusing the face-matching scores in a unified multi-biometric system. In this chapter, a parallel architecture is considered, which allows users a high degree of flexibility to use different approaches for encoding the facial features depending on the required security level and the users' convenience.

3. Discriminative training schemes equipped with a number of training strategies (e.g., Dropout method, AdaGrad algorithm, Data augmentation, etc.) are also employed in order to prevent overfitting during the learning process and increase the generalization ability of the proposed DNNs.

4. Another contribution of this work is to further examine and validate the current findings for improving the generalization ability and reducing the computational complexity of the DNNs (e.g., CNN and DBN) by assigning the handcrafted features (e.g., feature representations obtained from the FDT approach) to the visible units of the DNNs instead of the pixel intensity representations to learn additional and complementary representations. We demonstrate that the proposed *FaceDeepNet* system can represent large-scale face datasets, with less time required to obtain the final trained model compared to the direct use of the raw data. In addition, the proposed *FaceDeepNet* system is able to efficiently handle nonlinear variations of face images and is unlikely to overfit the training data due to the high non-linearity of the proposed DNNs.

5. The performance of the proposed approaches has been tested and evaluated in both face identification and verification tasks by conducting a number of extensive experiments on four large-scale unconstrained face datasets: SDUMLA-HMT, FRGC V 2.0, UFI, and LFW. Recognition rates comparable to those of the state-of-the-art methods have been achieved using only the proposed FDT-DRBM approach. New state-of-the-art recognition rates were achieved using the *FaceDeepNet* system on all the employed datasets, in particular when it is trained on top of the output feature representations of the FDT approach.

4.2 RELATED WORK

Generally, a number of approaches have been proposed and developed to overcome all/some of the aforementioned drawbacks and problems in the face recognition system. However, very few of them were capable of working in fully unconstrained environments. These approaches can be broadly divided into two categories: hand-crafted-descriptor approaches, and learning-based approaches [17]. The next subsections are devoted to outlining these approaches and their usage to address the face recognition problem, before discussing their respective strengths and limitations.

4.2.1 HANDCRAFTED-DESCRIPTOR APPROACHES

The majority of previous face recognition systems depend on feature representations produced by local/global handcrafted-descriptor approaches, such as scale-invariant feature transform (SIFT), local binary patterns (LBP), 2D Gabor Wavelet, etc. Handcrafted-descriptor approaches can be further divided into three groups: feature-based, holistic-based, and hybrid-based approaches. In the first category, a vector of geometrical facial features is extracted by measuring and calculating the locations and geometric relationships among the facial features (e.g., the mouth, eyes, and nose). For instance, Biswas et al. [18] developed a face recognition system based on feature representations obtained using SIFT-based descriptors [19] at fiducial locations on the face image, and then the SIFT features of all facial landmarks were formed into a single vector representing the face. A best Rank-1 identification rate of 91% was achieved on the CMU Multi-PIE face database containing a total of 337 subjects. Unlike the feature-based approaches, the holistic-based approaches (e.g., Eigenface methods) operate on the whole face image instead of measuring the local geometrical features. The major limitation of these approaches is caused by the need to perform the matching process in very high dimensionality space. Thus, a number of face recognition algorithms have been developed that employ different dimensionality reduction methods to handle the issue of high dimensionality before the matching process can take place. The first well-known face recognition system based on the Eigenface methods was proposed by Sirovich and Kirby [20].

The hybrid-based approaches use a combination of both feature and holistic approaches to address the face recognition problems. For example, one can argue that hybrid approaches that use both local and global features (e.g., obtained from both feature and holistic approaches, respectively) can be an efficient way to make use of complementary information, reduce the complexity of classifiers, and increase their generalization ability. However, the main challenging factors that can influence the performance of such approaches lie in the identification of the features to be fused, and the fusion method [21]. For example, local features are very sensitive to the illumination changes, while expression changes have more influence on global features. Some of the recent examples belonging to this category were proposed by Fischer et al. [22], Berg and Belhumeur [23], Annan et al. [24], Zhu et al. [25], and Al-Waisy [26, 27], and other examples can be found in Meethongjan and Mohamad [21]. Although previous research has demonstrated the efficiency of handcrafted-descriptor approaches used as robust and discriminative feature detectors to solve the face recognition problem, their performance declines dramatically in unconstrained environments. The decline in performance is caused by the fact that constructed facial feature representations are very sensitive to highly nonlinear intrapersonal variations such as illumination, expression, pose, and occlusion [28]. In addition, this approach usually requires significantly more effort and domain knowledge from the researcher to find the best compact and discriminative feature set for a given problem. Therefore, they are considered to be challenging and time demanding.

4.2.2 Learning-Based Approaches

Inspired by their ability to learn more discriminative feature representations, considerable attention has been paid recently to deep learning approaches, such as CNN. The main advantage of the deep learning approaches over handcrafted-descriptor approaches is their ability to learn from experience and their robustness in handling nonlinear variations of face images caused by pose variations, self-occlusion, etc. [29]. Moreover, in contrast to handcrafted-descriptor approaches, the applications making use of deep learning approaches can generalize well to other new fields. Some recent examples belonging to this category were proposed by Sun et al. A face recognition system, called Deep Identification-verification features (DeepID2), is presented in [28]. In this work, two supervisory signals (e.g., the face identification and verification signals) were combined simultaneously to boost the discriminative power of extracted features using CNNs. A verification rate of 99.15% was achieved using the "Unrestricted, Labeled Outside Data" evaluation protocol. The same group of researchers developed the deep ConvNet-RBM model based on a combination of a set of CNNs and a restricted Boltzmann machine (RBM) model for face verification in the wild environment [30]. Verification rates of 91.75% and 92.52% were achieved on the LFW and CelebFaces databases, respectively.

Taigman et al. [31] proposed the DeepFace framework for face verification using the frontal face images generated from a 3D shape model of a large-scale face dataset. They proposed to use the Softmax function on the top of CNN as the supervisory signal to train the proposed framework. Experiments on the LFW database using the "Unrestricted, Labeled Outside Data" setting yielded a verification rate of 97.35%. In the literature, several metric learning approaches have been proposed to address the face verification problem, including Multimodal Deep Face Representation (MM-DFR) [32], Information-Theoretic Metric Learning (ITML) [33], Discriminative Deep Metric Learning (DDML) [34], Pairwise-constrained Multiple Metric Learning (PMML) [35], Logistic Discriminant-based Metric Learning (LDML) [36], Large Scale Metric Learning (LSML) [37], and Cosine Similarity Metric Learning (CSML) [38]. The key limitations of the deep learning approaches include requiring a large amount of training data to avoid the overfitting problem and increase the generalization ability of the neural network, and the number of hyper-parameters that need to be set. The DNNs mainly depend on a tedious iterative optimization procedure that can be computationally expensive for large-scale databases. However, these issues can be effectively alleviated using high-performance computing systems, equipped with graphics processing units (GPUs) [29].

Finally, a number of approaches have been proposed to address the problem of the multi-biometric face recognition system. For instance, Peng et al. [39] have proposed a novel graphical representation-based HFR method (G-HFR) to address the heterogeneous face recognition (HFR) problem. The HFR problem refers to identifying the person's identity using face images captured from different sources (e.g., different sensors or different wavelengths). The same research group has also addressed another face recognition task, known as cross-modality face recognition. This task aims to match face images across different modalities, include matching sketches with real photos, low-resolution face images with high-resolution images, and near-infrared images with visual lighting images [40, 41]. Gao et al. [42] have also addressed the problem of the insufficient labeled samples by proposing the Semi-Supervised Sparse Representation-based Classification method. This method has been tested on four different datasets, including AR, Multi-PIE, CAS-PEAL, and LFW. For more information on the most significant challenges, achievements, and future directions related to the face recognition task, the reader is referred to [43].

4.3 THE PROPOSED APPROACHES

This section provides a brief description of the proposed approaches, including the FDT and DRBM approaches, which are used in the proposed local facial feature-based extraction approach. In addition, a brief description of the proposed deep learning approaches is given, including the CNN and DBN. The main goal here is to review and identify their strengths and shortcomings to empower the proposal of a novel multi-biometric face recognition system that consolidates the strengths of these approaches.

4.3.1 FRACTAL DIMENSION

The term "fractal dimension" was first introduced by the mathematician Benoit Mandelbrot as a geometrical quantity to describe the complexity of objects that show self-similarity at different scales [44]. The most important characteristic of the fractal dimension is self-similarity, which means that an object has a similar representation to the original under different magnifications. This property has been efficiently used in reflecting the roughness and fluctuation of the surface of an image, where increasing the scale of magnification provides more and more details of the imaged surface. Moreover, the non-integer value of the fractal dimension provides a quantitative measure of objects that have complex geometry, and cannot be well described by an integral dimension (e.g., the length of a coastline) [45–46]. The fractal dimension has been widely employed to resolve many problems in image processing and computer vision (e.g., texture segmentation, texture classification, medical imaging, etc.), due to its simplicity and robustness in reflecting the roughness and fluctuations of the imaged surface. However, not much work has been done to explore and address the potential of using the fractal dimension to resolve pattern recognition problems. Lin et al. [47] proposed an algorithm for human eye detection by exploiting the fractal dimension as an efficient approach for representing the texture of facial features. Farhan et al. [48] developed a personal identification system based on fingerprint images using the fractal dimension as a feature extraction method. Therefore, it appears that the texture of the facial image can be efficiently described by using the fractal dimension. Various methods have been proposed in the literature to estimate fractal dimension [44], e.g., box counting (BC), differential box counting (DBC), fractional Brownian motion (FBM), and others. In this work, the fractal dimension for each pixel over the whole face image is calculated using an improved differential box counting (IDBC) algorithm, which could be used as a powerful edge enhancement technique to enhance the edge representations for the face image without increasing the noise level.

4.3.2 RESTRICTED BOLTZMANN MACHINE

A restricted Boltzmann machine (RBM) is an energy-based bipartite graphical model composed of two fully connected layers via symmetric undirected edges, but there are no connections between units of the same layer. The first layer consists of m visible units $v = (v_1, v_2,..., v_m)$, which represent observed data, while the second layer consists of n hidden units $h = (h_1, h_2,..., h_n)$, which can be viewed as nonlinear feature detectors to capture higher-order correlations in the observed data. In addition, $W = \{w_{11}, w_{12},..., w_{1n},..., w_{mn}\}$ is the connecting weights matrix between the visible and hidden units. A typical RBM structure is shown in Figure 4.2 (a). The standard RBM was designed to be used only with binary stochastic visible units, and is termed Bernoulli RBM (BRBM). However, using binary units is not suitable for real-valued data (e.g., pixel intensity values in images). Therefore, a new model has been developed called the Gaussian RBM (GRBM) to address this limitation of the standard RBM [49]. The energy function of the GRBM is defined as follows:

$$E(v,h) = -\sum_{i=1}^{m}\sum_{j=1}^{n} w_{i,j} h_j \frac{v_i}{\sigma_i} - \sum_{i=1}^{m} \frac{(v_i - b_i)^2}{2\sigma_i^2} - \sum_{j=1}^{n} c_j h_j \qquad (4.1)$$

Here, σ_i is the standard deviation of the Gaussian noise for the visible unit v_i, w_{ij} represents the weights for the visible unit v_i and the hidden unit h_j, and b_i and c_j are biases for the visible and hidden units, respectively. The conditional probabilities for

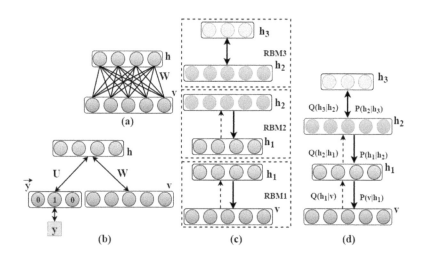

FIGURE 4.2 (a) A typical RBM structure. (b) A discriminative RBM modeling the joint distribution of input variables v and target classes y (represented as a one-hot vector by $\vec{y}$). (c) A greedy layer-wise training algorithm for the DBN composed of three stacked RBMs. (d) Three layers of the DBN as a generative model, where the top-down generative path is represented by the **P** distributions (solid arcs), and bottom-up inference and the training path are represented by the **Q** distributions (dashed arcs).

the visible units, given hidden units, and vice versa for the hidden units are defined as follows:

$$p(v_i = 1 \mid h) = N\left(v \mid b_i + \sum_j w_{i,j} h_j, \sigma_i^2\right)$$ (4.2)

$$p(h_j = 1 \mid v) = f\left(c_j + \sum_i w_{i,j} \frac{v_i}{\sigma_i^2}\right)$$ (4.3)

Here, $N(\cdot \mid \mu, \sigma^2)$ refers to the Gaussian probability density function with mean μ and standard deviation σ. $f(x)$ is a sigmoid function. During the training process, the log-likelihood of the training data is maximized using stochastic gradient descent, and the update rules for the parameters are defined as follows:

$$\Delta w_{i,j} = \epsilon\left(\left\langle \frac{1}{\sigma_i^2} v_i h_j \right\rangle_{data} - \left\langle \frac{1}{\sigma_i^2} v_i h_j \right\rangle_{model}\right)$$ (4.4)

$$\Delta b_i = \epsilon\left(\left\langle \frac{1}{\sigma_i^2} v_i \right\rangle_{data} - \left\langle \frac{1}{\sigma_i^2} v_i \right\rangle_{model}\right)$$ (4.5)

$$\Delta c_i = \epsilon\left(\left\langle h_j \right\rangle_{data} - \left\langle h_j \right\rangle_{model}\right)$$ (4.6)

Here, ϵ is the learning rate and $\langle \cdot \rangle_{data}$ and $\langle \cdot \rangle_{model}$ represent the expectations under the distribution specified by the input data (*Positive phase*) and the internal representations of the RBM model (*Negative phase*), respectively. Finally, b_i and c_i are bias terms for visible and hidden units, respectively. More details on the GRBM model can be found in [49]. Typically, RBMs can be used in two different ways: either as generative models or as discriminative models, as shown in Figure 4.2 (a, b). The generative models use a layer of hidden units to model a distribution over the visible units, as described above. Such models are usually trained in an unsupervised way and used as feature extractors to model only the inputs for another learning algorithm. On the other hand, discriminative models can also model the joint distribution of the input data and associated target classes. The discriminative models aim to train a joint density model using an RBM that has two layers of visible units. The first represents the input data and the second is the *Softmax* label layer that represents the target classes. A discriminative RBM with n hidden units is a parametric model of the joint distribution between a layer of hidden units (referred to as features) $h = (h_1, h_2, ..., h_n)$ and visible units made of the input data $v = (v_1, v_2, ..., v_m)$ and targets $y \in \{1,2,...C\}$, which can be defined as follows:

$$p(y, v, h) \propto e^{-E(y,v,h)}$$ (4.7)

where

$$E(y,v,h) = -h^T W v - b^T v - c^T h - d^T \vec{y} - h^T U \vec{y} \tag{4.8}$$

Here, (W, b, c, d, U) refer to the model parameters and $\vec{y} = \left(1_{y=i}\right)_{i=1}^{C}$ to the C classes. More details about the discriminative RBM model can be found in [50].

4.3.3   DEEP BELIEF NETWORKS

A deep belief network (DBN) is a generative probabilistic model that differs from conventional discriminative neural networks. DBNs are composed of one visible layer (observed data) and many hidden layers that have the ability to learn the statistical relationships between the units in the previous layer. They model the joint probability distribution over the input data (observations) and labels, which facilitates the estimation of both *P(Observations |Labels)* and *P(Labels |Observations)*, while conventional neural networks are limited only to the latter [51]. DBN has been proposed to address issues encountered when applying the Back-propagation algorithm to very deep neural networks, including:

1. The constraint of having a labeled dataset in the training phase.
2. The long time required to converge (slow learning process).
3. The increased number of free parameters that get trapped in poor local optima.

As depicted in Figure 4.2 (c), a DBN can be viewed as a composition of bipartite undirected graphical models, each of which is an RBM. Therefore, DBNs can be efficiently trained using an unsupervised greedy layer-wise algorithm, in which the stacked RBMs are trained one at a time in a bottom-to-top manner. For instance, consider training a DBN composed of three hidden layers, as shown in Figure 4.2(c).

According to the greedy layer-wise training algorithm proposed by Hinton et al. [52], the first RBM is trained using the CD algorithm to learn a layer (h_1) of feature representations from the visible units, as described in Section 4.1.2. Then, the hidden layer units (h_1) of the first RBM are used as visible units to train the second RBM. The whole DBN is trained when the learning of the final hidden layer is completed. A DBN with l layers can model the joint distribution between the observed data vector v and l hidden layers h_k as follows:

$$P\left(v, h^1, \ldots, h^l\right) = \left(\prod_{k=0}^{l-2} P\left(h^k \mid h^{k+1}\right)\right) P\left(h^{l-1}, h^l\right) \tag{4.9}$$

Here, $v = h^0$, $P\left(h^k \mid h^{k+1}\right)$ is the conditional distribution for the visible units given hidden units of the RBM associated with level k of the DBN, and $P\left(h^{l-1}, h^l\right)$ is the visible-hidden joint distribution in the top-level RBM. An example of a three-layer DBN as a generative model is shown in Figure 4.2 (d), where the symbol Q is introduced for exact or approximate posteriors of that model, which are used for bottom-up inference. During the bottom-up inference, the Q posteriors are all approximate except for the top-level $P(h^l \mid h^{l-1})$, which is formed as an RBM and then the exact inference is possible. Although the DBN is considered as one of the most popular unsupervised deep learning approaches, and has been successfully applied in a wide range of applications (e.g., face recognition [13], speech recognition [14], audio classification [15], etc.), one of the main issues of the DBN is that the feature representations it learns are very sensitive to the local translations of the input image, in particular when the pixel intensity values (raw data) are assigned directly to the visible units. This can cause the local facial features of the input image, which are known to be important for face recognition, to be disregarded.

4.3.4 Convolutional Neural Network

A convolutional neural network (CNN) is a feed-forward multi-layer neural network, which differs from traditional fully connected neural networks by combining a number of locally connected layers aimed at automated feature recognition, followed by a number of fully connected layers aimed at classification [51]. As shown in Figure 4.3, the CNN architecture is composed of several distinct layers, including sets of locally connected convolutional layers (each layer contains a number of learnable kernels), sub-sampling pooling layers, and one or more fully connected layers. The internal structure of the CNN is associated with three architectural concepts, which make the CNN successful in resolving different problems in image processing, computer vision, and pattern recognition. The first concept is applied in both convolutional and pooling layers, in which each neuron receives input from a small region of the previous layer called the local receptive field, which is equal in size to a convolution kernel [53]. This local connectivity scheme ensures that the trained CNN produces strong responses to capture local dependencies and extracts elementary features in the input image (e.g., edges, curves, etc.), which can play a significant role in minimizing the intra-class variations and inter-class similarities and hence increasing the discriminative power of the learned facial feature representations. Secondly, the convolutional layer applies the sharing parameters (weights) scheme in order to control the model capacity and

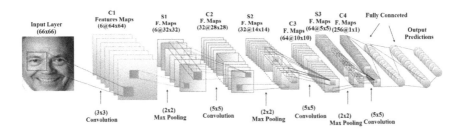

FIGURE 4.3 An illustration of the CNN architecture, where the gray and green squares refer to the activation maps and the learnable convolution kernels, respectively. The crossed lines between the last two layers show the fully connected neurons [55].

reduce its complexity. At this point, a form of translational invariance is obtained using the same convolution kernel to detect a specific feature at different locations in the face image [54]. Finally, the nonlinear down-sampling applied in the pooling layers reduces the spatial size of the convolutional layer's output and reduces the model's number of free parameters. Together, these characteristics make the CNN very robust and efficient at handling image deformations and other geometric transformations, such as translation, rotation, and scaling [51]. More details on the CNN can be found in [55].

4.4   THE PROPOSED METHODOLOGY

As depicted in Figure 4.4, a novel multi-biometric face recognition system named *FaceDeepNet* system is proposed to learn high-level facial feature representations by training the proposed deep learning approaches (e.g., CNN and DBN) on top of local facial feature representations obtained from the FDT approach instead of the pixel intensity representations. Firstly, the main stage of the proposed FDT-DRBM approach is described in detail. This is followed by describing how to learn additional and complementary representations by applying a *FaceDeepNet* system on top of existing local feature representations.

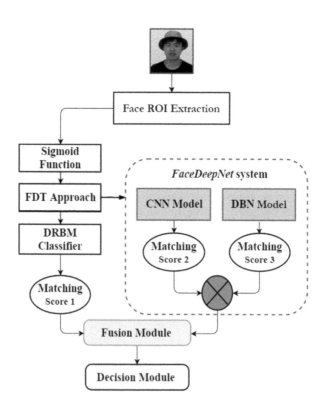

FIGURE 4.4   Illustration of the proposed multi-biometric face recognition system.

4.4.1   THE PROPOSED **FDT-DRBM** APPROACH

The proposed face recognition algorithm starts by detecting the face region using a Viola-Jones face detector [56]. Then, after rescaling the input image $f(x, y)$ to the range of $[0, 1]$, a simple pre-processing algorithm using a sigmoid function is applied as follows:

$$g(x,y) = \frac{1}{1 + e^{\left(c*(Th - f(x,y))\right)}} \tag{4.10}$$

Here, $g(x, y)$ is the enhanced image. The contrast factor (c) and the threshold value (Th) are empirically set to be **5** and **0.3**, respectively. As shown in Figure 4.5, the advantage of the sigmoid function is to reduce the effect of lighting changes by expanding and compressing the range of values of the dark and bright pixels in the face image, respectively. This is followed by estimating the fractal dimension value of each pixel in the enhanced face image using the proposed FDT approach. As mentioned above, the fractal dimension has many important characteristics, for instance its ability to reflect the roughness and fluctuations of a face image's surface, and to represent the facial features under different environmental conditions (e.g., illumination changes).

However, fractal estimation approaches are very time consuming and cannot meet real-time requirements [7]. Thus, the face image is rescaled to (64×64) pixel after detecting the face region in order to speed up the experiments and meet the real-time system demands. The result is then reshaped into a row feature vector FDT_{Vector} which is used as an input to a nonlinear classifier using DRBM to efficiently model the joint distribution of inputs and target classes. In this chapter, the

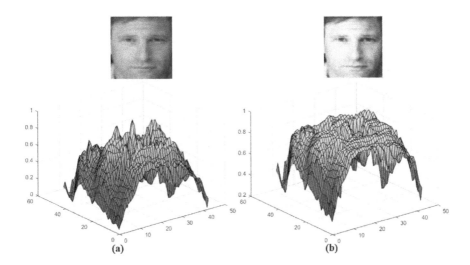

FIGURE 4.5   Output of the image contrast enhancement procedure. (a) The original face image. (b) The enhanced face image. p_{max} and p_{min} refer to maximum and minimum pixel values in the enhanced face image.

number of the produced matching scores and predicted classes using the DRBM classifier is varied according to the applied recognition task. For instance, the goal of the identification system is to determine the true identity (I) of the query trait based on these N matching scores, where $I \in \{I_1, I_2, I_3, \cdots, I_N, I_{N+1}\}$. Here, $(I_1, I_2, I_3, \cdots, I_N)$ correspond to the identities of the N persons enrolled in the dataset, while (I_{N+1}) point to the "reject" option, which is generated when no associated identity can be determined for the given query trait; this is known as open set identification. However, when the biometric system assumes that the given query trait is among the templates enrolled in the dataset, this is referred to as closed set identification [57]. In this proposed system, the open set identification protocol is adopted, because of its efficiency with large-scale, real-world applications (e.g., surveillance and watch list scenarios) [58]. On the other hand, in the verification task the DRBM classifier is used as a binary classifier to determine whether a pair of face images belong to the same person or not. The next two subsections describe in more detail the FDT and the DRBM methods mentioned above.

4.4.1.1   Fractal Dimension Transformation

Using the proposed FDT approach, the face image of size $I(M{\times}N)$ is transformed to its fractal dimension form using the differential box-counting (DBC) algorithm where a kernel function $f(p, q)$ of size $(n{\times}n)$ pixel is applied on the entire face image to produce a 3D matrix that represents the number of boxes necessary to overlay the whole face image at each pixel as follows:

$$N_d\left(x,y,d\right) = \sum_{p=-a}^{a}\sum_{q=-b}^{b} f\left(p,q\right) I\left(x+p, y+q\right)\left(\frac{r_{max}}{r}\right)^2 \qquad (4.11)$$

Here, $d = \left(1, 2, 3, \ldots, r_{max} - 1\right)$ is the third dimension of matrix $N_d\left(x, y, d\right)$, which represents the number of the produced feature maps obtained from applying the kernel function $f(p, q)$ along with different values of the scaling factor (r). Here, the values of the scaling factor (r) are empirically chosen to be in the range between 2 and 9. The kernel function $f(p, q)$ operates by block processing on ($7{\times}7$) neighboring pixels of the face image, and calculating the fractal dimension value of each pixel from its surrounding neighbors, as follows:

$$f(p,q) = \sum_{p=-a}^{a}\sum_{q=-b}^{b} \mathbf{floor}\left[\frac{p_{max} - p_{min}}{r}\right] + 1 \qquad (4.12)$$

Here, a and b are non-negative integer variables, which are used to center the kernel function on each pixel in the face image, and are defined as a and $b = ceil\left((n - 1)/2\right)$. p_{max} and p_{min} are the highest- and lowest-intensity values of neighboring pixels in the processing block. The size of the kernel function was determined empirically, noting that increasing its size can affect the accuracy of the calculated fractal dimension, causing the obtained image to become less distinct, while decreasing its size can result in an insufficient number of surrounding pixels to calculate the fractal dimension value accurately. Once the $N_d\left(x, y, d\right)$ matrix is obtained, each element from each array in $N_d(x, y, d)$ will be stored in a new row vector (V). In other words, the first element in all arrays of $N_d(x, y, d)$ will compose vector (V_1), and all second elements will compose (V_2), etc., as follows:

$$
\begin{pmatrix} V_1 \\ V_2 \\ \vdots \\ V_{M \times N} \end{pmatrix} = \begin{pmatrix} p_{111} & p_{112} & \cdots & \cdots & p_{11r_{max}} \\ p_{121} & p_{122} & \cdots & \cdots & p_{12r_{max}} \\ \vdots & \vdots & \ddots & & \vdots \\ \vdots & \vdots & & \ddots & \vdots \\ p_{MN1} & p_{MN2} & \cdots & \cdots & p_{MNr_{max}} \end{pmatrix}
\tag{4.13}
$$

Finally, the fractal dimension value at each pixel (x, y) is computed as the fractal slope of the least square linear regression line as follows:

$$
FD(x, y) = \frac{S_v(x, y)}{S_r}
\tag{4.14}
$$

Here, (S_v) and (S_r) are the sums of squares and can be computed as follows:

$$
S_v(x, y) = \sum_{r=1}^{j} \log(r) \left(v_{(x,y),r} \right) - \frac{\left(\sum_{r=1}^{j} \log(r) \right) \left(\sum_{r=1}^{j} v_{(x,y),r} \right)}{j}
\tag{4.15}
$$

$$
S_r(x, y) = \sum_{r=1}^{j} (\log(r))^2 - \frac{\left(\sum_{r=1}^{j} \log(r) \right)^2}{j}
\tag{4.16}
$$

Here, $j = r_{max} - 1$ and a fractal transformed face image is illustrated in Figure 4.6.

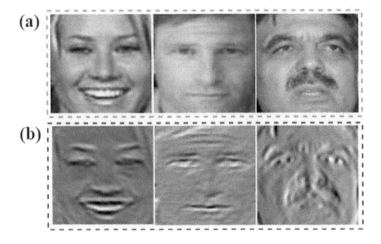

FIGURE 4.6 (a) Original face images. (b) FDT fractal-transformed images.

4.4.1.2 Discriminative Restricted Boltzmann Machines

In the classification stage, a single RBM that has two sets of visible units is trained as a nonlinear classifier to efficiently model the joint distribution of inputs and target classes. In addition to the visible units that represent a data vector, there is a Softmax label unit that represents the predicted classes. After training, each possible label is tried in turn and the one that gives lowest free energy is chosen as the most likely class. As reported in the literature, it is intractable to compute the $v_i h_{j\text{model}}$ in Eq. 4.4. Therefore, the contrastive divergence (CD) algorithm [59] has become the standard learning method to update the RBM parameters by sampling k steps from the RBM distribution to approximate the second term in Eq. 4.4. Efficient training requires optimized training parameters (e.g., learning rate, the weight-cost, the momentum, the initial values of the weights, the size of each mini-batch, the number of hidden units, etc.). Other decisions need to be made, including the types of hidden units to be used, how many times to update the binary states of the hidden units for each training epoch, how to monitor the progress of learning, and when to terminate the training process. One step of a CD algorithm for a single sample can be implemented as follows:

1. Visible units (v_i) are initialized using training data, and the probabilities of hidden units are computed with Eq. 4.2. Then, a hidden activation vector (h_j) is sampled from this probability distribution.
2. Compute the outer product of (v_i) and (h_j), which refers to the positive phase.
3. Sample a reconstruction (v_i') of the visible units from (h_j) with Eq. 4.3, and then from (v_i') resample the hidden units' activations (h_i'). (One Gibbs sampling step.)
4. Compute the outer product of (v_i') and (h_i'), which refers to the negative phase.
5. Update weights matrix and biases with Eq. 4.4, Eq. 4.5, and Eq. 4.6.

The computation steps of the CD-1 algorithm are graphically shown in Figure 4.7. In the CD learning algorithm, k is usually set to 1 for many applications. In this work, the weights were initialized with small random values sampled from a zero-mean normal distribution and standard deviation of 0.02. In the positive phase, the binary states of the hidden units are determined by computing the probabilities of weights

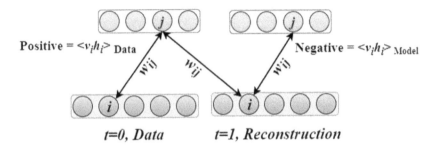

$t=0$, *Data* $t=1$, *Reconstruction*

FIGURE 4.7   A graphic illustration of a single step of the CD algorithm.

and visible units. Since the probability of training data is increased in this phase, it is called positive phase. Meanwhile, the probability of samples generated by the model is decreased during the negative phase. A complete positive-negative phase is considered as one training epoch, and the error between produced samples from the model and actual data vector is computed at the end of each epoch. Finally, all the weights are updated by taking the derivative of the probability of visible units with respect to weights, which is the expectation of the difference between positive phase contribution and negative phase contribution. Algorithm 1 shows pseudo-code of the procedure proposed to train the DRBM classifier using the CD algorithm.

Algorithm 1. Find the optimal DRBM classifier.

Input: Labeled dataset: **Training Set** (x_i, y_i) and **Testing Set** (x_i, y_i).
Output: *Optimal_h* = The best number of hidden units (h).
Initialization: Max_Accuracy= 0, **LR** = 10^{-2}, **WD** = 0.0002, **MM**=0.9, and **Epoch**=300;
for h_now = 1000 : 100 : 5000
    for *Epoch* = 1 : 300
- Train the DRBM classifier using the CD learning algorithm with one step of Gibbs sampling.
- Apply the 10-fold cross-validation evaluation procedure on the **Training Set**.
- Calculate the AER over 10-fold using Eq. 4.17.
- Calculate the Validation_Accuracy = 1-AER;
- **if** Validation_Accuracy>Max_Accuracy
      Max_Accuracy = Validation_Accuracy ;
      Optimal_h = h_now:
    end
    end
end
- Train the DRBM classifier on the whole **Training Set** using the *Optimal_h*.
- Evaluate the generalization ability of the final trained DRBM classifier using the **Testing Set**.

4.4.2 THE PROPOSED FACEDEEPNET SYSTEM

An efficient deep learning system is employed. This system, whose architecture is based on a combination of a CNN and a DBN to extract discriminative facial features from the input image of the same person, will be referred to as *FaceDeepNet*. To the best of our knowledge, this is the first work that investigates the potential use the CNN and the DBN together in a unified multi-biometric face recognition system operating in both the identification and verification modes. Here, we argue that using the local feature representations obtained from a powerful handcrafted-descriptor instead of the pixel intensity representations as an input to the proposed deep learning approaches can significantly improve the ability of the proposed *FaceDeepNet* system to learn more discriminating facial features, with a shorter training time required to obtain the final trained model. As depicted in Figure 4.4, the local facial features are firstly extracted using the proposed FDT approach. Then, the extracted local features are assigned to the feature extraction units of the proposed deep learning approaches (e.g., CNN and DBN) to learn additional and complementary feature representations.

In this work, the architecture of the proposed CNN involves a combination of convolutional layers and sub-sampling max-pooling. The top layers in the proposed CNN are two fully connected layers for the multi-class classification purpose. Then, the output of the last fully connected layer is fed into the Softmax classifier, which produces a probability distribution over the N class labels. Finally, a cross-entropy loss function, a suitable loss function for the classification task, is used to quantify the agreement between the predicted class scores and the target labels, and to calculate the cost value for different configurations of CNN. On the other hand, the architecture of the proposed DBN stacks a number of RBMs (as hidden layers). The first $N-1$ layers of the network can be seen as feature detection layers. The last layer of the network is trained as a discriminative model associated with Softmax classifier for the classification task. Finally, the hidden layers of the proposed DBN are trained using an unsupervised greedy layer-wise algorithm (e.g., the CD learning algorithm), in which the stacked RBMs are trained one at a time in a bottom-to-top manner. The proposed deep learning approaches (e.g., CNN and DBN) are trained using a powerful training methodology equipped with a number of training strategies and techniques (e.g., Back-propagation, AdaGrad algorithm, Dropout technique, Data augmentation, etc.), in order to avoid overfitting problems and increase the generalization ability of the proposed *FaceDeepNet* system. For more details on how both the proposed CNN and DBN are trained, the reader is referred to [55, 7]. Finally, as all the employed classifiers produce the same type of output (similarity score), and all of them have the same numeric range **[0, 1]**, no normalization procedure is needed before applying the score fusion method. In this work, the weighted sum rule is used to fuse the matching scores produced from the proposed CNN and DBN.

4.5   EXPERIMENTAL RESULTS

In this section, the performances of the proposed unimodal approaches (e.g., using only FDT-DRBM, DBN, and CNN) and multi-biometric approaches (e.g., using *FaceDeepNet*) are tested on the most challenging face datasets currently available in the public domain in order to demonstrate their effectiveness and compare their performances with other existing face recognition approaches. A brief description of the face datasets used in these experiments is given. Then, an elaborated evaluation and comparison with the state-of-the-art face recognition approaches in both face identification and verification tasks are presented. In addition, some insights and findings relevant to learning additional and complementary facial feature representations by training the proposed deep learning approaches on top of local feature representations are presented. The system code was written to run in MATLAB R2018a and later versions. The recognition time of the proposed approaches was measured by implementing them on a personal computer with the Windows 10 operating system, a 3.60 GHz Core i7-4510U CPU, and 16 GB of RAM. It should be noted that the recognition times of the proposed approaches are proportional to the number of registered subjects and their images in the database. However, the test time is less than 4 seconds on average, which is fast enough to meet the demands of real-world applications. Finally, the matching scores obtained from the multi-biometric face recognition systems are combined using different fusion methods at the score level

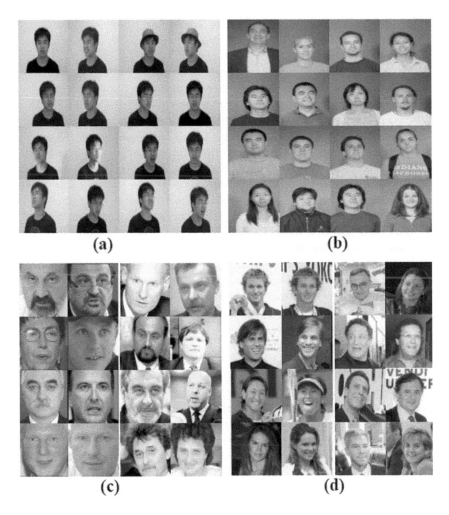

FIGURE 4.8 Examples of face images in four face datasets: (a) SDUMLA-HMT, (b) FRGC V 2.0, (c) UFI, and (d) LFW.

(e.g., using product, sum, weighted sum, max, and min rule) and rank level (e.g., using highest rank, Borda count, and logistic regression).

4.5.1 DESCRIPTION OF FACE DATASETS

All the experiments were conducted on four large-scale unconstrained face datasets: SDUMLA-HMT [60], Face Recognition Grand Challenge Version 2.0 (FRGC V 2.0) [61], Unconstrained Facial Images (UFI) [62], and Labelled Faces in the Wild (LFW) [63]. Some examples of face images from each dataset are shown in Figure 4.8.

- **SDUMLA-HMT Face dataset** [60]: This includes 106 subjects; each one has 84 face images taken from 7 viewing angles, and under different experimental conditions including facial expressions, accessories, poses, and illumination. The main purpose of this dataset is to simulate real-world conditions during face image acquisition. The image size is (640×480) pixels.

- **FRGC V 2.0 dataset** [61]: This is a large-scale challenging dataset collected at the University of Notre Dame and is composed of two main partitions. The first part is the training partition; it was collected in the 2002–2003 academic year and is used for training still face recognition approaches. This partition consists of 12,776 images from 222 subjects, with 6,388 controlled still images and 6,388 uncontrolled still images. This large training partition consists of 9 to 16 sessions per subject. The second part is the validation partition; it was collected during the 2003–2004 academic year and contains images from 466 subjects collected in 4,007 subject sessions. This partition consists of 1 to 22 subject sessions per subject. The image size is either (1704×2272) pixels or (1200×1600) pixels, stored in JPEG format.

- **Unconstrained Facial Images (UFI) dataset** [62]: It contains real face images obtained from a large set of images owned by the Czech News Agency (CTK). The UFI dataset is used mainly for evaluating face identification approaches; however, it is possible to be used in other related tasks (e.g., face detection, verification, etc.). The UFI dataset is composed of two partitions: cropped images (CI) and large images (LI). The CI partition contains images of 605 subjects with an average of 7.1 images per subject in the training set and one in the test set. The images are cropped to a size of (128×128) pixels, and Figure 4.8 (c) shows some examples from the CI partition. The LI partition contains a total of 530 subjects with an average of 8.2 images per subject. The size of images in this partition is (384×384) pixels, and some examples of this partition are shown in Figure 4.8 (c). The main goal of the large images partition is to report the final performance of the proposed face recognition system and compare it with other existing systems.

- **LFW dataset** [63]: This contains a total of 13,233 images taken from 5,749 subjects, where 1,680 subjects appear in two or more images. In the LFW dataset, all images were collected from Yahoo! News articles on the web, with a high degree of intrapersonal variations in facial expression, illumination conditions, occlusion from wearing hats and glasses, etc. It has been used to address the problem of unconstrained face verification tasks in recent years. The image size is (250×250) pixels, and some examples of the matched pairs and mismatched pairs are shown in Figure 4.7 (d).

4.5.2 FACE IDENTIFICATION EXPERIMENTS

Initially, the robustness and effectiveness of the proposed face recognition approaches were tested to address the unconstrained face identification problem using large-scale unconstrained face datasets: SDUMLA-HMT, FRGC V 2.0, and UFI dataset. The SDUMLA-HMT dataset was employed as the main dataset to fine-tune the hyper-parameters of the proposed FDT-DRBM approach (e.g., number of hidden units in the RBM) as well as the hyper-parameters of the proposed deep learning approaches used in the *FaceDeepNet* system (e.g., number of layers, number of filters per each layer, learning rate, etc.), as it has more images per person in its image gallery than the other datasets. This allowed more flexibility in dividing the face images into training, validation, and testing sets.

4.5.2.1 Parameter Settings of the FDT-DRBM Approach

The most important hyper-parameter in the proposed FDT-DRBM approach is the number of the hidden units in the DRBM classifier. The other hyper-parameters (e.g., learning rate, the number of epochs, etc.) were determined based on domain knowledge from the literature to be a learning rate of 10^{-2}, a weight decay of 0.0002, momentum of 0.9, and 300 epochs, as in [7]. The weights were initialized with small random values sampled from a zero-mean normal distribution and standard deviation of 0.02. In this work, the number of the hidden units was determined empirically by varying its values from 1,000 units to 5,000 units in steps of 100 units, using the CD learning algorithm with one step of Gibbs sampling to train the DRBM as a nonlinear classifier, as described in Section 4.1.2. Hence, 41 experiments were conducted using 60% randomly selected samples per person for the training set, and the remaining 40% for the testing set. In these experiments, the parameter optimization process is performed on the training set using the 10-fold cross-validation procedure that divides the training set into k subsets of equal size. Consecutively, one subset is used to assess the performance of the DRBM classifier trained on the remaining $k-1$ subsets. Then, the average error rate (AER) over 10 trials is computed as follows:

$$AER = \frac{1}{K}\sum_{i=1}^{k}Error_i \qquad (4.17)$$

Here, **Error**$_i$ refers to the error rate per trial. After finding the best number of hidden units with higher validation accuracy rate (VAR), the DRBM classifier was trained using the whole training set, and its performance in predicting unseen data properly was then evaluated, using the testing set. Figure 4.9 shows the VAR produced throughout 41 experiments. One can see that the higher VAR was obtained when the 3,000 units were employed in the hidden layer of the DRBM classifier. To verify the proposed FDT-DRBM approach, its performance was compared with two current state-of-the-art face recognition approaches, namely the Curvelet-Fractal approach [7] and the Curvelet Transform-Fractional Brownian Motion (CT-FBM) approach [64], using the same dataset. Figure 4.10 shows the cumulative match characteristic (CMC) curve to visualize their performance. It can be seen in Figure 4.10 that the proposed DFT-DRBM approach has the potential to outperform the CT-FBM approach, as the Rank-1 identification rate has dramatically increased from 0.90 to 0.95 using the CT-FBM to more than 0.9513 to 1.0 using the DFT-DRBM approach. On the other hand, despite the Curvelet-Fractal approach achieving slightly a higher Rank-1 identification rate, it is observed that after Rank-10 identification rate the accuracy of the Curvelet-Fractal approach drops compared to the FDT-DRBM approach.

As mentioned previously, the main structure of the *FaceDeepNet* system is based on building deep learning representations and fusing the matching scores obtained

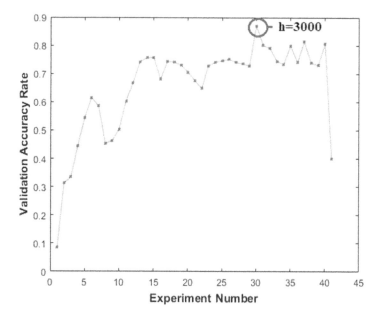

FIGURE 4.9   The VAR produced throughout 41 experiments for finding the best number of hidden units.

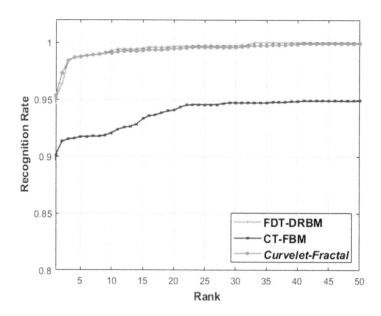

FIGURE 4.10   Performance comparison between the FDT-DRBM, Curvelet-Fractal, and CT-FBM approaches on the SDUMLA-HMT dataset.

from two discriminative deep learning approaches (e.g., CNN and DBN). It is well known in the literature that the major challenge of using deep learning approaches is the number of model architectures and hyper-parameters that need to be tested, such as the number of layers, the number of units per layer, the number of epochs, etc. In this section, a number of extensive experiments are performed to find the best model architecture for both CNN and DBN, along with studying and analyzing the influence of the different training parameters on the performance of the proposed deep learning approaches. All these experiments were conducted on the SDUMLA-HMT face dataset using the same sizes of the training and testing sets described in Section 5.2.1, and the parameters with the best performance (e.g., highest VAR and best generalization ability) were kept to be used later in finding the best network architecture.

For an initial CNN architecture, the architecture of our *IrisConvNet* system as described in Al-Waisy et al. [55] was used to test its generalization ability and adaptation to another new application (e.g., face recognition). This initial CNN architecture will be denoted as *FaceConvNet-A* through the subsequent experiments. The main structure of the *IrisConvNet* system is shown in Figure 4.3. Initially, the *FaceConvNet-A* system was trained using the same training methodology along with the same hyper-parameters described in [55]. However, it was observed that when 500 epochs were evaluated, the obtained model started overfitting the training data, and poor results were obtained on the validation set. Thus, a number of experiments were conducted to fine-tune this parameter. Firstly, an initial number of epochs was set to 100 epochs, and then larger numbers of epochs were also investigated, including 200, 300, and 400. Figure 4.11 (a) shows the CMC curves used to visualize the performance of the final obtained model on the validation set. It can be seen that as long as the number of epochs is increased, the performance of the last model gets better and the highest accuracy rate was obtained when 300 epochs were evaluated. Furthermore, it was also observed that a better performance can be obtained by adding a new convolution layer (C0= 10@64×64) before the first one (C1= 6@64×64), using the same filter size and increasing the number of filters per these two layers from 6 to

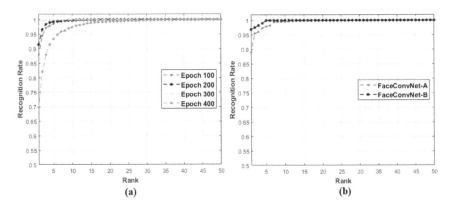

FIGURE 4.11 CMC curves for (a) epoch number parameter evaluation of CNN and (b) performance comparison between *FaceConvNet-A* and *FaceConvNet-B* on the testing set of the SDUMLA-HMT dataset.

10 learnable filters. This new CNN architecture was denoted as the *FaceConvNet-B* system, and it was used in our assessment procedure for all remaining experiments instead of the *FaceConvNet-A* system. The CMC curves of the *FaceConvNet-A* and *FaceConvNet-B* systems are shown in Figure 4.11 (b) to visualize their performances on the testing set of the SDUMLA-HMT dataset. Table 4.1 summarizes the details of the *FaceConvNet-B* system and its hyper-parameters.

Following, the training methodology described in [7], the DBN approach was greedily trained using input data acquired from the FDT-DRBM approach. Once the training of a given RBM (hidden layer) is completed, its weights matrix is frozen, and its activations are served as input to train the next RBM (next hidden layer) in the stack. Unlike [7], in which the DBN models are composed of only 3 layers, here larger and deeper DBM models have been investigated by stacking five RBM models. As shown in Table 4.2, five different 5-layer DBN models were greedily trained in a bottom-up manner using different numbers of hidden units only in the last three layers. For the first four hidden layers, each one was trained separately as an RBM model in an unsupervised way, using the CD learning algorithm with one step of Gibbs sampling. Each individual RBM was trained for 100 epochs with the learning rate set to be 0.01, a momentum value of 0.95, a weight-decay value of 0.0005, and a mini-batch size of 100. The weights of each model were initialized with small random values sampled from a zero-mean normal distribution and standard deviation of 0.02. The last RBM model was trained in a supervised way as a nonlinear classifier associated with Softmax units for the multi-class classification purpose. In this supervised phase, the last RBM model was trained using the same values of the hyper-parameters used to train the first four models. Finally, in the fine-tuning phase, the whole DBN model was trained in a top-down manner using the Back-propagation algorithm equipped with the Dropout technique, to find the optimized parameters and to avoid overfitting. The Dropout ratio was set to 0.5 and the number of epochs through the training set was determined using an early stopping procedure, in which

TABLE 4.1

Details of Hyper-Parameters for the Proposed Deep Learning Approaches (e.g., CNN and DBN)

FaceConvNet-B		5-Layer DBN Model	
Hyper-parameter	Value	Hyper-parameter	Value
No. of Conv. Layers	5	CD Learning Algorithm	1 step of Gibbs sampling
No. of Max-Pooling Layers	3	No. of Layers	5 RBMs
Optimization Method	Adagrad	No. of Epochs for each RBM	100
Momentum	0.9	Momentum	0.95
Weight-Decay	0.0005	Weight-Decay	0.0005
Dropout	0.5	Dropout	0.5
Batch Size	100	Batch Size	100
Learning Rate	0.01	Learning Rate	0.01
Total No. of Epochs	300	Total No. of Epochs (BW)	1000

TABLE 4.2

Rank-1 Identification Rates Obtained for Different DBN Architectures Using Validation Set

DBN Model	Accuracy Rate
4096-4096-2048-1024-512	0.92
4096-4096-2048-1024-1000	0.94
4096-4096-2048-2000-1000	0.95
4096-4096-3000-2000-1000	0.94
4096-4096-3000-2000-2000	0.97

the training process is stopped as soon as the classification error on the validation set starts to rise again. As can be seen in Table 4.2, the last DBN model provided significantly better recognition rate compared to the other hidden layers investigated. More information on the FaceConvNet-B and its hyper-parameters are presented in Table 4.1.

Moreover, a few experiments were also carried out using the SDUMLA-HMT dataset to find out how the performance of the proposed deep learning approaches can be greatly improved by assigning the local feature representations to the visible units of the DNNs instead of the raw data, as a way of guiding the neural network to learn only the useful facial features. As shown in Figure 4.4, the local facial features are firstly extracted using the proposed FDT-DRBM approach. Then, these extracted local facial features are assigned to the feature extraction units of the CNN and DBN to learn additional and complementary feature representations. As expected and as shown in Table 4.3, our experiments demonstrate that training the proposed deep learning approaches on the top of the FDT-DRBM approach's output can significantly increase the ability of the DNNs to learn more discriminating facial features with a shorter training time. Generally, the CNN produced the highest recognition rate compared to the DBN, due to its ability to capture the local dependencies and extract elementary features in the input image (e.g., edges, curves, etc.), which are proven to be important for face recognition. Thus, in the fusion phase when the

TABLE 4.3

Rank-1 Identification Rates Obtained for the Proposed Deep Learning Approaches (CNN and DBN) Using Different Types of Input Data

	Raw Data		Local Facial Features	
Approach	Accuracy Rate	Training Time	Accuracy Rate	Training Time
CNN	0.89	12 h & 40 min	0.96	4 h & 10 min
DBN	0.85	15 h & 25 min	0.94	4 h & 40 min

weighted sum rule is applied, a higher weight was assigned to the proposed CNN compared to the DBN approach, such that the maximum recognition rate is achieved.

4.5.2.2 Testing the Generalization Ability of the Proposed Approaches

To further examine the robustness of the proposed face recognition approaches, a number of comprehensive experiments were conducted on the FRGC V 2.0, and UFI datasets, and the results obtained were compared with the current state-of-the-art approaches. For a fair comparison, the performance of the FDT-DRBM approach was evaluated using the standard evaluation protocols of the FRGC V 2.0 and UFI datasets, described in [61, 62], respectively. It is worth noting that DNNs need to be trained on a massive amount of training data to achieve satisfactory prediction and prevent overfitting. Thus, to increase the generalization ability of the proposed deep learning approaches (e.g., CNN and DBN), the data augmentation procedure as described in [7] was applied to the gallery set of these two datasets. Then, their performances during the learning process were observed on a separate validation set taken from the fully augmented gallery set. According to the standard evaluation protocol, the FRGC V 2.0 dataset consists of six experiments. The experiments used to evaluate the performance of the 3D face recognition systems are not relevant to the objectives of this work. Hence, the focus will be on experiments 1, 2, and 4. In experiment 1, both the gallery and testing sets consist of a single controlled still image of each person. Experiment 2 addresses the effect of using multiple still images per person on performance. The gallery in this experiment consists of four images for each person, taken in the same subject session. Likewise, a testing set is composed of four images per person. Experiment 4 addresses the problem of uncontrolled face recognition. In experiment 4, the gallery is composed of a single controlled still image, and the testing set is composed of a single uncontrolled still image. In the standard UFI evaluation protocol, both partitions (CI and LI) are divided into training and testing sets. For comparison purposes, the face images in the cropped images were kept at the same size: (128×128) pixels.

Firstly, the matching scores are combined using different fusion methods at the score level (e.g., using product, sum, weighted sum, max, and min rule) and rank level (e.g., using highest rank, Borda count, and logistic regression). As shown in Figure 4.4, three matching scores are produced when the face trait is submitted by the user at identification point. One matching score is obtained from the FDT-DRBM approach and two matching scores from the *FaceDeepNet* system. In the implementation of weighted fusion methods, a higher weight was assigned to the proposed *FaceDeepNet* system compared to the FDT-DRBM approach, to ensure that the maximum recognition rate is achieved. The Rank-1 identification rates, using different fusion methods at the score-level and rank-level fusion, are listed in Tables 4.4 and 4.5, respectively. In general, the results obtained show a noticeable improvement in the performance of the proposed multi-biometric face recognition system, compared to applying only the FDT-DRBM approach or the *FaceDeepNet* system, by achieving higher Rank-1 identification rates in most cases. From Tables 4.4 and 4.5, one can see that the highest Rank-1 identification rates on all the testing sets of all the face datasets were achieved by employing the weighted sum rule and the highest rank method at the score-level and rank-level fusion, respectively. In this work, the

TABLE 4.4

Rank-1 Identification Rate of the Proposed Multi-Biometric Face Recognition System on Three Different Face Datasets Using Score-Level Fusion SR and PR are referred to the Sum Rule and Product Rule, respectively

Dataset		FDT-DRBM	*FaceDeepNet*	Score Fusion Method				
				SR	WSR	PR	Max	Min
FRGC V 2.0	Exp. 1	94.11	97.62	95.12	97.34	94.92	95.34	95.21
	Exp. 2	95.10	99.54	97.33	100	95.54	97.82	96.32
	Exp. 4	96.21	98.33	97.13	99.12	97.13	97.12	97.12
UFI	CI	92.23	93.60	94.32	96.66	92.12	92.31	93.21
	LI	93.12	95.29	96.12	97.78	93.64	95.12	94.12
SDUMLA-HMT		95.13	95.13	98.89	100	97.53	96.23	96.21

TABLE 4.5

Rank-1 Identification Rate of the Proposed Multi-Biometric Face Recognition System on Three Different Face Datasets Using Rank-Level Fusion HR, BC, and LR are referred to Highest Ranking, Borda Count, and Logistic regression, respectively

Dataset		FDT-DRBM	*FaceDeepNet*	Rank Fusion Method		
				HR	BC	LR
FRGC V 2.0	Exp. 1	94.11	97.62	97.12	96.33	99.45
	Exp. 2	95.10	99.54	98.31	97.1	100
	Exp. 4	96.21	98.33	97.23	97.67	100
UFI	CI	92.23	93.60	95.45	95.54	97.34
	LI	93.12	95.29	96.23	96.11	98.88
SDUMLA-HMT		95.13	98.89	98.89	96.42	100

highest-ranking approach was adopted for comparing the performance of the proposed multi-biometric face recognition system with other existing approaches, due to its efficiency compared to other fusion methods in exploiting the strength of each classifier effectively and breaking the ties between the subjects in the final ranking list. Table 4.6 compares the Rank-1 identification rates of the proposed approaches and the current state-of-the-art face recognition approaches using the three experiments with the FRGC V 2.0 dataset.

Although some approaches, such as the Partial Least Squares (PLS) approach [65] achieved a slightly higher identification rate in experiments 1 and 2, they obtained inferior results in the remaining experiments with the FRGC V 2.0 dataset. In addition, the *FaceDeepNet* system and the fusion results obtained from the proposed approaches achieved higher identification rates on all the experiments with the FRGC V 2.0 dataset. Table 4.7 compares the Rank-1 identification rates of the proposed approaches and the current state-of-the-art face recognition approaches

TABLE 4.6

Comparison of the Proposed
Multi-Biometric Face Recognition
System with the State-of-the-Art
Approaches on FRGC V 2.0 Dataset

	FRGC V 2.0		
Approach	Exp.1	Exp.2	Exp.4
Deep C2D-CNN [67]	–	–	97.86
PLS [65]	97.5	99.4	78.2
LBP+HDM[68]	–	–	63.7
FDT-DRBM	94.11	95.10	96.21
FaceDeepNet	97.62	99.54	98.33
Fusion (LR)	99.45	100	100

TABLE 4.7

Comparison of the Proposed
Multi-Biometric Face
Recognition System with the
State-of-the-Art Approaches
on UFI Dataset

	UFI	
Approach	CI	LI
E-LBP [69]	65.28	–
MGM+LBP [70]	51.07	–
POEMmask-50 [71]	68.93	–
M- BNCC [72]	74.55	–
LBPHS [62]	55.04	31.89
LDPHS [62]	50.25	29.43
POEMHS [62]	67.11	33.96
FS-LBP [62]	63.31	43.21
FDT-DRBM	92.23	93.12
FaceDeepNet	93.60	95.29
Fusion (LR)	97.34	98.88

using these two partitions of the UFI dataset. It can be seen that we were able to achieve a higher Rank-1 identification rate using only the FDT-DRBM approach compared with other existing approaches, and better results were also achieved using the *FaceDeepNet* system and the fusion framework on both partitions of the UFI dataset. Finally, the authors in [66] have proposed a multimodal biometric system

TABLE 4.8

Comparison of the Proposed Multi-Biometric Face Recognition System with the State-of-the-Art Approaches on SDUMLA-HMT Dataset

Approach	Accuracy Rate
MLP [66]	93.35
CLVQ [66]	96.54
CRBF[66]	92.25
FDT-DRBM	95.13
FaceDeepNet	98.89
Fusion (LR)	100

using only the face trait in the SDUMLA-HMT dataset. Therefore, for the purpose of comparison, the best results obtained on the SDUMLA-HMT dataset using the highest-ranking approach were employed, and the results are listed in Table 4.8. It can be seen that the accuracy rate of 100% achieved using the proposed system is higher than the best results reported in [66], which is 96.54 % using the CLVQ approach.

4.5.3 FACE VERIFICATION EXPERIMENTS

In this section, LFW datasets were used to evaluate the robustness and effectiveness of the proposed face recognition approaches to address the unconstrained face verification problem. The face images in the LFW dataset were separated into two distinct views. "View-1" is used for selecting and fine-tuning the parameters of the face recognition model, while "View-2" is used to report the final performance of the selected recognition model. In "View-2," the face images are paired into 6,000 pairs, with 3,000 pairs labeled as positive pairs, and the rest as negative pairs. As described in [63], the final performance is reported by calculating the mean accuracy rate ($\hat{\mu}$) and the standard error of the mean accuracy (SE) over 10-fold cross-validation, with 300 positive and 300 negative image pairs per fold. For a fair comparison between all face recognition approaches, six evaluation protocols were pre-defined by the creators of the LFW dataset, as described in [73].

In these experiments, the *"Image-Restricted, Label-Free Outside Data"* protocol is followed, where only the outside data is used to train the proposed deep learning approaches. Additionally, the aligned LFW-a* dataset is used, and the face images were resized to (64×64) pixels after the face region has been detected using pre-trained Viola-Jones[†] face detector. No data augmentation or outside data (e.g.,

---

* http://www.openu.ac.il/home/hassner/data/lfwa/
[†] Incorrect face detection results were assessed manually to ensure that all the subjects were included in the subsequent evaluation of the proposed approaches.

creating additional positive/negative pairs from any other sources) was used in the training phase of the proposed FDT-DRBM approach. The general flow of information, when the proposed multi-biometric face recognition system operates in the verification mode, is shown in Figure 4.12. The feature representations, f_x and f_y, of a pair of two images, I_x and I_y, are obtained firstly by applying FDT approach, and then a feature vector F for this pair is formed using element-wise multiplication ($F = f_x \odot f_x$). Finally, this vector F (e.g., extracted from pairs of images) is used as input data to the DBN and DRBM to learn additional feature representations and perform face verification in the last layer, respectively. The final performance is reported in Table 4.9 by calculating the mean accuracy rate ($\hat{\mu}$) and the standard error of the mean accuracy (SE) over 10-fold cross-validation using different score-level fusion methods, and the corresponding receiver operating characteristic (ROC) curves are shown in Figure 4.13. From Table 4.9 and Figure 4.13, the highest accuracy rate was obtained using the weighted sum rule as a fusion mothed in the score level, where a higher weight was assigned to the proposed *FaceDeepNet* system compared to the FDT-DRBM approach. The accuracy rate has been improved by 4.5% and 0.98% compared to the FDT-DRBM approach and the *FaceDeepNet* system, respectively.

The proposed FDT-DRBM approach was compared to the current state-of-the-art approaches on the LFW dataset, such as Curvelet-Fractal [7], DDML [34], LBP, Gabor [74], and MSBSIF-SIEDA [75] using the same evaluation protocol (e.g., Restricted), as shown in Table 4.10. It can be seen that the Curvelet-Fractal [7] has achieved slightly a higher accuracy rate, but the proposed FDT-DRBM approach was able to achieve competitive results with the other state-of-the-art face verification results reported on the LFW dataset. The performance of the proposed face recognition approaches were also compared with the current state-of-the-art deep learning approaches, such as DeepFace [31], DeepID [76], ConvNet-RBM [77], ConvolutionalDBN [78], MDFR Framework [7], and DDML [34]. The first three approaches were mainly trained using a different evaluation protocol, named the *"Unrestricted, Labeled Outside Data"* protocol,

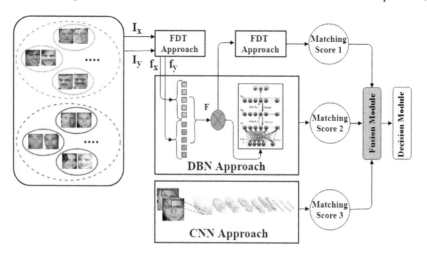

FIGURE 4.12 Illustration of the proposed multi-biometric face recognition system operating in the verification mode.

TABLE 4.9

Performance Comparison between the Proposed Face Recognition Approaches on LFW Dataset Using Different Score Fusion Methods

		Acc. ($\hat{\mu} \pm S_E$)				
FDT-DRBM	*FaceDeepNet*	SR	WSR	PR	Max	Min
0.9542 ± 0.0192	0.9899 ± 0.0210	0.9918 ± 0.0110	0.9992 ± 0.0232	0.9837± 0.0164	0.9758 ± 0.0177	0.9880 ± 0.0211

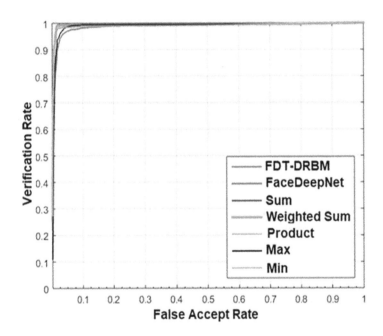

FIGURE 4.13   ROC curves using different score-level fusion methods.

in which a private dataset consisting of a large number of training samples (>100K) is used. As shown in Table 4.10, the accuracy rate has been further improved by 0.16% and 1.09% compared to the next highest results reported by MDFR Framework [7] using the *FaceDeepNet* system and the whole multi-biometric face recognition system (e.g., using the weighted sum rule (WSR)). These promising results demonstrate the good generalization ability of the proposed face recognition approaches, and their feasibility for deployment in real-life applications.

4.6   CONCLUSIONS

In this chapter, a multi-biometric face recognition approach based on multimodal deep learning termed FDT-DRBM is proposed. The approach is based on integrating the advantages of the fractal dimension and the DRBM classifier depending on

TABLE 4.10

Performance Comparison between the Proposed Approaches and the State-of-the-Art Approaches on LFW Dataset under Different Evaluation Protocols

Approach	Acc. ($\hat{\mu} \pm S_E$)	Protocol
DeepFace [31]	0.9735 ±0.0025	Unrestricted
DeepID [76]	0.9745 ± 0.0026	Unrestricted
ConvNet-RBM[77]	0.9252 ± 0.0038	Unrestricted
VGG Net [79]	98.95	Unrestricted
ConvolutionalDBN [78]	0.8777 ± 0.0062	Restricted
DDML [34]	0.9068 ± 0.0141	Restricted
VMRS [80]	0.9110 ± 0.0059	Restricted
HPEN+HD-LBP+ DDML [74]	0.9257 ± 0.0036	Restricted
HPEN+HD-Gabor+ DDML [74]	0.9280 ± 0.0047	Restricted
Sub-SML+Hybrid+LFW3D [81]	0.9165 ± 0.0104	Restricted
MSBSIF-SIEDA [75]	0.9463 ± 0.0095	Restricted
3-layers DBN [7]	0.9353 ± 0.0165	Restricted
Curvelet-Fractal [7]	0.9622 ± 0.0272	Restricted
MDFR Framework [7]	0.9883 ± 0.0121	Restricted
FDT-DRBM	**0.9542 ± 0.0192**	Restricted
FaceDeepNet	**0.9899 ± 0.0210**	Restricted
Fusion (WSR)	**0.9992 ± 0.0232**	Restricted

extracting the face texture roughness and fluctuations in the surface under unconstrained environmental conditions and efficiently modeling the joint distribution of inputs and target classes, respectively. Furthermore, a novel *FaceDeepNet* system is proposed to learn additional and complementary facial feature representations by training two discriminative deep learning approaches (i.e., CNN and DBN) on top of the local feature representations obtained from the FDT-DRBM approach. The proposed approach was tested on four large-scale unconstrained face datasets (i.e., SDUMLA-HMT, FRGC V 2.0, UFI, and LFW) with high diversity in facial expressions, illumination conditions, pose, noise, etc. A number of extensive experiments were conducted, and a new state-of-the-art accuracy rate was achieved for both the face identification and verification tasks by applying the proposed *FaceDeepNet* system and the whole multi-biometric face recognition system (e.g., using the WSR) on all the employed datasets. The obtained results demonstrate the reliability and efficiency of the FDT-DRBM approach by achieving competitive results with the current state-of-the-art face recognition approaches (e.g., CLVQ, Curvelet-Fractal, DeepID, etc.). For future work, it would be necessary to validate further the efficiency and reliability of the proposed multi-biometric system using a larger multi-modal dataset, containing more individuals with face images captured under more challenging conditions.

BIBLIOGRAPHY

1. K. Nandakumar, A. Ross, and A. K. Jain, "Introduction to multibiometrics," in *Appeared in Proc of the 15th European Signal Processing Conference (EUSIPCO) (Poznan Poland)*, 2007, pp. 271–292.
2. A. S. Al-Waisy, "Ear identification system based on multi-model approach," *International Journal of Electronics Communication and Computer Engineering*, vol. 3, no. 5, pp. 2278–4209, 2012.
3. M. S. Al-ani and A. S. Al-Waisy, "Milti-view face datection based on kernel principal component analysis and kernel," *International Journal on Soft Computing (IJSC)*, vol. 2, no. 2, pp. 1–13, 2011.
4. A. S. Al-Waisy, R. Qahwaji, S. Ipson, and S. Al-Fahdawi, "A fast and accurate Iris localization technique for healthcare security system," in *IEEE International Conference on Computer and Information Technology; Ubiquitous Computing and Communications; Dependable, Autonomic and Secure Computing; Pervasive Intelligence and Computing*, 2015, pp. 1028–1034.
5. R. Jafri and H. R. Arabnia, "A survey of face recognition techniques," *Journal of Information Processing Systems*, vol. 5, no. 2, pp. 41–68, 2009.
6. C. Ding, J. Choi, D. Tao, and L. S. Davis, "Multi-directional multi-level dual-cross patterns for robust face recognition," *IEEE Transactions on Pattern Analysis and Machine Intelligence*, vol. 38, no. 3, pp. 518–531, 2016.
7. A. S. Al-Waisy, R. Qahwaji, S. Ipson, and S. Al-Fahdawi, "A multimodal deep learning framework using local feature representations for face recognition," *Machine Vision and Applications*, vol. 29, no. 1, pp. 35–54, 2017.
8. A. S. Al-Waisy, R. Qahwaji, S. Ipson, and S. Al-fahdawi, "A multimodal biometric system for personal identification based on deep learning approaches," in *Seventh International Conference on Emerging Security Technologies (EST)*, 2017, pp. 163–168.
9. A. Ross, K. Nandakumar, and J. K. Anil, "Handbook of multibiometrics," *Journal of Chemical Information and Modeling*, vol. 53, no. 9, pp. 1689–1699, 2006.
10. A. Ross and A. K. Jain, "Multimodal biometrics : An overview," in *12th European Signal Processing Conference IEEE*, 2004, pp. 1221–1224.
11. A. Lumini and L. Nanni, "Overview of the combination of biometric matchers," *Information Fusion*, vol. 33, pp. 71–85, 2017.
12. L. Deng and D. Yu, "Deep learning methods and applications," *Signal Processing*, vol. 28, no. 3, pp. 198–387, 2013.
13. J. Liu, C. Fang, and C. Wu, "A fusion face recognition approach based on 7-layer deep learning neural network," *Journal of Electrical and Computer Engineering*, vol. 2016, pp. 1–7, 2016.
14. P. Fousek, S. Rennie, P. Dognin, and V. Goel, "Direct product based deep belief networks for automatic speech recognition," in *Proceedings of the IEEE International Conference on Acoustics, Speech and Signal Processing, ICASSP*, 2013, pp. 3148–3152.
15. H. Lee, P. Pham, Y. Largman, and A. Ng, "Unsupervised feature learning for audio classification using convolutional deep belief networks," in *Advances in Neural Information Processing Systems Conference*, 2009, pp. 1096–1104.
16. R. Sarikaya, G. E. Hinton, and A. Deoras, "Application of deep belief networks for natural language understanding," *IEEE Transactions on Audio, Speech and Language Process.*, vol. 22, no. 4, pp. 778–784, 2014.
17. C. Ding and D. Tao, "A comprehensive survey on pose-invariant face recognition," *ACM Transactions on Intelligent Systems and Technology (TIST)*, vol. 7, no. 3, pp. 1–40, 2016.

18. S. Biswas, G. Aggarwal, P. J. Flynn, and K. W. Bowyer, "Pose-robust recognition of low-resolution face images," *IEEE Transactions on Pattern Analysis and Machine Intelligence*, vol. 35, no. 12, pp. 3037–3049, 2013.

19. D. G. Lowe, "Distinctive image features from scale-invariant keypoints," *International Journal of Computer Vision*, vol. 60, no. 2, pp. 91–110, 2004.

20. L. Sirovich and M. Kirby, "Low-dimensional procedure for the characterization of human faces.," *Journal of the Optical Society of America. A, Optics and Image Science*, vol. 4, no. 3. pp. 519–524, 1987.

21. K. Meethongjan and D. Mohamad, "A summary of literature review : Face recognition," *Review Literature and Arts of the Americas*, vol. 2007, no. July, pp. 1–12, 2007.

22. M. Fischer, H. K. Ekenel, and R. Stiefelhagen, "Analysis of partial least squares for pose-invariant face recognition," *in IEEE Fifth International Conference on Biometrics: Theory, Applications and Systems (BTAS)*, 2012, pp. 331–338.

23. T. Berg and P. Belhumeur, "Tom-vs-Pete classifiers and identity-preserving alignment for face verification," in *Proceedings of the British Machine Vision Conference*, 2012, pp. 129.1–129.11.

24. A. Li, S. Shan, and W. Gao, "Coupled bias – variance tradeoff for cross-pose face recognition," *IEEE Transactions on Image Processing*, vol. 21, no. 1, pp. 305–315, 2012.

25. Z. Zhu, P. Luo, X. Wang, and X. Tang, "Multi-view perceptron : A deep model for learning face identity and view representations," *Advances in Neural Information Processing Systems*, pp. 1–9, 2014.

26. A. S. Al-Waisy, "Detection and recognition of human faces based on hybrid techniques," *International Journal of Applied Computing (IJAC)*, vol. 5, no. 2, pp. 115–126, 2012.

27. M. S. Al-Ani and A. S. Al-waisy, "Face recognition approach based on wavelet-curvelet technique," *Signal & Image Processing: An International Journal (SIPIJ)*, vol. 3, no. 2, pp. 21–31, 2012.

28. Y. Sun, X. Wang, and X. Tang, "Deep learning face representation by joint identification-verification," *Advances in Neural Information Processing Systems*, pp. 1988–1996, 2014.

29. L. Deng, "Three classes of deep learning architectures and their applications: A tutorial survey," *APSIPA Transactions on Signal and Information Processing*, 2013.

30. Y. Sun, X. Wang, and X. Tang, "Hybrid deep learning for face verification," in *Proceedings of the IEEE International Conference on Computer Vision*, 2013, pp. 1489–1496.

31. Y. Taigman, M. A. Ranzato, T. Aviv, and M. Park, "DeepFace: Closing the gap to human-level performance in face verification," in *Proceedings of the IEEE Conference on Computer Vision and Pattern Recognition*, 2014, pp. 1701–1708.

32. C. Ding and D. Tao, "Robust face recognition via multimodal deep face representation," *IEEE Transactions on Multimedia*, vol. 17, no. 11, pp. 2049–2058, 2015.

33. J. V. Davis, B. Kulis, P. Jain, S. Sra, and I. S. Dhillon, "Information-theoretic metric learning," in *in Proceedings of the 24th International Conference on Machine Learning*, 2007, pp. 209–216.

34. J. Hu, J. Lu, and Y. P. Tan, "Discriminative deep metric learning for face verification in the wild," in *Proceedings of the IEEE Computer Society Conference on Computer Vision and Pattern Recognition*, 2014, pp. 1875–1882.

35. Z. Cui, W. Li, D. Xu, S. Shan, and X. Chen, "Fusing robust face region descriptors via multiple metric learning for face recognition in the wild," in *Proceedings of the IEEE Computer Society Conference on Computer Vision and Pattern Recognition*, 2013, pp. 3554–3561.

36. M. Guillaumin, J. Verbeek, C. Schmid, M. Guillaumin, J. Verbeek, C. Schmid, M. Guillaumin, J. Verbeek, C. Schmid, and L. J. Kuntzmann, "Is that you ? Metric learning approaches for face identification To cite this version," in *IEEE 12th International Conference on Computer Vision*, 2009, pp. 498–505.

37. M. Kostinger, M. Hirzer, P. Wohlhart, P. M. Roth, and H. Bischof, "Large scale metric learning from equivalence constraints," in *Proceedings of the IEEE Computer Society Conference on Computer Vision and Pattern Recognition*, 2012, pp. 2288–2295.
38. H. V Nguyen and L. Bai, "Cosine Similarity Metric Learning for Face Verification," in *Asian Conference on Computer Vision*. Springer Berlin Heidelberg, 2011, pp. 1–12.
39. C. Peng, X. Gao, N. Wang, and J. Li, "Graphical representation for heterogeneous face recognition," *IEEE Transactions on Pattern Analysis and Machine Intelligence*, vol. 39, no. 2, pp. 301–312, 2017.
40. C. Peng, N. Wang, J. Li, and X. Gao, "DLFace: Deep local descriptor for cross-modality face recognition," *Pattern Recognition*, vol. 90, pp. 161–171, 2019.
41. C. Peng, X. Gao, S. Member, N. Wang, and D. Tao, "Multiple representations-based face sketch – photo synthesis," *IEEE Transactions on Neural Networks and Learning Systems*, vol. 27, no. 11, pp. 2201–2215, 2016.
42. Y. Gao, J. Ma, and A. L. Yuille, "Semi-supervised sparse representation based classification for face recognition with insufficient labeled samples," *IEEE Transactions on Image Processing*, vol. 26, no. 5, pp. 2545–2560, 2017.
43. M. Hassaballah and S. Aly, "Face recognition: Challenges, achievements and future directions," *IET Computer Vision*, vol. 9, no. 4, pp. 614–626, 2015.
44. R. Lopes and N. Betrouni, "Fractal and multifractal analysis: A review, *Medical Image Analysis*, vol. 13, no. 4, pp. 634–649, 2009.
45. B. Mandelbrot, *The Fractal Geometry of Nature*. Library of Congress Cataloging in Publication Data, United States of America, 1983.
46. B. Mandelbrot, "Self-affinity and fractal dimension," *Physica Scripta*, vol. 32, 1985, pp. 257–260.
47. K. Lin, K. Lam, and W. Siu, "Locating the human eye using fractal dimensions," in *Proceedings of International Conference on Image Processing*, 2001, pp. 1079–1082.
48. M. H. Farhan, L. E. George, and A. T. Hussein, "Fingerprint identification using fractal geometry," *International Journal of Advanced Research in Computer Science and Software Engineering*, vol. 4, no. 1, pp. 52–61, 2014.
49. G. Hinton, "A practical guide to training restricted boltzmann machines a practical guide to training restricted boltzmann machines," *Neural Networks: Tricks of the Trade*. Springer, Berlin, Heidelberg, 2010, pp. 599–619.
50. H. Larochelle and Y. Bengio, "Classification using discriminative restricted Boltzmann machines," in *Proceedings of the 25th International Conference on Machine Learning*, 2008, pp. 536–543.
51. B. Abibullaev, J. An, S. H. Jin, S. H. Lee, and J. Il Moon, "Deep machine learning a new frontier in artificial intelligence research," *Medical Engineering and Physics*, vol. 35, no. 12, pp. 1811–1818, 2013.
52. G. E. Hinton, S. Osindero, and Y.-W. Teh, "A fast learning algorithm for deep belief nets," *Neural Computation*, vol. 18, no. 7, pp. 1527–1554, 2006.
53. H. Khalajzadeh, M. Mansouri, and M. Teshnehlab, "Face recognition using convolutional neural network and simple logistic classifier," *Soft Computing in Industrial Applications*. Springer International Publishing, pp. 197–207, 2014.
54. Y. Bengio, "Learning deep architectures for AI," *Foundations and Trends® in Machine Learning,* vol. 2, no. 1, pp. 1–127, 2009.
55. A. S. Al-Waisy, R. Qahwaji, S. Ipson, S. Al-Fahdawi, and T. A. M. Nagem, "A multi-biometric iris recognition system based on a deep learning approach," *Pattern Analysis and Applications*, vol. 21, no. 3, pp. 783–802, 2018.
56. P. Viola, O. M. Way, and M. J. Jones, "Robust real-time face detection," *International Journal of Computer Vision*, vol. 57, no. 2, pp. 137–154, 2004.

57. A. R. Chowdhury, T.-Y. Lin, S. Maji, and E. Learned-Miller, "One-to-many face recognition with bilinear CNNs," in *IEEE Winter Conference on Applications of Computer Vision (WACV)*, 2016, pp. 1–9.

58. D. Wang, C. Otto, and A. K. Jain, "Face search at scale: 80 million gallery," arXiv Prepr. arXiv1507.07242, pp. 1–14, 2015.

59. G. E. Hinton, "Training products of experts by minimizing contrastive divergence," *Neural Computation*, vol. 14, no. 8, pp. 1771–1800, 2002.

60. Y. Yin, L. Liu, and X. Sun, "SDUMLA-HMT: A multimodal biometric database," in *Chinese Conference Biometric Recognition*, Springer-Verlag Berlin Heidelberg, pp. 260–268, 2011.

61. P. J. Phillips, P. J. Flynn, T. Scruggs, K. W. Bowyer, J. Chang, K. Hoffman, J. Marques, J. Min, and W. Worek, "Overview of the face recognition grand challenge," in *Proceedings IEEE Computer Society Conference on Computer Vision and Pattern Recognition, CVPR*, 2005, vol. I, pp. 947–954.

62. L. Lenc and P. Král, "Unconstrained facial images: Database for face recognition under real-world conditions," *Lecture Notes in Computer Science (including subseries Lecture Notes in Artificial Intelligence and Lecture Notes in Bioinformatics)*, vol. 9414, pp. 349–361, 2015.

63. G. B. Huang, M. Mattar, T. Berg, and E. Learned-miller, "Labeled faces in the wild: A database for studying face recognition in unconstrained environments," in *Technical Report 07-49, University of Massachusetts, Amherst*, 2007, pp. 1–14.

64. A. S. Al-Waisy, R. Qahwaji, S. Ipson, and S. Al-Fahdawi, "A robust face recognition system based on curvelet and fractal dimension transforms," in *IEEE International Conference on Computer and Information Technology; Ubiquitous Computing and Communications; Dependable, Autonomic and Secure Computing; Pervasive Intelligence and Computing*, 2015, pp. 548–555.

65. W. R. Schwartz, H. Guo, and L. S. Davis, "A robust and scalable approach to face identification," in *European Conference on Computer Vision*, 2010, pp. 476–489.

66. M. Y. Shams, A. S. Tolba, and S. H. Sarhan, "A vision system for multi-view face recognition," *International Journal of Circuits, Systems and Signal Processing*, vol. 10, pp. 455–461, 2016.

67. J. Li, T. Qiu, C. Wen, K. Xie, and F.-Q. Wen, "Robust face recognition using the deep C2D-CNN model based on decision-level fusion," *Sensors*, vol. 18, no. 7, pp. 1–27, 2018.

68. J. Holappa, T. Ahonen, and M. Pietikäinen, "An optimized illumination normalization method for face recognition," in *IEEE Second International Conference on Biometrics: Theory, Applications and Systems*, 2008, pp. 1–6.

69. P. Král, L. Lenc, and A. Vrba, "Enhanced local binary patterns for automatic face recognition," arXiv Prepr. arXiv1702.03349, 2017.

70. J. Lin and C.-T. Chiu, "Lbp edge-mapped descriptor using mgm interest points for face recognition," in *IEEE International Conference on Acoustics, Speech and Signal Processing (ICASSP)*, 2017, pp. 1183–1187.

71. L. Lenc, "Genetic algorithm for weight optimization in descriptor based face recognition methods," in *Proceedings of the 8th International Conference on Agents and Artificial Intelligence.* SCITEPRESS-Science and Technology Publications, 2016, pp. 330–336.

72. J. Gaston, J. Ming, and D. Crookes, "Unconstrained face identification with multi-scale block-based correlation," in *IEEE International Conference on Acoustics, Speech and Signal Processing (ICASSP)*, 2017, pp. 1477–1481.

73. G. B. Huang and E. Learned-miller, "Labeled faces in the wild : Updates and new reporting procedures," *Department of Computer Science, University of Massachusetts Amherst*, Amherst, MA, USA, Technical Report, pp. 1–14, 2014.

74. X. Zhu, Z. Lei, J. Yan, D. Yi, and S. Z. Li, "High-fidelity pose and expression normalization for face recognition in the wild," in *Proceedings of the IEEE Conference on Computer Vision and Pattern Recognition*, 2015, pp. 787–796.

75. A. Ouamane, M. Bengherabi, A. Hadid, and M. Cheriet, "Side-information based exponential discriminant analysis for face verification in the wild," in *11th IEEE International Conference and Workshops on Automatic Face and Gesture Recognition (FG)*, 2015, pp. 1–6.

76. Y. Sun, X. Wang, and X. Tang, "Deep learning face representation from predicting 10,000 classes," in *Proceedings of the IEEE Computer Society Conference on Computer Vision and Pattern Recognition*, 2014, pp. 1891–1898.

77. Y. Sun, X. Wang, and X. Tang, "Hybrid deep learning for face verification," *IEEE Transactions on Pattern Analysis and Machine Intelligence*, vol. 38, no. 10, pp. 1997–2009, 2016.

78. G. B. Huang, H. Lee, and E. Learned-Miller, "Learning hierarchical representations for face verification with convolutional deep belief networks," in *Proceedings of the IEEE Computer Society Conference on Computer Vision and Pattern Recognition*, 2012, pp. 2518–2525.

79. O. M. Parkhi, A. Vedaldi, and A. Zisserman, "Deep face recognition," in *British Machine Vision Conference*, 2015, pp. 1–12.

80. O. Barkan, J. Weill, L. Wolf, and H. Aronowitz, "Fast high dimensional vector multiplication face recognition," in *Proceedings of the IEEE International Conference on Computer Vision*, 2013, pp. 1960–1967.

81. T. Hassner, S. Harel, E. Paz, and R. Enbar, "Effective face frontalization in unconstrained images," in *IEEE Conference on Computer Vision and Pattern Recognition (CVPR)*, 2015, pp. 4295–4304.

5 Deep LSTM-Based Sequence Learning Approaches for Action and Activity Recognition

Amin Ullah, Khan Muhammad, Tanveer Hussain, Miyoung Lee, and Sung Wook Baik

CONTENTS

5.1 INTRODUCTION

Human action and activity recognition is one of the challenging areas in computer vision due to the increasing number of cameras installed everywhere, which enables its numerous applications in various domains. The enormous amount of data generated through these cameras requires human monitoring for identification of different activities and events. Smarter surveillance, through which normal and abnormal activities can be automatically identified using artificial intelligence and computer vision technologies, is needed in this era. It has vast demand due to its importance in real-life problems like detection of abnormal and suspicious activities, intelligent video surveillance, elderly care, patient monitoring in healthcare centers, and video retrieval based on different actions [1, 2]. In the literature, many researchers have analyzed action and activity recognition through various sensors data including RGB, grayscale, night vision, Kinect depth, and skeleton data [3].

Smartphone and smartwatch sensors are also utilized by some researchers for activity analysis and fitness applications [4]. Categorizations of action and activity recognition methods using available sensors are given in Figure 5.1. In this chapter, we will study only vision RGB-based action and activity recognition methods in which long short-term memory (LSTM) has been deployed for sequence learning. Processing visual contents of video data for activity recognition is very challenging due to the inter-similarity of visual and motion patterns, changes in viewpoint, camera motion in sports videos, scale, action poses, and different illumination conditions [5].

Recently, convolutional neural network (CNN) outperformed state-of-the-art techniques in various domains such as image classification, object detection, and localization. However, an activity is recorded in sequences of images, so processing a single image surely cannot give us good results to capture the whole idea of the activity. Therefore, in videos, the whole activity can be recognized by analyzing the motions and other visual changes of human body parts like hands and legs in a sequence of frames. For instance, jumping for a headshot in football and skipping rope have the same action pose in the initial frame. However, the difference between these actions can be easily recognized in a sequence of frames. Another challenge for analyzing long-term activities in video data is that each activity has parts that overlap with other activities; for example, diving for swimming is one activity, but whenever the swimmer dives into the pool he also jumps in the air, which is another activity. So, when two activities like jumping and diving appear in the same sequence, it can lead to false predictions because the sequence is interrupted. To address these issues, many researchers [6, 7] have introduced different kinds of techniques to represent sequences and learn those sequences for activity recognition. For instance, the current literature on human activity recognition highlights some popular methods

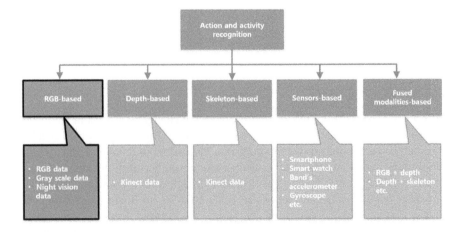

FIGURE 5.1   Categorization of action and activity recognition using available sensor data.

including volume motion templates [6], spatio-temporal features [7], motion history volume [8], interest point descriptors [9], bag-of-words [10], optical flow trajectories [11], and CNN-based learned features [12, 13] for sequence representation followed by some machine learning algorithms such as support vector machine (SVM), decision tree, recurrent neural network (RNN), and LSTM. However, these methods are not enough for sequence representation and learning. Recent studies prove that the LSTM is one of the most effective sources of sequence learning. Therefore, in this chapter, we will explore sequence learning concepts integrated with LSTM for action and activity recognition.

5.1.1 RECURRENT NEURAL NETWORK (RNN)

RNNs are a type of artificial neural network that has turned out to be more popular in recent years because of its power to learn time series data. The RNN is a special network that consists of recurrent connections, unlike traditional feedforward neural networks. The major benefit is that with its network of recurrent connections, it is able to process arbitrary time series sequences of input data. RNNs are presented for investigating hidden sequential patterns in both temporal sequential and spatial sequential data [14]. A standard RNN is shown in Figure 5.2 (a). It processes data by taking input at different time steps over time. At each time step, it has two inputs: the first input comes from the hidden state of the previous RNN, and the second is from the time series data.

From Figure 5.2 (a) it can be seen that RNN is taking input at each time step and giving output as well. This process of taking input at each recurrent step disturbs the patterns for sequences in time series data, because the neural network performs multiplication of weights with data and addition with bias terms. Therefore, multiplication of data many times with several hidden states causes the effect of early sequential data in the series to be forgotten. This issue, known as the vanishing gradient problem, is shown in Figure 5.2 (b), where the effect of data is shown as gray shading, which decreases when the sequence is going long. In practice, the range of sequential information that standard RNNs access is limited to approximately ten time steps between the relevant input and target output. Video data is also sequential where changes in visual contents are represented in many frames such

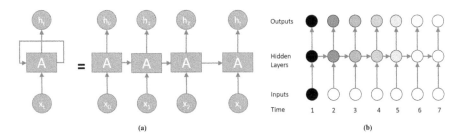

FIGURE 5.2 (a) A standard unrolled RNN network; (b) vanishing gradient problem in RNN.

that a sequence of frames helps in understanding the context of an action. In long-term sequences, RNNs can interpret such sequences but forget the earlier inputs of the sequence.

5.1.2 LONG SHORT-TERM MEMORY (LSTM)

The vanishing gradient problem can be solved by utilizing a special type of RNN known as LSTM [15]. It is capable of learning long-term sequential dependencies. Its special structure, with input, output, and forget gates, controls the long-term sequential pattern identification in time series data. The gates are adjusted by a sigmoid function connected to each unit, which learns during training those states in which it needs to be open and those in which it should be closed. Eq. (1.1) to Eq. (1.7) [16] explain the operations performed in the LSTM unit, where x_t is the input at time t, where input is fed in the form of chunks from the long sequence. f_t is the forget gate at time t, which clears information from the memory cell when needed and keeps a record of the previous frame, whose information needs to be cleared from the memory. The output gate o_t keeps information about the upcoming step, where g is the recurrent unit with activation function "tanh" and is computed from the input of the current frame and the state of the previous frame s_{t-1}. The hidden state of an RNN step is calculated through tanh activation and memory cell c_t. The output of each state can be used in other tasks; however, action recognition does not need the intermediate output of the LSTM, so the final decision is made by applying the Softmax classifier to the final state of the LSTM network, which gives probabilities for the predicted activity.

$$i_t = \sigma\left(\left(x_t + s_{t-1}\right)W^i + b_i\right) \tag{5.1}$$

$$f_t = \sigma\left(\left(x_t + s_{t-1}\right)W^f + b_f\right) \tag{5.2}$$

$$o_t = \sigma\left(\left(x_t + s_{t-1}\right)W^o + b_o\right) \tag{5.3}$$

$$g = \tanh\left(\left(x_t + s_{t-1}\right)W^g + b_g\right) \tag{5.4}$$

$$c_t = c_{t-1} \cdot f_t + g \cdot i_t \tag{5.5}$$

$$s_t = \tanh(c_t) \cdot o_t \tag{5.6}$$

$$\text{Predictions} = \text{soft max}(Vs_t) \tag{5.7}$$

Recent studies have proven [17–21] that huge amounts of data increase the accuracy of a machine learning task. Therefore, learning complex sequential patterns in video data of large-scale datasets is very challenging, and a single LSTM is not effective enough to do it. Therefore, researchers have recently utilized multi-layer LSTM, which is constructed by stacking multiple LSTM cells to learn long-term dependencies in video data.

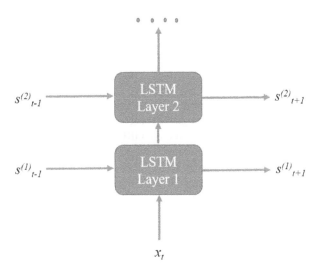

FIGURE 5.3 Two-layer LSTM network [22].

5.1.3 MULTI-LAYER LSTM

The performance of most deep neural networks can be boosted in terms of accuracy by increasing the number of layers in the neural network models. The same strategy is followed in LSTM by stacking multiple LSTM layers to the network. By adding deep layers, it captures a higher level of sequence information [16]. In standard RNN, data is fed to a single layer for activation and processing before output, but in time sequence problems, we need to process data on several layers to capture hierarchical patterns. By stacking LSTM layers, each layer in the network is a hierarchy that receives the hidden state of the previous layer as input. Figure 5.3 shows a two-layer LSTM. Layer 1 receives input from data x_t, while the input of layer 2 is from its previous time step $s_{t-1}^{(2)}$, the output of the current time step of layer 1, $s_t^{(1)}$. The computation of the multi-layer LSTM cell is same as simple LSTM cell, but only each layer's information is added to the superscript of each $i_t, f_t, o_t, c_t,$ and s_t. The procedure for calculating the state of the layer is as follows:

$$s_t^l = \tanh\left(c_t^l\right) \cdot o_t^l \qquad (5.8)$$

5.1.4 BIDIRECTIONAL LSTM

In bidirectional LSTM, the output at time t is dependent not only on the previous frames in the sequence, but also on the upcoming frames [23]. Bidirectional LSTMs are quite simple, having two recurrent units, one stacked on top of the other. One recurrent unit goes in the forward direction and the other goes in the backward direction. The combined output is then computed based on the hidden state of both recurrent units. The input data is fed to the bidirectional LSTM, and the hidden states of forward pass and backward pass are combined in the output layer. The cost function is adjusted after the output layer, weights, and biases are adjusted through back-propagation.

5.1.5   GATED RECURRENT UNIT

Two dominant variants of RNNs have been introduced by researchers: LSTM [22] and GRU [24]. The LSTM contains gated recurrent units, including the input, forget, output gates, and memory cell, while the GRU consists of the reset gate, update gate, and activation unit. The structure of the LSTM is more complex and contains gates and memory cell, resulting in more computation to process a sequence. On the other hand, GRU involves only two gates, which makes it applicable for processing real-time sequential data. However, many researchers have reported LSTM as more effective than GRU [24].

5.2   ACTION AND ACTIVITY RECOGNITION METHODS

Over the last decade, researchers have presented many handcrafted and deep-nets-based approaches for action recognition. The earlier work was based on handcrafted features for non-realistic actions, where an actor used to perform some actions in a scene with a simple background. Such systems extract low-level features from the video data and then feed them to a classifier such as SVM, decision tree, and k-Nearest Neighbors (KNN) for action recognition. However, current methods utilize realistic video data for action and activity recognition tasks. For instance, an action recognition framework is proposed by Ullah et al. [22], where they used CNN features followed by deep bidirectional LSTM (DB-LSTM) for sequence learning. They extracted features from every six frames of the input video to reduce the time complexity and redundancy. The sequential information extracted from these frames is inputted to the DB-LSTM network. Multiple layers in the DB-LSTM are stacked together in both forward and backward pass in order to increase the network depth and make it capable of learning long-term sequences. The framework of DB-LSTM is shown in Figure 5.4.

Another recent method proposed by Ullah et al. [1] is for activity recognition in industrial surveillance environments. Their framework is visualized in Figure 5.5. First, they divided the continuous surveillance video stream into important shots using the proposed CNN-based human saliency features. Next, the temporal features of activity in the sequence of frames are extracted by utilizing the convolutional layers of a FlowNet2 CNN model. Finally, a multi-layer LSTM is presented for learning long-term sequences in the temporal optical flow features for activity recognition.

The method proposed by Li et al. [25] selected discriminative parts in spatial dimension, as visualized in Figure 5.6. They split the videos into multiple clips with fixed-length frames in the temporal direction and used the sliding windows to generate local cubes in the spatial direction. The cubes are extracted using the C3D network. Then, they designed the part selection method to select useful cubes and utilized the max-pooling function to generate the representation of clips. Finally, the representation of clips was fed into two layers of the LSTM network to train the multiple LSTM and the Softmax layer, which yielded the prediction by using the feature vector generated from the LSTM network.

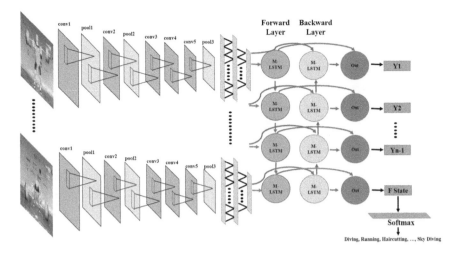

FIGURE 5.4   The framework for action recognition in videos using CNN and deep bidirectional LSTM [22].

An adaptive recurrent-convolutional hybrid network for long-term action recognition is introduced by Xin et al. [7]. Xin et al. introduced a framework learning from static, non-static, and sequential mixed features to aim at different problems such as domain variation and temporal diversities. They represented human actions by first reducing data based on cognitive techniques and a hybrid "network-over-network" architecture. Li et al. in [4] present end-to-end sequence learning for action recognition based on video LSTM. To exploit the spatial correlations, they hardwired convolutions in soft-Attention LSTM architecture. They also introduced motion-based attention, as motion informs about action contents. The authors performed experiments on UCF101, HMDB51, and THUMOS13 datasets to present the better accuracy of their system. All of the statistics of the exiting LSTM-based action and activity recognition models are briefly given in Table 5.1. A group activity recognition framework using temporal dynamics is presented by Ibrahim et al. [26], where they used an LSTM-based deep model to capture these dynamics. Their framework is shown in Figure 5.7. As dynamics of group activity can be inferred on the basis of individual persons' activities, the authors presented a two-stage deep temporal model to recognize the group activity occurring in video sequences. The LSTM model in their method is designed especially to represent action dynamics of individual people in a sequence and the second LSTM model take the average of person-level information in order to understand the group activity.

Ma et al. [33] presented a framework to improve accuracy by advancing the training process of temporal deep learning models for activity detection and recognition tasks. Their proposed framework is shown in Figure 5.8. Conventional methods based on RNNs and LSTMs consider only classification errors. But in this method, they studied two problems: activity detection and then activity recognition. To detect activity, the authors recognized the category of the activity and its initial and final end points and then classified the activity as visualized in Figure 5.8. A novel

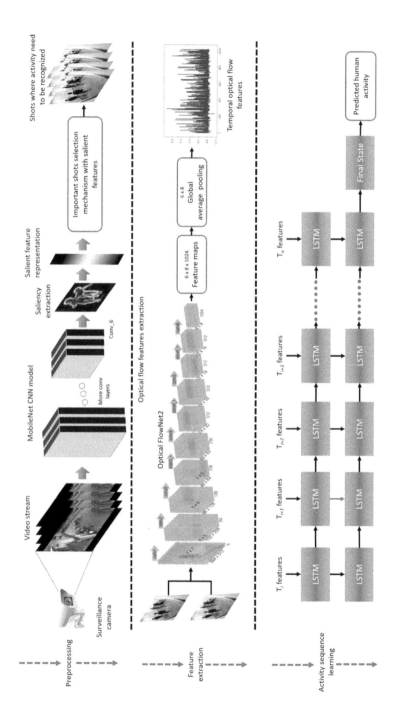

FIGURE 5.5 The framework of activity recognition in industrial systems. In the first stage, the surveillance stream is preprocessed using CNN-based salient feature extraction mechanism for shot segmentation. The second stage extracts CNN-based temporal optical flow features from the sequence of frames. Finally, multi-layer LSTM is presented for sequence learning and the recognition of activities [1].

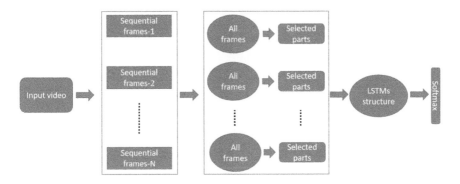

FIGURE 5.6 The framework of selected spatio-temporal features via bidirectional LSTM for action recognition [25].

approach based on the convolutional features of ResNet-101-v2 model is proposed by Chen et al. [29], where the input video is divided into segments of equal duration. Their proposed framework is shown in Figure 5.9. Given an input video, they divided it into T segments (T = 3 in this case for succinct illustration) of equal duration and randomly sampled one short snippet from its corresponding segment. The short snippet is either a single RGB image (spatial stream) or a sequence of 10 optical flow fields (temporal stream). After feeding the selected short snippets through the ResNet-101-v2 conv1 layer to the ResNet-101-v2 conv5 2 layer, they had two separated branches.

The local branch (top dotted box) captures local appearance and short-term motion features from snippets and generates a video-level feature representation using temporal average pooling. The non-local branch (bottom dotted box) is for relation reasoning using the relational LSTM module. Eventually, they obtained an overall video-level feature representation by concatenating feature vectors generated by the two branches, added one fully connected layer, and optimized it using a standard Softmax with cross entropy classification loss. Recently, encoder-decoder frameworks based on attention mechanisms have been applied to action recognition and have produced results comparable to those obtained with state-of-the-art methods, but they have always faced a bottleneck when it comes to distinguishing similar actions. To tackle this challenge, Yang et al. [36] proposed a novel framework based on recurrent attention CNN (RACNN). RACNN integrates CNNs and LSTMs along with an attention mechanism.

Figure 5.10 contains four major components: (1) convolutional feature extraction; (2) bidirectional LSTM sequence model and spatial attention model; (3) temporal attention model; and (4) sequence recognition. CNNs encode the input video frames into feature sequences. AttLSTM means hierarchical LSTM with spatial attention model. The output of bidirectional AttLSTM is designed to be integrated and sent to the temporal attention model. After refining the input, the architecture takes the output of the last LSTM unit and calculates the resulting features extracted, which are encoded and forward propagated to the second module to seek spatial attention, followed by another module for temporal attention, which is advanced to decoder

TABLE 5.1

Description of LSTM-Based Approaches with Details about the Nature of the Activity

Method	Sequence representation		Sequence learning				Nature of activity	
	CNN/Deep features	Handcrafted features	RNN	LSTM	Bidirectional LSTM	Multi-layer LSTM	Individual action/activity	Group action/activity
Ibrahim et al. [26]	✓	–	–	–	–	✓	–	✓
Ullah et al. [22]	✓	–	–	–	✓	–	✓	–
Ullah et al. [1]	✓	–	–	✓	–	–	✓	–
Xin et al. [7]	✓	–	✓	–	–	–	✓	–
Li et al. [27]	✓	–	–	✓	–	–	✓	–
Ma et al. [28]	✓	–	✓	–	–	–	✓	–
Chen et al. [29]	✓	✓	–	–	–	✓	✓	–
Du et al. [30]	–	–	–	✓	–	–	✓	–
Donahue et al. [31]	✓	–	–	✓	–	–	✓	–
Huang et al. [32]	✓	–	–	✓	–	–	✓	–
Ma et al. [33]	✓	–	–	–	–	✓	✓	–
Sun et al. [34]	✓	–	–	–	–	✓	✓	–
Li et al. [25]	✓	✓	✓	–	–	–	✓	–
Veeriah et al. [35]	–	✓	✓	–	–	–	✓	–
Yang et al. [36]	✓	–	–	–	✓	–	✓	–

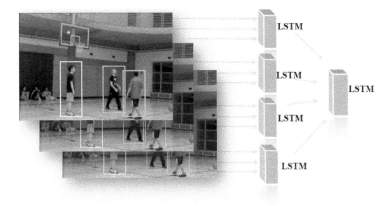

FIGURE 5.7 Group activity recognition via a hierarchical model [26]. Each person in a scene is modeled using a temporal model that captures his/her dynamics; these models are integrated into a higher-level model that captures scene-level activity.

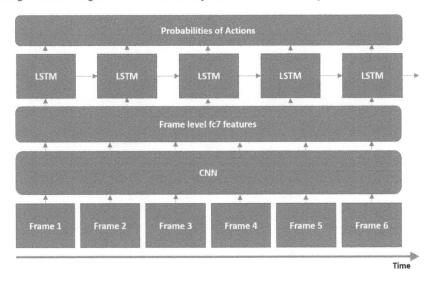

FIGURE 5.8 Framework of early activity detection [33]. At each video frame, their model first computes deep CNN features (named as fc7), and then the features are fed into the LSTM to compute detection scores of activities.

and sequence recognition for final action prediction. The LSTM used by the authors is bidirectional with hierarchical structure, and they employ spatial-temporal attention in their method. Finally, they find the integrated spatio-temporal features that are extracted using three-dimensional CNNs (3DCNNs) to advance the relationship mined in each frame.

Similarly, there are many other methods that utilize sequence learning based on RNN or LSTM with different configurations to recognize different actions and activities from video data. Furthermore, researchers (see [25] and [29–36]) have

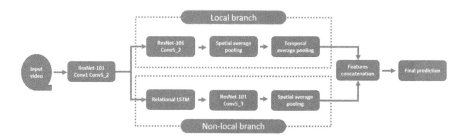

FIGURE 5.9   Network architecture for relational LSTM for video action recognition [29].

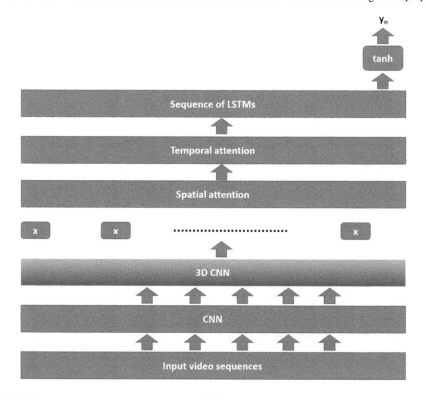

FIGURE 5.10   The architecture of the RACNN model for action recognition.

introduced different types of LSTM and RNN configurations for precise action and activity recognition. Mainstream LSTM-based action recognition literature follows convolutional features for the representation of sequences and extracts deep features as a prerequisite step for different versions of LSTM or RNN. The sequence learning step for action recognition based on LSTM uses different flavors of it, such as bidirectional, multi-layer, and simple single layer. The majority of the action recognition literature uses a simple LSTM structure. The discussed LSTM-based methods are limited to individual action or activity recognition and localization in video frames. They process all the frames instead of taking different clues from individual in frame based on simple LSTM to predict the activity for a group.

5.3   RESULTS AND DISCUSSION

In this section, we compare the results achieved by different action and activity recognition methods on benchmark datasets. Detailed comparisons of state-of-the-art techniques are given in Table 5.2. The listed LSTM-based approaches are given in chronological order, starting with the most recent first. As can be seen from Table 5.2, the latest approach [36] used recurrent attention CNN, incorporating CNNs and LSTM with an attention mechanism. Using this method, the authors obtained 94.8% and 71.9% accuracy scores on UCF-101 [37] and HMDB-51 [38], respectively. The CNN-based temporal optical flow features followed by multi-layer LSTM [1] achieved accuracy scores of 94.45% and 72.1% on UCF-101 and HMDB51 datasets, respectively. This method improved the computational performance of the activity recognition system by utilizing the lightweight FlowNet2 CNN model for feature extraction. Another approach [29] uses relational LSTM, which captures long- and short-range relations among the objects in the video data. This work obtained accuracy scores of 94.8% and 71.9% on UCF-101 and HMBD-51, respectively. A list of popular action and activity datasets is given in Table 5.3, and sample categories are given in Figure 5.11. These datasets contain human actions collected from YouTube and different other resources, presenting real-life action categories.

Ullah et al. [1] used the AlexNet CNN model for frame-level feature extraction followed by DB-LSTM and obtained 91.21%, 87.64%, and 92.84% accuracy for UCF-101, HMDB51, and YouTube action datasets, respectively. This method achieved higher accuracy for the HMDB51 dataset. The differential RNN (dRNN) [35] used a differential gating scheme with the LSTM network in order to consider the impact of spatio-temporal correspondences to salient motion patterns for action recognition. This method was evaluated on a very old and simple dataset, KTH2 [39], obtaining 92.12% accuracy. An end-to-end long-term RNN [31] reported 65% accuracy on

TABLE 5.2

Comparison of Different LSTM-Based Action and Activity Recognition Methods on Different Challenging Benchmark Datasets

Method	UCF-101	YouTube	HMDB-51
3DCNNs and bidirectional hierarchical LSTM [36]	94.8%	–	71.9%
Temporal optical flow with multi-layer LSTM [1]	94.45	95.8%	72.21
TS-LSTM and temporal inception [28]	91.1%	–	69.0%
Relational LSTM [29]	94.8%	–	71.4%
Deep bidirectional LSTM [22]	91.21%	92.84%	87.64%
Video LSTM [27]	92.2%	–	73.3%
RSTAN [30]	95.1%	–	76.9%
ARMA [32]	85.2%	–	
Bidirectional LSTM [25]	94.2%	–	70.4%
Lattice-LSTM [34]	93.6%	–	66.2%
ARCH [7]	85.3%	–	58.2%
Long-term RNN [31]	65.4%	–	–

TABLE 5.3
List of Popular Benchmark Action and Activity Recognition Datasets

Dataset	Number of videos	Number of action/ activity classes
UCF101 [37]	13,320	101
HMDB51 [38]	7,000	51
Hollywood2 [40]	3,669	12
ActivityNet [41]	28,000	203
Kinetics [42]	500,000	600
AVA [43]	57,600	80
YouTube [44]	8,000,000	4,716
Sports-1M [45]	1,133,158	487
Moments in Time [46]	1,000,000	339

FIGURE 5.11   Sample activities from UCF-101, HMDB51, and YouTube action datasets [22].

UCF-101. After this technique, an improvement of 20% on UCF-101 was achieved by Xin et al. [7], which solves the problem of the spatial and temporal domains using sequential and dynamic features. The accuracy scores obtained were 85%, 95.2%, and 58.2% on UCF-101, KTH2, and HMDB51, respectively. Apart from all these methods, other approaches presented in Table 5.2 proposed various procedures using LSTMs and RNNs to precisely recognize the activities. The majority of these methods challenged the performance over the UCF-101 and HMDB51 datasets and improved the accuracy by 3% to 5%. The highest accuracy on the UCF-101 dataset till now is 95.1%, achieved by RSTAN [30], and a high score of 87% was obtained for the HMDB51 dataset by DB-LSTM [22].

The confusion matrix, also known as the error matrix, is a popular evaluation matrix for recognition tasks. Usually, it is a table layout where the diagonal cells represent the true position predictions. However, in recognition tasks for large-scale categories, making a large table is not an attractive representation. Therefore, for this purpose, a color-coded confusion matrix is used to show the error matrix of the algorithm: values near 100% are brighter and lower values are darker. The confusion matrices for the DB-LSTM approach on the HMDB51 and UCF-101 datasets are given in Figure 5.12, where the intensity of true positives (diagonal) is high for each category, proving the efficiency of this method. A recent approach by Ullah et al. [1] evaluated their method on five benchmark datasets. Confusion matrices from their method for UCF-101 and HMDB51 datasets are shown in Figure 5.13. They achieved high accuracy for UCF-101 and UCF-50 datasets; therefore, from Figure 5.13 (a) and Figure 5.13 (d), it can be seen that for each class the true positive is very high. Similarly, for two challenging datasets, namely HMDB51 and Hollywood2, they achieved average performance; therefore, in Figure 5.13 (b) and Figure 5.13 (c) the diagonals are not as bright in the confusion matrix. Figure 5.13 (e) is the confusion matrix for a YouTube action dataset that contains only 11 action categories; therefore the results are good for each class and the true positives are very bright in the confusion matrix.

Category-wise accuracy is another metric for large-scale dataset evolution. It shows the overall performance of the algorithm for all categories in the dataset. We have compared the class-wise accuracies obtained by DB-LSTM [22] and temporal optical flow with multi-layer LSTM [1] methods for the UCF-101 dataset in Figure 5.14 (a) and Figure 5.14 (b). The horizontal axis shows classes, and the vertical axis represents percentage accuracies for the corresponding classes for test sets of the UCF-101 dataset. It is evident from Figure 5.14 (a) and Figure 5.14 (b) that the best accuracies are above 80%, approaching 100% for some categories.

However, in the results for DB-LSTM shown in Figure 5.14 (a), many categories are less than 50%, which shows that the performance of the method is not effective for this dataset. On the other hand, the temporal optical flow method with multi-layer LSTM [1] given in Figure 5.14 (b) achieves higher than 70% accuracy for each class, which shows the robustness of this approach for different kinds of activity recognition in real-world scenarios. The class-wise accuracies achieved by DB-LSTM [22]

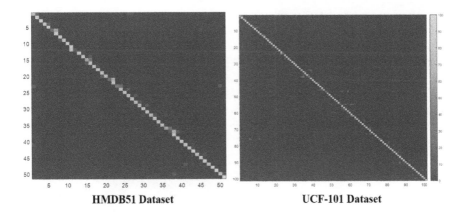

HMDB51 Dataset   **UCF-101 Dataset**

FIGURE 5.12 Confusion matrices of HMDB51 and UCF-101 datasets obtained by deep features of AlexNet model for frame-level representation followed by DB-LSTM for action recognition [22].

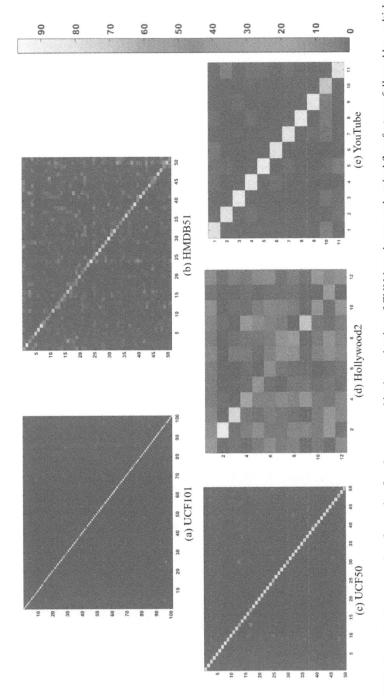

FIGURE 5.13 Confusion matrices for testing five datasets used in the evaluation of CNN-based temporal optical flow features followed by multi-layer LSTM. (a) UCF101 dataset, (b) HMDB51 dataset, (c) Hollywood2 dataset, (d) UCF50 dataset, and (e) YouTube dataset [1, 22].

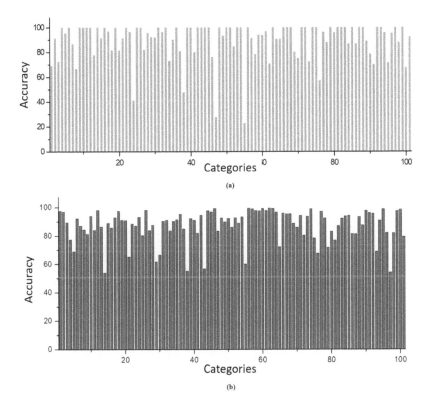

FIGURE 5.14 Comparison of category-wise accuracy on the UCF101 dataset for DB-LSTM and temporal optical flow with multi-layer LSTM approaches [1, 22].

and temporal optical flow with multi-layer LSTM [1] approach for HMDB51 dataset are shown in Figure 5.15 (a) and Figure 5.15 (b). respectively.

HMDB51 is one of the most challenging datasets; for this dataset, the method of [1] achieved 71% and the method of [22] obtained 87% overall accuracy. However, the class-wise accuracy for the method shown in Figure 5.15 (a) the method of [22] is better for maximum class, but a few of them are still less than 10%, while from the method of [1], shown in Figure 5.15 (b), we can see that a few category scores are above 90%, but most of them are within the range of 70% to 90%. The accuracy for all classes is greater than 45% except kicking a football, picking, and swinging a baseball bat. This is due to the overlapping of visual contents and similarity of motion information with other classes. Therefore, predictions of these classes are confused with others.

5.4 FUTURE RESEARCH DIRECTIONS

It has been observed that many researchers have addressed the problem of action and activity recognition by applying different strategies in LSTM according to the need in a particular domain. The current literature shows that many researchers have

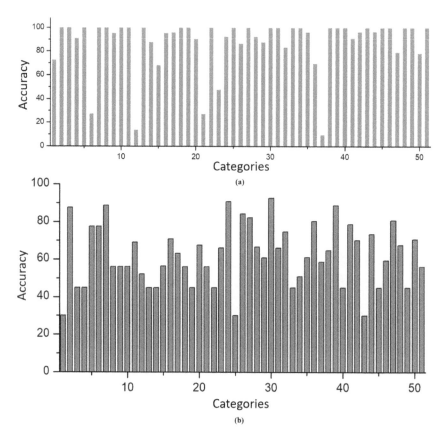

FIGURE 5.15 Comparison of category-wise accuracy on the HMDB51 dataset for DB-LSTM and temporal optical flow with multi-layer LSTM approaches [1, 22].

used different kinds of features for sequence representations followed by RNN or LSTM. For instance, some researchers also used deep features for sequence representation; however, the time complexity of those methods is very high. So, there is a need for state-of-the-art techniques that can make tradeoffs between accuracy and time complexity. Recommendations and future directions about actions and activity recognition are as follows.

5.4.1 Benefiting from Image Models

Image classification has achieved state-of-the-art results using CNN and beaten human error rates in recent deep models [47]. These existing image models can be utilized for frame-level representation in sequential video data. There are many lightweight CNN models, such as MobileNet [48], SqueezeNet [49], and ShuffleNet [50], that can help in real-time processing for action and activity recognition. These CNN models have already achieved state-of-the-art results in other tasks such as object detection and localization [51, 52], image and video retrieval [53], video summarization [54], etc. The dominance of these models is that they are trained on large-scale image datasets, which helps to learn discriminative hidden local and global representations from visual data.

5.4.2 MULTIPLE ACTIONS AND ACTIVITY RECOGNITION

Several action recognition methods exist, but we can see that these methods are applicable for processing single-person action data effectively. The existing methods are trained on UCF101, UCF sports, and HMDB51 datasets; in these datasets, only one person is performing an activity in sample videos. From Table 5.1, we can see that only one LSTM-based method addresses the problem of multiple activity recognition in video data. In CCTV, multiple activities are performed in a single scene. So, in the future, this task needs to be tackled by detecting and localizing the individual in the video data and then further processing it for activity recognition. Localization and tracking are very challenging tasks in group activity recognition, for which we can use existing deep learning-based object detection and localization models such as YOLO [55], faster RCNN [56], MobileNet SSD [48], etc.

5.4.3 UTILIZATION OF LARGE-SCALE DATASETS

The existing datasets utilized by researchers are very small and are not sufficient for training deep learning models from scratch, although there exist several large-scale video action and activity recognition datasets that can be utilized for effective recognition. Table 5.3 shows some of the popular datasets in the action and activity recognition literature. The first three datasets are already used by many researchers and have achieved good results, as shown in Table 5.2. However, by utilizing ActivityNet [41], Kinetics [42], AVA [43], YouTube [44], Sports-1M [45], and Moments in Time [46], we can recognize a larger number of activities and actions through a single recognition model.

5.5 CONCLUSION

Human action and activity recognition is the problem of analyzing and predicting the movements of human body parts in a sequence of video frames. In this chapter, we have described the concept of sequence learning for action and activity recognition using LSTM and its variants such as multi-layer or deep LSTM and bidirectional LSTM networks. Moreover, recent LSTM-based methods for action and activity recognition are surveyed, and its achievements and drawbacks are discussed. We also explained the working of traditional RNNs and their limitations, along with why LSTM is better than RNN. Finally, the chapter concluded with recommendations for future directions that may help overcome the present challenges of action and activity recognition using LSTM.

ACKNOWLEDGMENT

This work was supported by the National Research Foundation of Korea (NRF) grant funded by the Korea Government (MSIP) (No. 2019R1A2B5B01070067).

BIBLIOGRAPHY

1. A. Ullah, K. Muhammad, J. Del Ser, S. W. Baik, and V. Albuquerque, "Activity recognition using temporal optical flow convolutional features and multi-layer LSTM," *IEEE Transactions on Industrial Electronics*, vol. 66, no. 12, pp. 9692–9702, 2019.
2. I. U. Haq, K. Muhammad, A. Ullah, and S. W. Baik, "DeepStar: Detecting starring characters in movies," *IEEE Access*, vol. 7, pp. 9265–9272, 2019.
3. Y. Liu, L. Nie, L. Han, L. Zhang, and D. S. Rosenblum, "Action2Activity: Recognizing complex activities from sensor data," in *IJCAI*, 2015, pp. 1617–1623.
4. Y. Liu, L. Nie, L. Liu, and D. S. Rosenblum, "From action to activity: Sensor-based activity recognition," *Neurocomputing*, vol. 181, pp. 108–115, 2016.
5. A. Ullah, K. Muhammad, I. U. Haq, and S. W. Baik, "Action recognition using optimized deep autoencoder and CNN for surveillance data streams of non-stationary environments," *Future Generation Computer Systems*, vol. 96, pp. 386–397, 2019.
6. M.-C. Roh, H.-K. Shin, and S.-W. Lee, "View-independent human action recognition with volume motion template on single stereo camera," *Pattern Recognition Letters*, vol. 31, pp. 639–647, 2010.
7. M. Xin, H. Zhang, H. Wang, M. Sun, and D. Yuan, "Arch: Adaptive recurrent-convolutional hybrid networks for long-term action recognition," *Neurocomputing*, vol. 178, pp. 87–102, 2016.
8. D. Weinland, R. Ronfard, and E. Boyer, "Free viewpoint action recognition using motion history volumes," *Computer Vision and Image Understanding*, vol. 104, pp. 249–257, 2006.
9. M. Baccouche, F. Mamalet, C. Wolf, C. Garcia, and A. Baskurt, "Action classification in soccer videos with long short-term memory recurrent neural networks," in *International Conference on Artificial Neural Networks*, 2010, pp. 154–159.
10. A. Kovashka, and K. Grauman, "Learning a hierarchy of discriminative space-time neighborhood features for human action recognition," in *2010 IEEE Conference on Computer Vision and Pattern Recognition (CVPR)*, 2010, pp. 2046–2053.
11. M. Sekma, M. Mejdoub, and C. B. Amar, "Human action recognition based on multi-layer fisher vector encoding method," *Pattern Recognition Letters*, vol. 65, pp. 37–43, 2015.
12. J. Hou, X. Wu, Y. Sun, and Y. Jia, "Content-attention representation by factorized action-scene network for action recognition," *IEEE Transactions on Multimedia*, vol. 20, pp. 1537–1547, 2018.
13. F. U. M. Ullah, A. Ullah, K. Muhammad, I. U. Haq, and S. W. Baik, "Violence detection using spatiotemporal features with 3D convolutional neural network," *Sensors*, vol. 19, p. 2472, 2019.
14. K.-i. Funahashi, and Y. Nakamura, "Approximation of dynamical systems by continuous time recurrent neural networks," *Neural Networks*, vol. 6, pp. 801–806, 1993.
15. S. Hochreiter, and J. Schmidhuber, "Long short-term memory," *Neural Computation*, vol. 9, pp. 1735–1780, 1997.
16. H. Sak, A. Senior, and F. Beaufays, "Long short-term memory recurrent neural network architectures for large scale acoustic modeling," in *Fifteenth Annual Conference of the International Speech Communication Association*, 2014.
17. M. Sajjad, S. Khan, K. Muhammad, W. Wu, A. Ullah, and S. W. Baik, "Multi-grade brain tumor classification using deep CNN with extensive data augmentation," *Journal of Computational Science*, vol. 30, pp. 174–182, 2019.
18. J. Ahmad, K. Muhammad, S. Bakshi, and S. W. Baik, "Object-oriented convolutional features for fine-grained image retrieval in large surveillance datasets," *Future Generation Computer Systems*, vol. 81, pp. 314–330, 2018/04/01/ 2018.

19. K. Muhammad, J. Ahmad, Z. Lv, P. Bellavista, P. Yang, and S. W. Baik, "Efficient deep CNN- based fire detection and localization in video surveillance applications," *IEEE Transactions on Systems, Man, and Cybernetics: Systems*, vol. 49, no. 7, pp. 1419–1434, 2019.

20. K. Muhammad, R. Hamza, J. Ahmad, J. Lloret, H. H. G. Wang, and S. W. Baik, "Secure surveillance framework for IoT systems using probabilistic image encryption," *IEEE Transactions on Industrial Informatics*, vol. 14, no. 8, pp. 3679–3689, 2018.

21. M. Sajjad, A. Ullah, J. Ahmad, N. Abbas, S. Rho, and S. W. Baik, "Integrating salient colors with rotational invariant texture features for image representation in retrieval systems," *Multimedia Tools and Applications*, vol. 77, pp. 4769–4789, 2018.

22. A. Ullah, J. Ahmad, K. Muhammad, M. Sajjad, and S. W. Baik, "Action recognition in video sequences using deep Bi-directional LSTM with CNN features," *IEEE Access*, vol. 6, pp. 1155–1166, 2018.

23. A. Ogawa, and T. Hori, "Error detection and accuracy estimation in automatic speech recognition using deep bidirectional recurrent neural networks," *Speech Communication*, vol. 89, pp. 70–83, 2017.

24. J. Chung, C. Gulcehre, K. Cho, and Y. Bengio, "Empirical evaluation of gated recurrent neural networks on sequence modeling," arXiv preprint arXiv:1412.3555, 2014.

25. W. Li, W. Nie, and Y. Su, "Human action recognition based on selected spatio-temporal features via bidirectional LSTM," *IEEE Access*, vol. 6, pp. 44211–44220, 2018.

26. M. S. Ibrahim, S. Muralidharan, Z. Deng, A. Vahdat, and G. Mori, "A hierarchical deep temporal model for group activity recognition," In *Proceedings of the IEEE Conference on Computer Vision and Pattern Recognition*, pp. 1971–1980.

27. Z. Li, K. Gavrilyuk, E. Gavves, M. Jain, and C. G. Snoek, "VideoLSTM convolves, attends and flows for action recognition," *Computer Vision and Image Understanding*, vol. 166, pp. 41–50, 2018.

28. C.-Y. Ma, M.-H. Chen, Z. Kira, and G. AlRegib, "TS-LSTM and temporal-inception: Exploiting spatiotemporal dynamics for activity recognition," *Signal Processing: Image Communication*, vol. 71, pp. 76–87, 2019.

29. Z. Chen, B. Ramachandra, T. Wu, and R. R. Vatsavai, "Relational long short-term memory for video action recognition," arXiv preprint arXiv:1811.07059, 2018.

30. W. Du, Y. Wang, and Y. Qiao, "Recurrent spatial-temporal attention network for action recognition in videos," *IEEE Transactions on Image Processing*, vol. 27, pp. 1347–1360, 2018.

31. J. Donahue, L. Anne Hendricks, S. Guadarrama, M. Rohrbach, S. Venugopalan, K. Saenko, et al., "Long-term recurrent convolutional networks for visual recognition and description," in *Proceedings of the IEEE Conference on Computer Vision and Pattern Recognition*, 2015, pp. 2625–2634.

32. Y. Huang, X. Cao, Q. Wang, B. Zhang, X. Zhen, and X. Li, "Long-short term features for dynamic scene classification," *IEEE Transactions on Circuits and Systems for Video Technology*, vol. 29, no. 4, pp. 1038–1047, 2019.

33. S. Ma, L. Sigal, and S. Sclaroff, "Learning activity progression in lstms for activity detection and early detection," in *Proceedings of the IEEE Conference on Computer Vision and Pattern Recognition*, 2016, pp. 1942–1950.

34. L. Sun, K. Jia, K. Chen, D.-Y. Yeung, B. E. Shi, and S. Savarese, "Lattice long short-term memory for human action recognition," in *ICCV*, 2017, pp. 2166–2175.

35. V. Veeriah, N. Zhuang, and G.-J. Qi, "Differential recurrent neural networks for action recognition," in *Proceedings of the IEEE International Conference on Computer Vision*, 2015, pp. 4041–4049.

36. H. Yang, J. Zhang, S. Li, and T. Luo, "Bi-direction hierarchical LSTM with spatial-temporal attention for action recognition," *Journal of Intelligent & Fuzzy Systems*, vol. 36, no. 1, pp. 775–786, 2019.

37. K. Soomro, A. R. Zamir, and M. Shah, "UCF101: A dataset of 101 human actions classes from videos in the wild," arXiv preprint arXiv:1212.0402, 2012.

38. H. Kuehne, H. Jhuang, E. Garrote, T. Poggio, and T. Serre, "HMDB: A large video database for human motion recognition," in *2011 IEEE International Conference on Computer Vision (ICCV)*, 2011, pp. 2556–2563.

39. C. Schuldt, I. Laptev, and B. Caputo, "Recognizing human actions: A local SVM approach," in *Proceedings of the 17th International Conference on Pattern Recognition, ICPR 2004*, 2004, pp. 32–36.

40. M. Marszalek, I. Laptev, and C. Schmid, "Actions in context," in *IEEE Conference on Computer Vision and Pattern Recognition, CVPR 2009*, 2009, pp. 2929–2936.

41. F. Caba Heilbron, V. Escorcia, B. Ghanem, and J. Carlos Niebles, "Activitynet: A large-scale video benchmark for human activity understanding," in *Proceedings of the IEEE Conference on Computer Vision and Pattern Recognition*, 2015, pp. 961–970.

42. W. Kay, J. Carreira, K. Simonyan, B. Zhang, C. Hillier, S. Vijayanarasimhan, et al., "The kinetics human action video dataset," arXiv preprint arXiv:1705.06950, 2017.

43. C. Gu, C. Sun, S. Vijayanarasimhan, C. Pantofaru, D. A. Ross, G. Toderici, et al., "AVA: A video dataset of spatio-temporally localized atomic visual actions," arXiv preprint arXiv:1705.08421, vol. 3, p. 6, 2017.

44. S. Abu-El-Haija, N. Kothari, J. Lee, P. Natsev, G. Toderici, B. Varadarajan, et al., "Youtube-8m: A large-scale video classification benchmark," arXiv preprint arXiv:1609.08675, 2016.

45. A. Karpathy, G. Toderici, S. Shetty, T. Leung, R. Sukthankar, and L. Fei-Fei, "Large-scale video classification with convolutional neural networks," in *Proceedings of the IEEE Conference on Computer Vision and Pattern Recognition*, 2014, pp. 1725–1732.

46. M. Monfort, B. Zhou, S. A. Bargal, A. Andonian, T. Yan, K. Ramakrishnan, et al., "Moments in time dataset: One million videos for event understanding," arXiv preprint arXiv:1801.03150, 2018.

47. S. Khan, K. Muhammad, S. Mumtaz, S. W. Baik, and V. H. C. de Albuquerque, "Energy-efficient deep CNN for smoke detection in foggy IoT environment," *IEEE Internet of Things Journal*, vol. 6, no. 6, pp. 9237–9245, 2019.

48. A. G. Howard, M. Zhu, B. Chen, D. Kalenichenko, W. Wang, T. Weyand, et al., "Mobilenets: Efficient convolutional neural networks for mobile vision applications," arXiv preprint arXiv:1704.04861, 2017.

49. F. N. Iandola, S. Han, M. W. Moskewicz, K. Ashraf, W. J. Dally, and K. Keutzer, "Squeezenet: Alexnet-level accuracy with 50x fewer parameters and< 0.5 mb model size," arXiv preprint arXiv:1602.07360, 2016.

50. N. Ma, X. Zhang, H.-T. Zheng, and J. Sun, "Shufflenet v2: Practical guidelines for efficient cnn architecture design," arXiv preprint arXiv:1807.11164, vol. 5, 2018.

51. M. Sajjad, M. Nasir, F. U. M. Ullah, K. Muhammad, A. K. Sangaiah, and S. W. Baik, "Raspberry Pi assisted facial expression recognition framework for smart security in law-enforcement services," *Information Sciences*, vol. 479, pp. 416–431, 2019.

52. M. Sajjad, S. Khan, T. Hussain, K. Muhammad, A. K. Sangaiah, A. Castiglione, et al., "CNN-based anti-spoofing two-tier multi-factor authentication system," *Pattern Recognition Letters*, vol 126, pp. 123–131, 2019.

53. J. Ahmad, K. Muhammad, J. Lloret, and S. W. Baik, "Efficient conversion of deep features to compact binary codes using Fourier decomposition for multimedia big data," *IEEE Transactions on Industrial Informatics*, vol. 14, pp. 3205–3215, July 2018.

54. K. Muhammad, T. Hussain, and S. W. Baik, "Efficient CNN based summarization of surveillance videos for resource-constrained devices," *Pattern Recognition Letters*, 2018.

55. J. Redmon, S. Divvala, R. Girshick, and A. Farhadi, "You only look once: Unified, real-time object detection," in *Proceedings of the IEEE Conference on Computer Vision and Pattern Recognition*, 2016, pp. 779–788.

56. S. Ren, K. He, R. Girshick, and J. Sun, "Faster r-cnn: Towards real-time object detection with region proposal networks," in *Advances in Neural Information Processing Systems*, 2015, pp. 91–99.

6 Deep Semantic Segmentation in Autonomous Driving

*Hazem Rashed, Senthil Yogamani,
Ahmad El-Sallab, Mahmoud Hassaballah, and
Mohamed ElHelw*

CONTENTS

6.1   INTRODUCTION

The first autonomous cars (ACs) developed by Carnegie Mellon University and ALV appeared in 1984 [1]. Later, in 1987, Mercedes-Benz collaborated with Bundeswehr University Munich in a project called the Eureka Prometheus Project [2], which was the largest R&D project ever in autonomous driving (AD). Since then, numerous car manufacturers have developed prototypes for their self-driving cars [3, 4]. The first truly autonomous car, VaMP, developed in the 1990s by Mercedes-Benz, was a 500 SEL Mercedes-Benz that had the capabilities to control the steering wheel, throttle, and brakes by computers in real time. The car relied only on four cameras for localization, as there was no GPS at that time. The vehicle was able to drive more than 1,000 km in normal traffic at speeds up to 130, kmh moving from lane to another and passing other cars autonomously after the approval of the safety driver [5]. Recently, AD has gained great attention with significant progress in computer vision algorithms [6, 7], and it is considered one of the highly trending technologies all over the globe. Within the next 5–10 years, AD is expected to be deployed commercially. Currently, most automotive original equipment manufacturers (OEMs) over the world, such as Volvo, Daimler, BMW, Audi, Ford, Nissan, and Volkswagen, are working on development projects focusing on the AD technology [8, 9].

In addition, every year large numbers of people die or are injured due to car accidents [10]. Autonomous cars (ACs) have great potential to significantly reduce these numbers. One of the potential benefits of ACs is the higher safety levels for both drivers and surrounding pedestrians [11]. It has been reported by consulting firm McKinsey & Company that deploying AD technology could eliminate 90% of accidents in the USA and save up to US 190 billion due to damage prevention annually. Besides, autonomous vehicles are predicted to have positive impact on traffic flow, with higher speed limits permitted for increased roadway capacity [12]. AD can also provide a higher level of satisfaction for customers, especially for the elderly and the disabled, introduce new business models such as autonomous taxis as a service, and reduce crime.

Additionally, when vehicles are self-aware, they will have the information needed to avoid numerous accidents [13]. For example, vehicles may recognize when they are being stolen or the driver is being threatened. Self-aware vehicles may recognize children being kidnapped or weapons being transported, or they may spot drunk drivers and consequently take proper action [14, 15]. Generally, there are five levels of autonomy [16]:

- **Level 0** is "no automation", where everything has to be controlled by the driver.
- **Level 1** is "driver assistance", where the driver controls everything; however, separate systems provide assistance for one or more tasks related to steering and acceleration/deceleration. This takes place on a condition that the system controls either steering or acceleration/deceleration, but not both together. A good example is the lane-keeping assist system, in which the car monitors the lanes ahead and constantly applies torque to the steering

system to keep the car in the center of the lane. Another example is adaptive cruise control (ACC), which controls the acceleration/deceleration according to the distance between the ego-vehicle and the car ahead with a predefined speed limit [17].

- **Level 2** is "partial automation", where the car can do maneuvers controlling both speed and steering together; however, the driver must keep his/ her hands on the steering wheel for any required intervention. Most of the current commercial autonomous driving systems are of Level 2.
- **Level 3** is "conditional automation", where the car manages almost all of the tasks but the system prompts the driver for intervention when there is a scenario it cannot navigate through. Unlike Level 2, the driver can remove his hands from the steering wheel; however, he/she should be available to take over at any time.
- **Level 4** is "high automation", where the car can operate without the driver's supervision but under conditional circumstances for the environment. Otherwise, the driver may take control. Google's Firefly prototype is an example for Level 4 autonomy; however, speed is limited to 25 mph.
- **Level 5** is "full automation", where the car can operate in any condition and the driver's role is only providing the destination. This level of autonomy has not been reached yet.

Nevertheless, like any new technology, AD faces various technical and non-technical challenges. The technical challenges include sensor accuracy compared to cost; that is, the sensors have to be reasonable in terms of price [18]. Therefore, efficient algorithms have to be implemented to deal with map generation errors. There are localization errors that take place, as data from GPS and intertial measurement units (IMUs) is not always stable due to connectivity issues and even odometry information has errors, especially when the car travels long distances. Path planning and navigation should not depend on imperfect inputs and should deal with various practical cases [19]. The complexity of a system must be acceptable for the purpose of producing commercial cars, which adds another limitation in terms of the hardware used for production. The non-technical challenges include the urgent need to take tough decisions into implementation [20]. For instance, the AC has to be equipped with a decision-making system for unavoidable accidents; it has to decide to choose hitting a woman or a child if it has to, and which one to protect the most, the child outside the car or the car occupant [21]. In this regard, a mobile robot must be able to perform four key tasks to be able to navigate autonomously [22, 23].

- **Environment perception and map creation** using different types of sensors. This task includes a lot of smaller subtasks, such as object recognition and classification, object tracking, criticality definition, etc. [24].
- **Localization** within the map where an accurate location for the robot must be calculated using sensors such as GPS and further algorithms utilized to improve accuracy and provide location even in areas where GPS signals are not available including tunnels and garages.

- **Path planning and decision making** according to both map and position information. In addition to path planning, there are some constraints and regulations to be followed especially for a self-driving car, e.g., observing traffic lights and following rules in special paths such as roundabouts, intersections, and U-turns.
- **Vehicle control** through sending signals to actuators over the vehicle's network for taking a desired action.

6.2 CONVOLUTIONAL NEURAL NETWORKS

Recently, convolutional neural networks (CNNs) and in particular deep learning have played a great role in computer vision, and they are exploited in various applications for AD. In this section, a brief introduction about deep CNN is presented, i.e., how neural networks are used in machine learning, and the idea behind it. Afterwards, semantic segmentation, which is one of the main applications of deep learning in AD, is discussed.

6.2.1 OVERVIEW OF DEEP CNN

Deep learning is a specialized form of machine learning. Generally, a machine learning algorithm starts after the phase of manual feature selection in a given input, whether it is an image or any type of input. The features are then used to create a model that performs the classification task according to the input data [25]. In the deep learning workflow, relevant features are automatically extracted from images. From this comes the power of deep learning, as it has the capability to extract hidden relationships and features that cannot be easily perceived by humans. In addition, deep learning performs end-to-end learning, where a network is given raw data and a task to perform, such as classification, and it learns how to do this automatically [26].

There are several types of deep neural networks; one of these, built to deal with input images, is called a CNN. In CNN algorithms, each level in the deep network learns to transform its input data into a slightly more abstract and composite representation [27]. For example, in an image recognition application, the raw input is a matrix of pixels. The first layer may abstract the pixels and encode edges; the second layer may encode arrangements of edges. The third layer may encode specific features like a nose or eyes, and the fourth layer may recognize that the image contains a face. The whole pipeline takes place without manual feature selection as opposed to conventional algorithms, where specific characteristics are being searched in order to recognize which given images contain faces. In fact, deep learning has gained great attention in the 2010s after the development of AlexNet [28] in the ImageNet challenge, where an error of 16% was achieved for the image classification task in the ImageNet challenge.

The CNNs are used for various tasks, like depth estimation, object detection, localization, classification, semantic segmentation, and several others. The goal of a CNN is to learn higher-order features in the data by using convolutions. In a conventional neural network, an input 3-RGB channel image of 300×300 pixels would create 270,000 connection weights per hidden neuron if it is fully connected, which is not practical from an implementation perspective. Additionally, unlike any other input vector, images are spatially correlated, where pixels belonging to the same object typically will have close pixel values. This piece of information is very important and can help the network to provide better outputs. Neural networks make use of this piece of information through full connectivity. CNNs, on the other hand, are able to utilize computation power better than fully connected neurons. With CNNs, one can arrange the neurons in a three-dimensional structure. These attributes are consistent with the image structure, as it has image width, image height, and 3 channels representing the RGB information. Neurons in a layer are connected to only a small region of neurons in the layer before it, where each layer transforms the 3D input volume from the previous layer into another 3D output volume of neuron activations with some differentiable function [29].

6.2.2 TYPES OF CNN LAYERS

The term "deep" in deep CNN refers to the number of hidden layers through which the data representation is transformed [30]. The common types of layers used in a typical deep CNN can be summarized as follows:

Input layers: This is where we load the raw input data of the image to be processed by the network. The input is a 3D volume of image width, height, and depth. Usually the depth represents the three RGB layers of an image.

Convolutional layers: These layers are the core building blocks of any CNN architecture. Convolutional layers transform the input data by using a patch of locally connected neurons from the previous layer. Generally, convolution is defined as a mathematical operation describing how to merge two sets of information. It computes a dot product between the region of neurons in the previous layer and the weights to which they are locally connected in the output layer. Figure 6.1 shows a visualization for neuron activation function, where each neuron holds a set of parameters that form a filter to be multiplied with the input volume as a dot product and

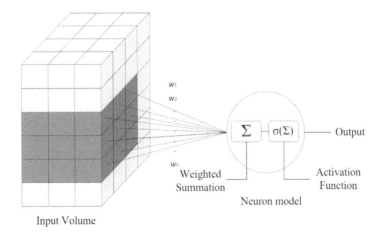

FIGURE 6.1 Visualization of the CNN activation function.

added to bias. The convolution input can be raw pixels or an output feature map from the previous layer. The convolution is known in deep learning as a feature detector for the CNN where the values in kernels define an operation for feature detection. A kernel with ones in the middle columns is used to detect vertical edges within the input patch. The output of each convolution operation is called a *feature map*. Stacking the output feature maps for different kernels generates a 3D output volume.

Convolution filters define a small bounded region to output activation maps from the input volume through connection with only a subset of the input volume. This allows us to have high-quality feature extraction utilizing the spatial correlation between pixels besides reduction the number of parameters per layer compared to fully connected neurons [31]. Since the depth of the input volume is 3, the convolution layer as well should have a depth of 3. Convolution kernels connect to the bounded region on the image in terms of width and height; however, the depth always has to be the same. CNNs use parameter sharing to control the total number of parameters, where the same weights being learned using a certain filter are used while the filter scans the whole image or feature map, instead of defining new parameters. There are several parameters to control using a filter in CNNs such as receptive field, output depth, stride, and zero-padding.

Receptive field: This term refers to the spatial filter size, in terms of width and height. A filter with receptive field of $5 \times 5 \times 3$ means that it has a 5-pixel width, a 5-pixel height, and 3 channels of depth. Usually, filters of the first convolutional layer have depths of 3 to accept an RGB image input volume of 3 channels.

Output depth: The neuron count in the convolutional layer connected to the same region of the input volume, meaning that it refers to the number of filters used in the convolutional layer.

Stride: This refers to the step used by the filter to scan through the input volume. A stride of 1 means that the filter moves horizontally and vertically with one-pixel steps, resulting in a dense output volume of a large number of depth columns, as there is a large overlap between receptive fields.

Zero-padding: This parameter is used to control the spatial size of the output volume. This is very helpful if the output volume is required to be same as the input volume. Additionally, if convolutions are done without zero-padding and the size is reduced, the information at boundaries will be lost quickly.

Batch normalization layer: This layer is used in deep learning to speed up the training by making normalization part of the network. For example, if input values are in a range from 1 to 1,000 and from 1 to 10, the change from 1 to 2 in the second input represents a 10% change in the input value, but actually represents only a 0.1% change in the first input. Normalizing the inputs will help the network learn faster and more accurately without biasing certain dimensions over the others. Batch normalization uses a very similar approach, however, for the hidden layers changing the outputs all the time. Since the activations of a previous layer are the inputs of the next layer, each layer in the neural network faces a problem due to input distribution change with each step. This problem is termed "covariate shift". The basic idea behind batch normalization is to limit covariate shift by normalizing the activations of each layer through transforming the inputs to be mean 0 and unit variance. This

allows each layer to learn on a more stable distribution of inputs, and would accelerate the training of the network. Limiting the covariate shift helps as well in avoiding the vanishing gradient problem, where gradients become very small when the
input is large in a sigmoid activation function, for example, as the distribution keeps
changing during training and might be large enough to create the vanishing gradient
problem [32].

Pooling layers: These layers are periodically inserted after convolutional layer
in a typical ConvNet architecture. Their function is to progressively reduce the spatial size of the input representation providing multiple advantages: (1) Obtaining an
abstract representation of the input data where only the most important features are
being considered; hence, this helps to avoid overfitting the model to the input data.
(2) Reducing the computational cost as the number of learnable parameters becomes
less than those for corresponding networks not using pooling. (3) Providing basic
translation invariance to the input representation. The pooling layer operates independently on each depth slice of the input representation. The most common form of
pooling is called max-pooling, where a 2×2 window scans the image with a stride
of 2, keeping the max value of that window only, resulting in keeping only 25% of
the input volume [33].

Fully connected layers: Neurons in a fully connected layer have connections
to all output activations from the previous layer, which is always good to be able to
model complex representations. The fully connected layer provides all the possible
combinations of the features in the previous layers, as the convolutional layer relies
only on the local spatial correlation between pixels. However, fully connected layers
contribute strongly to the network's complexity [34].

Output layer: In a normal classification task where the objects belong to one
class only, the sigmoid layer is used to finally generate the output. Let's say we
are trying to classify an object among five classes. The Softmax function provides an output probability for the likelihood that this object belongs to each class.
These output probabilities will sum up to 1, which is very helpful because the next
step would be classification of the object according to the maximum probability
generated.

6.3 SEMANTIC SEGMENTATION

Semantic image segmentation has witnessed tremendous progress recently, especially with the appearance of deep learning. It provides dense pixel-wise labeling
of the image, which leads to scene understanding. Semantic segmentation has been
used in many applications such as robotics [35–37], virtual reality [38], and more
importantly autonomous driving [39, 40], which is one of the main applications using
semantic segmentation. Figure 6.2 shows an example of the output of a deep convolutional network for an automated driving scene.

Pixel-wise annotation of large datasets is very hard and labor intensive. Thus,
there is a lot of research with the aim of reducing the complexity of segmentation networks by using knowledge from other domains and incorporating other
cues wherever possible. To this end, most of the current semantic segmentation

FIGURE 6.2   Semantic segmentation for a typical automotive scene.

networks are implemented to use only appearance cues without considering several other cues that might be helpful to a specific application. In this context, deep learning has played a great role in semantic segmentation tasks, ideally because of the automatic feature selection advantage that deep learning has over conventional methods. There are three main areas explored in deep learning for semantic segmentation. The first one uses patch-wise training, where, the input image is fed into the Laplacian pyramid, each scale is forwarded through a three-stage network for hierarchical feature extraction, and patch-wise classification is used [41].

The second area is end-to-end training, in which the CNN network is trained to take raw images as input and directly output dense pixel-wise classification instead of patches classification. In [42], famous classification networks, namely AlexNet [43], VGG-net [44], and GoogleLeNet [45], are adapted into fully convolutional networks, and the learned knowledge is transferred to the semantic segmentation task. Skip connections between shallow and deep features are implemented to avoid losing resolution. The features learned within the network are upsampled using deconvolution to get dense predictions. In [46], deeper deconvolution is implemented, in which stacked deconvolution and unpooling layers are utilized. In [47], an encoder-decoder network is implemented where the decoder upsampling made use of the pooling indices computed in the max-pooling step of the corresponding encoder for nonlinear upsampling.

Recent works have focused on multiscale semantic segmentation [48]. In [41], the scale issue has been dealt with using multiple rescaled versions of the image; however, after the emergence of end-to-end learning, the skip connections implemented in [42] helped to merge feature maps from different resolutions, as loss of resolution takes place due to image downsampling. Dilated convolution has been introduced in [49], where the receptive field has been expanded with zero values in between depending on the dilation factor in order to avoid losing resolution. This helped for the multiple scales issue. In the attention model [50, 51], the multiscale image segmentation model is adapted so that it learns to softly weight the multiscale features for objects with different scales.

6.3.1 AUTOMOTIVE SCENE SEGMENTATION

Since this work mainly focuses on autonomous driving applications, we provide a summary of automotive environment conditions and information specific to these applications that can be utilized to enhance the performance of tasks on hand.

Modern cars are equipped with a multi-camera network, usually of 4 cameras but up to 10 cameras for future generations. Figure 6.3 shows an example of 360° surround-view system using four fisheye cameras installed for viewing all the surroundings. Using such a system, there is a spatio-temporal relationship between the outputs of the four cameras. For instance, a moving car will be shown in one camera only, then overlap with another one, and so on. There is still no significant work targeting automotive fisheye setup. However, in [52], a method is used to enhance the way convolution captures information from neighbor pixels. Restricted deformable convolution is utilized as learnable dilated convolution. The idea is to initialize a dilated convolution kernel, then learn offset parameters through training, which will change the kernel's center to enhance the learning of neighbor pixels. This technique is used to learn semantic segmentation of four fisheye cameras in automotive surround-view systems.

There is no semantic segmentation annotation provided for fisheye cameras. To solve this problem, a method for data augmentation named zoom augmentation is implemented in [53], and focal length has been changed to generate fisheye images to train CNNs for semantic segmentation. The problem of depth estimation for fisheye cameras using sparse LiDAR data is investigated for automotive surround-view

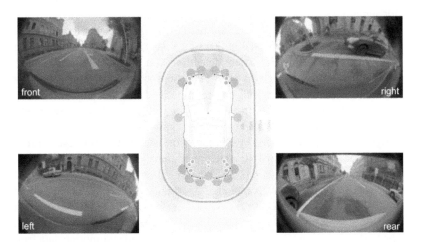

FIGURE 6.3   Sample images from the surround-view camera network demonstrating near-field sensing and wide field of view.

systems. The difference between viewpoints of both camera and LiDAR caused occluded points to be projected incorrectly to the image plane. Then, an occlusion correction method is presented to remove the incorrectly mapped points, and dense depth map for the fisheye camera is generated using the LiDAR depth information as supervised signals.

In a generic structure of automotive scenes, different types of prior information that are common in any automotive road scene can be utilized, e.g., spatial priors such as the road pixels usually taking almost all of the bottom half of the image. The lanes are seen as thick white lines expanding until the road pixels end. Other priors include the color priors for traffic lights, roads, and lanes. Additionally, there is strong geometric structure in the scene as the road is flat, and all the objects stand vertically above it. This is explicitly exploited in the formulation of a commonly used objects-depth representation, namely Stixels [54]. There are static and moving objects in addition to the vehicle ego-motion, which may provide good temporal information. Although there is a lot in the automotive scene structure that can be used as prior information, the system's complexity has to be taken into consideration in the automotive applications, given the computational bounds in memory and processing power of the embedded systems on board.

6.3.2 COMMON AUTOMOTIVE DATASETS

There are many datasets built specifically for testing autonomous driving applications. They are divided into two main categories: real and synthetic. The most commonly used automotive datasets are discussed below.

6.3.2.1 Real Datasets

- **CamVid** [55] is one of the first datasets released for automotive applications. It consists of 700 images from a 10-minute video sequence annotated for semantic segmentation.
- **KITTI** [56] (Karlsruhe Institute of Technology and Toyota Technological Institute) dataset was released in 2012 for several tasks like stereo, optical flow, visual odometry, 3D object detection, and tracking. Different annotations are provided such as GPS/IMU sensor readings for ego-motion odometry, 3D objects, tracking information, and LiDAR data. The dataset provides annotation for 200 images only for pixel-wise semantic segmentation training and another 200 images for testing.
- **Cityscapes** [57], introduced by Daimler, consists of 20,000 images of realistic urban road scenes with coarse annotations. Its images have high-resolution that reaches $2,048 \times 1,024$. A set of 5,000 images is provided with fine annotations. The labels include 30 semantic classes, of which only 19 are used in evaluation as per the official evaluation script provided on the website. Stereo images are provided with disparity map annotation using the SGM algorithm [58] besides GPS coordinates and ego-motion data from vehicle odometry. To obtain absolute disparity values, a method is provided with the dataset for such calculations.

- **Mapillary Vistas** [59] is a large-scale dataset containing 25,000 high-resolution images annotated into 66 object categories in addition to instance annotation. The dataset consists of images from all around the world for different weather conditions and seasons. It is built through collaborative work all over the globe and is the most diverse open dataset geographically.
- **ApolloScape** [60], provided by Baidu, is a large-scale dataset containing high-resolution images of $3,384 \times 2,710$ pixels. The scene parsing dataset includes 146,997 images with corresponding pixel-level poses, and depth map annotations. However, depth maps are provided only for static objects where the scene is scanned multiple times, the generated point clouds are aligned, and then points based on temporal consistency are removed.

6.3.2.2 Synthetic Datasets

- **Virtual KITTI** [61] consists of 50 high-resolution videos consisting of 21,260 image frames of $1,242 \times 375$ resolution, generated using a unity game engine in an urban environment. Different annotations are provided with the dataset like semantic and instance segmentation, dense optical flow (DOF), depth maps, and 2D and 3D object detection bounding boxes. The depth map is a PNG16 image of the same size of the input image. The value of each pixel represents the z coordinate of the point in camera coordinate space. Each pixel takes values from 0 to 65,535, which corresponds to 655.36 m from the camera image plane. Points at infinity, like sky, are clipped to a depth of 655.3 m. The DOF image is a representation of dense optical flow between two consecutive images. Semantic and instance information is given per frame as a unique color for each class. Fourteen classes are annotated: Building, Car, GuardRail, Misc, Pole, Road, Sky, Terrain, TrafficLight, TrafficSign, Tree, Truck, Van, and Vegetation.
- **SYNTHIA** [62] consists of 200,000 images from video sequences with a wide range of diversity regarding weather conditions and scene structure. $360°$ views are provided by simulating 8 RGB cameras. Depth maps are provided as well.
- **NUSCENES** [63] is a large-scale dataset providing the most sensor measurements. It contains 1,000 videos of 20 seconds each with several types of annotation data such as LiDAR, radar, camera, IMU, and GPS data. It also provides 3D bounding boxes over 25 classes of objects annotated at 2 Hz.
- **SYNSCAPES** [64] is very close to reality and provides both $1,440 \times 720$ and $2,048 \times 1,024$ resolution images, depth maps, semantic and instance segmentation labels. 2D and 3D bounding boxes are also provided. Each image of the 25,000 images has its own unique structure, which gives a very wide range of variations and features combinations.

6.4 MOTION AND DEPTH CUES

This section gives brief information about motion and depth cues, which are utilized to enhance semantic segmentation performance in our experiments.

6.4.1   Depth Estimation

Depth is a very important information cue that implies the distance between surrounding objects and the vehicle, and it is critical for autonomous driving applications. Knowing only semantics without depth is not enough for full environment perception [65]. In addition to depth sensors that have their own pros and cons, depth can be estimated using several approaches utilizing cameras only without further more expensive sensors (e.g., LiDAR). The most common method for camera-based depth estimation is based on stereo images using the SGM algorithm [58], where pixels from the left image are searched for in the right image across the epipolar lines. Disparity is calculated according to a cost function that also guarantees the smoothness of the output surface. Disparity is the difference between positions of a feature that is viewed from two different cameras looking at the same scene with different viewpoints. Depth has a nonlinear relationship with disparity and the absolute value can be calculated if camera calibration information is available. However, disparity maps using classical approaches are usually noisy; hence, depth estimation itself is topic of research and deep learning has been used extensively through multiple architectures for the purpose of providing better depth estimation.

Depth estimation using deep learning has been tackled through several approaches [66, 67], which can be classified as supervised, semi-supervised, and unsupervised methods. Each approach has its pros and cons. For example, supervised approaches generally provide better performance than unsupervised ones. Unsupervised approaches have the flexibility to be trained without explicit depth annotations; however, they usually provide less accuracy than supervised ones. Generally, supervised approaches require accurate depth annotation, which requires usage of synthetic datasets. In [68], ego-motion information is used to exploit temporal information. The depth network takes only the target view as input and a pixel-wise depth map is generated. The pose network takes the target view and the source views as input and generates relative camera poses. The outputs are used to inverse-warp the source views to reconstruct the target view in an unsupervised manner. View synthesis is used for that task, where a target image can be synthesized given a pixel-wise depth map of that image, in addition to pose and visibility in a nearby view. Hence, unsupervised training is performed. Godard et al. [69] formulated depth estimation as an unsupervised learning problem where epipolar geometry constraints are exploited to generate disparity images by training the network for image reconstruction loss. This approach has the advantage of using only stereo information for training in the unsupervised manner, while in cases of inference, only monocular images are needed.

Apart from color, depth is another dimension for which the impact on semantic segmentation has not yet been explored extensively. Most of the previous work using depth for semantic segmentations made use of RGB-D cameras for indoor scenes. Hazirbas et al. [70] investigated incorporating depth into semantic segmentation for indoor use only using RGB-D cameras on the SUN RGB-D dataset. In [71], depth and RGB were combined for multi-view semantic segmentation where depth was leveraged to rewarp different views. Lin et al. [72] used a fully convolutional network (FCN) with branch predictors to process layered discretized depth maps to match them with the corresponding regions in the RGB image. The approach is used for indoor scenes using the NYU dataset. Cao et al. [73] provided a way to learn depth map and empirically proved enhanced performance for 2% on the VOC2012 dataset.

6.4.2 OPTICAL FLOW ESTIMATION

In addition to geometric information, there is a very important information prior that can be used in autonomous driving applications (i.e., the motion information) [74]. In a normal driving scenario, there are a lot of moving objects including cars, buses, and pedestrians as well as the vehicle ego-motion. Optical flow provides a visualization for the motion vectors of features in both x and y directions of the image. Moreover, optical flow is already computed within the automotive vision platform and has been a standard module for more than ten years. Due to hardware restrictions, sparse optical flow was used, but recently it has been replaced by dense optical flow.

In classical approaches, the most common way to generate sparse optical flow is the Lucas-Kanade method [75], which assumes similar motion vectors on a small window of 3×3 size as an example for every pixel at hand to be able to solve the optical flow equation:

$$f_x u + f_y v + f_t = 0 \tag{6.1}$$

Where u and v denote the change in the pixels' x and y positions with respect to time, which implies the motion of a pixel in both directions; f_x and f_y denote image derivatives along x and y directions; and f_t is the image difference over time. The Lucas-Kanade method is computed over previously selected features where the final output will be sparse and hence provides computationally efficient solution. On the contrary, Farneback's method [76] computes optical flow, but for every pixel in the image, and provides dense output where every pixel in the source frame is associated with a 2D vector showing values for pixel motion in x and y directions called u and v. Both Lucas-Kanade and Farneback methods provide good performance only when there is little motion between corresponding pixels in both images, and they fail when the displacement is large.

On the other hand, deep learning has been used to formulate optical flow estimation as a learning problem where temporally sequential images are used as input to the network and the output is a dense optical flow. FlowNet [77] used a large-scale synthetic dataset, namely "Flying Chairs", as a ground truth for dense optical flow. Two architectures were proposed. The first one accepts an input of six channels for two temporally sequential images stacked together in a single frame. The second is a Mid-Fusion implementation in which each image is processed separately on one stream and 2D correlation is done to output a DOF map. This work has been later extended to FlowNet v2 [78] focusing on small displacement motion, where an additional specialized subnetwork is added to learn small displacements. Results are then fused with the original architecture using another network to learn the best fusion. In [79], the estimated flow is used to warp one of the two temporally sequential images to the other, and the loss function is implemented to minimize the error between the warped image and the original image. Junhwa and Stefan [80] implemented joint estimation of optical flow and semantic estimation such that each task benefits from the other.

6.5 PROPOSED FUSION ARCHITECTURES

The objective of this study is to experimentally investigate the impact of fusion of both depth and optical flow with RGB images for enhanced semantic segmentation of typical autonomous driving scenes. For this purpose, we implement three different segmentation network architectures: Single-Input network, Early-Fusion network, and Mid-Fusion network.

6.5.1 SINGLE-INPUT NETWORK

An encoder-decoder architecture is employed to perform semantic segmentation. For the encoder, the commonly used VGG-16 network [81] is used with the pretrained weights to initialize the encoder weights. A fully convolutional network architecture, FCN8s [42], is adopted for semantic segmentation, and the four fully connected layers of VGG16 are transformed to fully convolutional layers used to learn the upsampling weights and restore the original image size. The fully convolutional network is exploited for its flexibility with any input image size. In contrast to fully connected networks, the fully convolutional ones preserve spatial relationships between pixels at lower computational cost. Additionally, FCN8s skip connections combine deep semantic information with shallow ones to improve accuracy as reported in [42].

In the proposed architecture, the convolutional layers are used for feature extraction. Feature downsampling is performed using max-pooling layers, and layer activation is done using rectified linear unit (ReLU) activation function. Afterwards, the segmentation decoder follows the FCN architecture and 1×1 convolutional layers are used to adapt depth channels, followed by three transposed convolution layers to perform upsampling. In order to benefit from high-resolution features, skip connections are utilized to add features from the shallow layers to the partially upsampled deep ones to benefit from the high-resolution appearance. The architecture of this network illustrated in Figure 6.4 is used to conduct experiments for evaluation of a single information cue on semantic segmentation. The same architecture is used for a single input of RGB-only, depth-only, and DOF-only, which refers to dense optical flow.

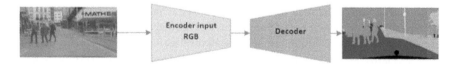

FIGURE 6.4 Single-Input network utilized for semantic segmentation given only one piece of information (e.g., RGB-only, depth-only, or DOF-only).

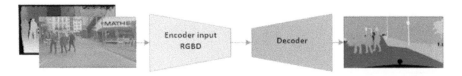

FIGURE 6.5 Early-Fusion Network: the pixel-level fusion is done on raw pixels before any processing or feature extraction.

6.5.2 EARLY-FUSION NETWORK

Using this network, we investigate early fusion for multiple cues. The term "early fusion" refers to pixel-level fusion, where the raw inputs are fused together before extracting any features using the deep network. The input to the network in this case is a 3D volume containing the original RGB image concatenated with the DOF image, or the depth map. The RGB image consists of 3 layers, and the DOF map consists of 1 layer. Concatenation is performed across the input volume's depth axis, so after concatenation we obtain an input 3D volume with the same image resolution, and with a depth of 4 layers. In this network, the same architecture of the Single-Input network is used and the VGG pre-trained weights are utilized to initialize the network. However, the first layer is adapted so that it accepts an input of four layers, and the corresponding weights are initialized randomly. Figure 6.5 shows the concatenation of the 3-layer RGB with a 1-layer depth map for a 4-layer-total 3D volume to be fed to the network. The same network is used to investigate the effect of early fusion with DOF, where several optical flow representations have been studied experimentally to find the best-performing input, namely, color wheel representation in 3 channels, magnitude and direction in 2 channels, and magnitude only in 1 channel.

6.5.3 MID-FUSION NETWORK

In this architecture, a two-stream network of two parallel VGG16 encoders for extracting information from two different cues is implemented inspired by the works [82–84]. The feature maps from both streams are fused together using a fusion junction producing encoded features of the same dimensions as the output from the Single-Input network encoder. The same decoder described in the Single-Input network, which consists of three transposed convolution layers, is used to perform upsampling. Figure 6.6 illustrates an example of a mid-level feature fusion approach where the information from both streams is fused together using a summation junction after feature vector extraction. Following the same approach as the single-stream network, skip connections are utilized to benefit from high-resolution feature maps. This approach provides more flexibility for each stream to learn separately. Accordingly, the outputs of both are then fused together, which has the advantage of the ability to initialize both encoders with the VGG pre-trained weights. However, the model is computationally more complex than the Early-Fusion single-stream network.

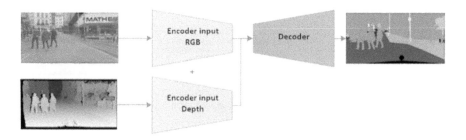

FIGURE 6.6   Mid-Fusion Network: information of both streams is fused together.

6.6 EXPERIMENTS

6.6.1 EXPERIMENTAL SETTINGS

This section discusses experimental settings. For all experiments, the Adam optimizer is used with a learning rate of value e^{-5}. The L2 regularization is used in the loss function, with a factor of value $5e^{-4}$ to overcome overfitting the training data. The VGG pre-trained weights are utilized for initialization. Dropout with probability 0.5 is utilized for 1×1 convolution layers. Intersection over Union (IoU) is used to evaluate class-wise accuracy, and the mean IoU is computed over all the classes. In this work, the TensorFlow—an open-source library developed by Google—is utilized. TensorFlow can run on CPUs as well as GPUs, and it is available for Windows, Linux, and macOS. In our experiments, we used TensorFlow v1.0.0 on a Python 2.7 platform under a Linux Ubuntu v14.04 operating system, and the implemented networks are executed on TitanX GPU, Maxwell architecture.

The datasets used in this work are chosen based on three factors: First, they have to be automotive datasets containing urban road scenes, as our application is mainly focused on autonomous driving. Second, the datasets should consist of video sequences to allow us to compute DOF and use the motion information for semantic segmentation. Third, there should be a way to calculate depth information from the chosen dataset. Two well-known automotive datasets fulfilling the requirements, namely, Virtual KITTI [61] and Cityscapes [57], are used in the experiments. For the Virtual KITTI, we shuffled the dataset randomly and split the frames to 80% training data, and 20% for test. In Cityscapes, we use the split provided by the dataset, which is 3,474 images for training and 1,525 images for testing. The optical flow annotation is not provided with the Cityscapes dataset; however, there are video sequences containing temporal information about the scenes. Thus, we exploit these sequences to generate different representations of dense optical flow using two methods. We train the network to output all of the classes provided in dataset; however, evaluation is done on 19 classes only, as mentioned by the Cityscapes team where the official evaluation script published for the dataset is exploited to report results.

6.6.2   RESULTS ANALYSIS AND DISCUSSION

6.6.2.1   Depth Experiments

Table 6.1 illustrates that depth augmentation consistently improves results. There is an improvement of 5.7% in mean IoU. Although the overall improvement is incremental, there is a large improvement for certain classes; for example, trucks, vans, buildings, and traffic lights show improvements of 32%, 28%, 9%, and 8% respectively. Table 6.2 shows quantitative evaluation for the Cityscapes dataset, and it had a relatively moderate improvement of 1% in IoU. Results show that even noisy depth maps with invalid values due to depth estimation errors can improve semantic segmentation. Results on Virtual KITTI show that using concatenated feature maps provides better results than using added feature maps. In Cityscapes, results for both approaches are quite close, although the results from the monoDepth [69] estimator using concatenated feature maps show slightly higher performance than those from added feature maps, which leads to the same conclusion of the KITTI dataset.

Figure 6.11 illustrates qualitative results for the Cityscapes dataset. Figure 6.9 (g) shows better detection of the white van in the background, which has a uniform texture; depth augmentation has helped to detect it better. Figure 6.11 (f) illustrates better detection of the bus compared to RGB-only detection and better detection of vegetation compared to depth-only, while showing degraded performance for the poles in the background, due to the noise shown in the depth map; the input image is shown in Figure 6.11 (b). Figure 6.11 (g) shows more immunity to noise regarding the background poles compared to Figure 6.11 (f) due to greater flexibility obtained from concatenated feature maps.

Depth-only network is reported to study the performance of depth cue alone on semantic segmentation. Surprisingly, depth provides good results especially for road, vegetation, vehicle, and pedestrians. This is also consistent with the results obtained by [70], when depth-only is tested for indoor scenes. Although the depth-only network shows good performance, the joint network shows only incremental improvement. It is likely that using a simple network architecture does not effectively combine depth and RGB, as they are completely different modalities with different properties. Another possible reason is that the network trained on RGB has the ability to learn geometric scene structure only implicitly.

6.6.2.2   Optical Flow Experiments

In this section, experiments were performed using different optical flow representations to help us decide which one gives the best performance. We found that results depend on the architecture used. In Cityscapes, we generated DOF using the Farneback [76] algorithm, which provides noisy optical flow. The experiments were conducted using FlowNet [78], the current state of the art for DOF computation, which is based on a deep network trained on virtual datasets. Using the Single-Input network, we attempted to investigate the relationship between motion cue alone and semantic segmentation. Using optical flow only, the network learned to segment the scene with good performance for moving objects. For static objects, the network

TABLE 6.1

Performance of Depth-Augmented Semantic Segmentation on the Virtual KITTI Dataset Using IoU Metric

Type	Mean	Truck	Car	Van	Road	Sky	Vegetation	Building	Guardrail	TrafficSign	TrafficLight	Pole
RGB	66.47	33.66	85.69	29.04	95.91	93.91	80.92	68.15	81.82	66.01	65.07	40.91
D (GT)	55	67.68	58.03	56.3	73.81	94.38	53.64	43.95	14.61	53.97	56.51	42.67
RGBD (GT)	66.76	65.34	91.74	56.93	95.46	94.41	79.17	54.91	73.42	60.21	46.09	30.46
RGB+D (GT-add)	68.6	43.38	91.59	29.19	69.01	94.32	85.17	77.6	80.13	69.54	72.73	32.09
RGB+D (GT-concat)	72.13	62.84	93.32	38.42	96.33	94.2	90.46	79.04	90.85	72.22	67.83	34.4
D (monoDepth)	46.1	36.05	75.46	33.2	77.3	87.3	39.3	32.3	6.8	42.14	45.9	15.9
RGB+D (monoDepth-add)	67.05	42.9	86.9	43.5	96.2	94.1	88.07	65.94	85.4	65.7	51.25	30.13
RGB+D (monoDepth-concat)	68.92	40.57	86.1	50.3	95.95	93.82	81.63	70.43	86.3	68.66	67.58	35.94

TABLE 6.2

Performance of Depth-Augmented Semantic Segmentation on the Cityscapes Dataset Using IoU Metric

Type	Mean	Bicycle	Person	Rider	Motorcycle	Bus	Car	Fence	Building	Road	Sidewalk	Sky	TrafficSign
RGB	62.47	63.52	67.93	40.49	29.96	62.13	89.16	44.53	87.86	96.22	74.98	89.79	59.88
D (SGM)	47.8	39.84	54.99	29.04	11.29	48.1	82.36	34.32	78.42	95.15	67.78	81.18	27.96
RGBD (SGM)	55.5	56.68	60.27	34.64	21.18	58	86.94	36.47	84.7	94.84	70.39	84.64	45.48
RGB+D (SGM-add)	63.48	66.46	67.85	42.31	41.37	63.1	89.77	46.28	88.1	96.38	75.66	90.23	60.78
RGB+D (SGM-concat)	63.13	65.32	67.79	39.14	37.27	69.71	90.06	42.75	87.44	96.6	76.35	91.06	59.44
D (monoDepth)	40.89	36.63	44.6	18.5	7.3	37.5	77.78	16.16	77.01	92.83	54.87	89.33	24.67
RGB+D (monoDepth-add)	61.39	66.23	67.33	39.9	44.01	55.7	89.1	40.2	87.34	69.47	75.7	88.7	57.7
RGB+D (monoDepth-concat)	63.03	65.85	67.44	41.33	46.24	66.5	89.7	33.6	87.25	96.01	73.5	90.3	59.8

learned only the generate the scene structure without any further details. This result motivated us toward the concept of fusion, which should help the network understand the scene semantics from the RGB image and enhance performance using motion information from optical flow. We used color wheel representation for this task, encoding both flow magnitude and direction.

Table 6.3 shows an evaluation of four architectures on the Virtual KITTI dataset. There is an improvement of 4% in mean IoU with the Mid-Fusion approach, with larger improvements in moving classes like Truck, Car, and Van (38%, 6%, and 27%, respectively), which demonstrates that the Mid-Fusion network provides the best capability of capturing the motion cues and thus the best semantic segmentation accuracy. The RGBF network provides very close performance to the RGB-alone network, but with significant improvement in moving classes such as Truck and Van. For all the results mentioned in Tables 6.3 and 6.4, we make use of a color wheel representation format for flow. Table 6.4 illustrates the performance on the Cityscapes dataset, indicating a marginal improvement of 0.1% in overall IoU. There is a significant improvement in moving object classes like motorcycle and train by 17% and 7%, even using noisy flow maps. The improvement obtained from moving objects is significant compared to overall IoU, which is dominated by sky and road pixels.

Quantitative comparisons between different DOF representations in the Early-Fusion network RGBF and the Mid-Fusion RGB+F architectures for both Virual KITTI and Cityscapes dataset are reported in Tables 6.5 and 6.6, respectively. Results show that the color wheel Mid-Fusion network provides the best performance, as color wheel representation encodes both magnitude and direction, while the Mid-Fusion network provides the capacity for each stream to learn separately. Hence, we obtained the best results using this approach. RGBF results are also interesting: they show that the network is incapable of providing good output segmentation when the input channels are increased to 6 layers including 3 for RGB and 3 for color wheel DOF. However, augmenting only one channel for magnitude DOF provides better results for moving classes such as Truck and Van in Virtual KITTI. This might be beneficial for an embedded system with limited resources. In the Cityscapes, the RGBF is incapable of providing better segmentation for several reasons, one of which is the simple network architecture used and the large number of classes in the Cityscapes dataset trained for. Additionally, the Cityscapes has only 3,475 images for training, while the Virtual KITTI has 17,008 frames. The Virtual KITTI results show the need for more research in that direction, especially for limited hardware resources. Figures 6.7 and 6.8 illustrate the qualitative results of the four architectures on Virtual KITTI and Cityscapes, respectively. Figure 6.7 (f) shows better detection of the van, which has a uniform texture and which flow cue has helped to detect more accurately. Figure 6.7 (h) shows that FlowNet-only results provides good segmentation for the moving van with the ability to capture generic scene structures like pavement, sky, and vegetation without color information. However, fusion in Figure 6.7 (i) still needs to be improved. Figure 6.8 (f) and (h) illustrate better detection of the bicycle and the cyclist after fusion with DOF. These examples visually verify the accuracy improvement shown in Tables 6.3 and 6.4.

TABLE 6.3

Performance of Flow-Augmented Semantic Segmentation Results on the Virtual KITTI Dataset Using IoU Metric

Type	Mean	Truck	Car	Van	Road	Sky	Vegetation	Building	Guardrail	TrafficSign	TrafficLight	Pole
RGB	66.47	33.66	85.69	29.04	95.91	93.91	80.92	68.15	81.82	66.01	65.07	40.91
F (GT)	42	36.2	55.2	20.7	62.6	93.9	19.54	34	15.23	51.5	33.2	29.3
RGBF (GT)	65.328	70.74	80.2	48.34	93.6	93.3	70.79	62.05	67.86	55.14	55.48	31.9
RGB+F (GT)	70.52	71.79	91.4	56.8	96.19	93.5	83.4	66.53	82.6	64.69	64.65	26.6
F (FlowNet)	28.6	24.6	47.8	14.3	57.9	68	13.4	4.9	0.8	31.8	18.5	6.6
RGB+F (FlowNet)	68.84	60.05	90.87	40.54	96.05	91.73	84.54	68.52	82.43	65.2	63.54	26.54

TABLE 6.4

Performance of Flow-Augmented Semantic Segmentation Results on the Cityscapes Dataset Using IoU Metric

Type	Mean	Bicycle	Person	Rider	Motorcycle	Bus	Car	Train	Building	Road	Truck	Sky	TrafficSign
RGB	62.47	63.52	67.93	40.49	29.96	62.13	89.16	44.19	87.86	96.22	48.54	89.79	59.88
F (Farneback)	34.7	34.48	37.9	12.7	7.39	31.4	74.3	11.35	72.77	91.2	19.42	79.6	11.4
RGBF (Farneback)	47.8	52.6	55.8	31.1	22.4	39.34	82.75	22.8	80.43	92.24	20.7	81.87	44.08
RGB+F (Farneback)	62.56	63.65	66.3	39.65	47.22	66.24	89.63	51.02	87.13	96.4	36.1	90.64	60.68
F (FlowNet)	36.8	32.9	50.9	26.8	5.12	25.99	75.29	15.1	65.16	90.75	25.46	50.16	29.14
RGBF (FlowNet)	52.3	54.9	58.9	34.8	26.1	53.7	83.6	40.7	79.4	94	28.1	79.4	45.5
RGB+F (FlowNet)	62.43	64.2	66.32	40.9	40.76	66.05	90.03	41.3	87.3	95.8	54.7	91.07	58.21

TABLE 6.5

Semantic Segmentation Results on the KITTI for Different DOF (GT) Representations

Type	Mean	Truck	Car	Van	Road	Sky	Vegetation	Building	Guardrail	TrafficSign	TrafficLight	Pole
RGBF (GT-color wheel)	59.88	41.7	84.44	40.74	93.76	93.6	66.3	49.43	52.18	62.21	49.61	21.52
RGBF (GT-Mag & Dir)	58.85	45.12	82.3	30.04	90.25	94.1	60.97	56.48	51.48	58.74	49.7	26.01
RGBF (GT-Mag only)	65.32	70.73	80.16	48.33	93.59	93.3	70.79	62.04	67.86	55.13	55.48	31.92
RGB+F (GT-3 layers Mag only)	67.88	35.7	91.02	24.78	96.47	94.06	88.72	74.4	84.5	69.48	68.95	34.28
RGB+F (GT-color wheel)	70.52	71.79	91.4	56.8	96.19	93.5	83.4	66.53	82.6	64.69	64.65	26.6

TABLE 6.6

Semantic Segmentation Results on the Cityscapes for Different DOF (Farneback) Representations

Type	Mean	Bicycle	Person	Rider	Motorcycle	Bus	Car	Train	Building	Road	Truck	Sky	TrafficSign
RGBF (Mag only)	47.8	52.63	55.82	31.08	22.38	39.34	82.75	22.8	80.43	92.24	20.7	81.87	44.08
RGBF (Mag & Dir)	54.6	57.28	58.63	33.56	18.49	56.44	84.6	41.15	84.41	95.5	31.8	87.86	44.26
RGBF (color wheel)	57.2	61.47	62.18	35.13	22.68	54.87	87.45	36.69	86.28	95.94	40.2	90.07	51.64
RGB+F (3 layers Mag only)	62.1	65.15	65.44	32.59	33.19	63.07	89.48	43.6	87.88	96.17	57.2	91.48	55.76
RGB+F (color wheel)	62.56	63.65	66.3	39.65	47.22	66.24	89.63	51.02	87.13	69.4	36.11	90.64	60.68

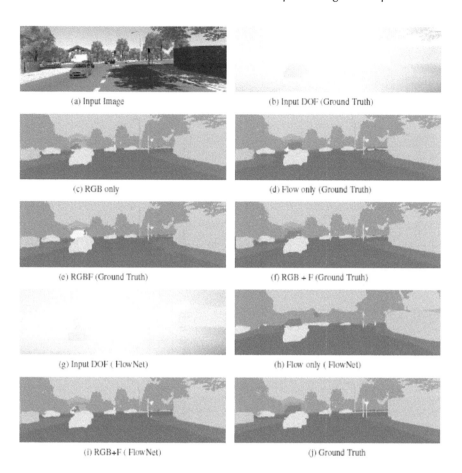

(a) Input Image   (b) Input DOF (Ground Truth)

(c) RGB only   (d) Flow only (Ground Truth)

(e) RGBF (Ground Truth)   (f) RGB + F (Ground Truth)

(g) Input DOF (FlowNet)   (h) Flow only (FlowNet)

(i) RGB+F (FlowNet)   (j) Ground Truth

FIGURE 6.7 Flow augmentation: qualitative comparison of semantic segmentation outputs from four architectures on the KITTI dataset.

Unexpectedly, Table 6.4 shows that Farneback and FlowNet v2 provide similar results; however, FlowNet shows better results for some classes like Truck. In Figure 6.8, fusion with DOF in a two-stream approach enhances results of semantic segmentation for both Farneback and FlowNet. Farneback DOF is noisy; however, the result Figure 6.8 (f) shows that the network has good immunity towards noise. Farneback has an advantage over FlowNet in fine details, where it is shown visually that the rider can be separated from the bicycle; however, FlowNet doesn't provide this level of detail. As a result, fusion with Farneback provides slightly higher performance than fusion with FlowNet. In both datasets, we observed some degradation in accuracy of static objects after fusion despite the improvement obtained for moving objects. This highlights the motivation towards multimodal fusion to combine RGB, motion, and depth together, where depth signal is expected to improve the accuracy of static objects. Moreover, there is a need for deeper investigation of fusion strategy to maximize the benefit from each modality without loss in accuracy of other classes (Tables 6.7 and 6.8).

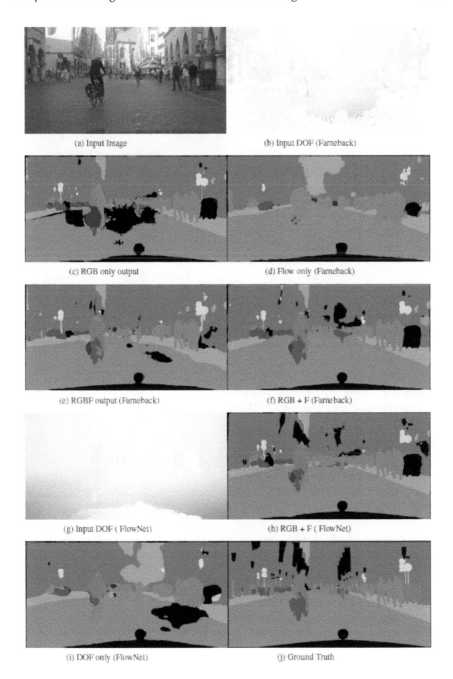

(a) Input Image   (b) Input DOF (Farneback)

(c) RGB only output   (d) Flow only (Farneback)

(e) RGBF output (Farneback)   (f) RGB + F (Farneback)

(g) Input DOF (FlowNet)   (h) RGB + F (FlowNet)

(i) DOF only (FlowNet)   (j) Ground Truth

FIGURE 6.8 Flow augmentation: qualitative comparison of semantic segmentation outputs from four architectures on the Cityscapes dataset.

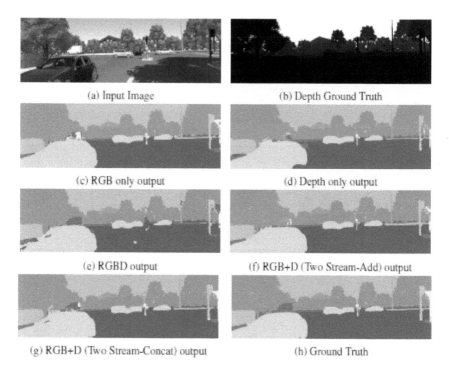

(a) Input Image   (b) Depth Ground Truth

(c) RGB only output   (d) Depth only output

(e) RGBD output   (f) RGB+D (Two Stream-Add) output

(g) RGB+D (Two Stream-Concat) output   (h) Ground Truth

FIGURE 6.9 Depth augmentation: qualitative comparison of semantic segmentation outputs from four architectures on the KITTI dataset.

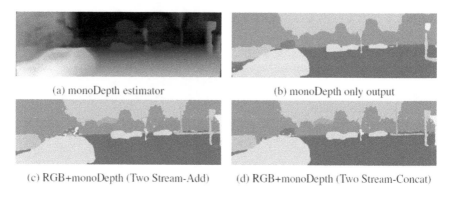

(a) monoDepth estimator   (b) monoDepth only output

(c) RGB+monoDepth (Two Stream-Add)   (d) RGB+monoDepth (Two Stream-Concat)

FIGURE 6.10 Depth augmentation: qualitative comparison of semantic segmentation outputs using the monoDepth estimator on the KITTI dataset.

6.7 CONCLUSION

This chapter focuses on computer vision tasks for autonomous driving applications. It explores the autonomous driving field and the role of recent computer vision algorithms in self-driving car development, such as deep learning CNN, which showed significant improvement in terms of accuracy for various tasks in different

(a) Input Image   (b) SGM Depth map

(c) RGB only output   (d) Depth only output

(e) RGBD output   (f) RGB + D (Two-Stream-Add) output

(g) RGB + D (Two-Stream-Concat) output   (h) Ground Truth

FIGURE 6.11 Depth augmentation: qualitative comparison of semantic segmentation outputs from four architectures on the Cityscapes dataset.

applications. The main goal is to construct simple architectures to demonstrate the benefit of flow and depth augmentation to standard CNN-based semantic segmentation networks. The impact of both motion and depth information on semantic segmentation was experimentally studied based on a simple network architecture. Results of experiments on two public datasets demonstrate reasonable improvement in overall accuracy. From the conducted experiments, it is shown that CNN can generalize well even when using noisy maps such as SGM and Farneback; however, the quality of the input map in terms of object structure largely affects the output. Furthermore, the obtained experimental results show that there is still a lot of

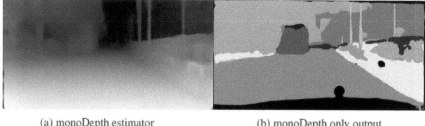

(a) monoDepth estimator   (b) monoDepth only output

(c) RGB+monoDepth (Two Stream-Add)   (d) RGB+monoDepth (Two Stream-Concat)

FIGURE 6.12   Depth augmentation: qualitative comparison of semantic segmentation outputs using the monoDepth estimator on the Cityscapes dataset.

scope for improvement, as depth, flow, and color are different modalities, so more research should be conducted to construct better fusion networks. Future research aims to

- Conduct deeper investigation to implement more robust multimodal networks to combine valuable pieces of information.
- Understand the effect of different network architectures with varying complexities in the multimodal fusion task.
- Incorporate sparsity invariant metrics to handle missing flow and depth estimates.
- Migrate from pixel-level to object-level classification and make use of depth per object, as it is more practical than pixel-wise depth, especially in an AD application where the objective is to avoid collision with full objects.
- Investigate multimodal fusion for different tasks such as moving and static object detection, especially those employing structural constraints [85] and compressive sensing techniques [86], which will be very helpful in the driving scenes.
- Train the network with fewer classes than the total number provided by the dataset. In a normal driving scenario, information about all classes is not critical for a vehicle to move. Dataset classes can be minimized to the ones that are needed only by an AC.
- Train on multiple real and synthetic datasets with fine-tuning to solve the problem of limited numbers of annotated images.

TABLE 6.7

Comparison between Baseline, Depth-Augmented, Flow-Augmented, and Depth-Flow-Augmented Semantic Segmentation Using Ground Truth

Type	Mean	Truck	Car	Van	Road	Sky	Vegetation	Building	Guardrail	TrafficSign	TrafficLight	Pole
RGB-only	66.47	33.66	85.69	29.04	95.91	93.91	80.92	68.15	81.82	66.01	65.07	40.91
RGB+D (GT)	72.13	62.84	93.32	38.42	96.33	94.2	90.46	79.04	90.85	72.22	67.83	34.4
RGB+F (GT)	70.52	71.79	91.4	56.8	96.19	93.5	83.4	66.53	82.6	64.69	64.65	26.6
RGB+D+F (GT)	71.88	71.688	92.08	61.44	95.85	94.83	84.32	71.86	83.42	64.69	60.67	31.08

TABLE 6.8

Comparison between Baseline, Depth-Augmented, Flow-Augmented, and Depth-Flow-Augmented Semantic Segmentation Using Classical Estimators

Type	Mean	Bicycle	Person	Rider	Motorcycle	Bus	Car	Train	Building	Road	Truck	Sky	TrafficSign
RGB-only	62.47	63.52	67.93	40.49	29.96	62.13	89.16	44.19	87.86	96.22	48.54	89.79	59.88
RGB+D (SGM)	63.48	66.46	67.85	42.31	41.37	63.1	89.77	46.28	88.1	96.38	75.66	90.23	60.78
RGB+F (Farneback)	62.56	63.65	66.3	39.65	47.22	66.24	89.63	51.02	87.13	96.4	36.1	90.64	60.68
RGB+D+F	62.58	65.46	68.18	43.09	37.5	64.4	88.34	57.13	86.86	96.13	41.45	87.96	59.23

BIBLIOGRAPHY

1. Richard S. Wallace, Anthony Stentz, Charles E. Thorpe, Hans P. Moravec, William Whittaker, and Takeo Kanade. First results in robot road-following. In *International Joint Conferences on Artificial Intelligence (IJCAI)*, pages 1089–1095. Citeseer, 1985.

2. Sebastian Thrun. Toward robotic cars. *Communications of the ACM*, 53(4): 99–106, 2010.

3. Matthew A. Turk, David G. Morgenthaler, Keith D. Gremban, and Martin Marra. VITS-A vision system for autonomous land vehicle navigation. *IEEE Transactions on Pattern Analysis and Machine Intelligence*, 10(3): 342–361, 1988.

4. Kichun Jo, Junsoo Kim, Dongchul Kim, Chulhoon Jang, and Myoungho Sunwoo. Development of autonomous car-Part II: A case study on the implementation of an autonomous driving system based on distributed architecture. *IEEE Transactions on Industrial Electronics*, 62(8): 5119–5132, 2015.

5. Ernst Dieter Dickmanns. *Dynamic Vision for Perception and Control of Motion.* Springer-Verlag, London, 2007.

6. Mahdi Rezaei and Reinhard Klette. *Computer Vision for Driver Assistance.* Springer, Cham Switzerland, 2017.

7. Mahmoud Hassaballah and Khalid M. Hosny. *Recent Advances in Computer Vision: Theories and Applications*, volume 804. Springer International Publishing, 2019.

8. David Michael Stavens. *Learning to Drive: Perception for Autonomous Cars.* Stanford University, 2011.

9. Yuna Ro and Youngwook Ha. A factor analysis of consumer expectations for autonomous cars. *Journal of Computer Information Systems*, 59(1): 52–60, 2019.

10. Adil Hashim, Tanya Saini, Hemant Bhardwaj, Adityan Jothi, and Ammannagari Vinay Kumar. Application of swarm intelligence in autonomous cars for obstacle avoidance. In *Integrated Intelligent Computing, Communication and Security*, pages 393–404. Springer, 2019.

11. Angelos Amanatiadis, Evangelos Karakasis, Loukas Bampis, Stylianos Ploumpis, and Antonios Gasteratos. ViPED: On-road vehicle passenger detection for autonomous vehicles. *Robotics and Autonomous Systems*, 112: 282–290, 2019.

12. Yusuf Artan, Orhan Bulan, Robert P. Loce, and Peter Paul. Passenger compartment violation detection in HOV/HOT lanes. *IEEE Transactions on Intelligent Transportation Systems*, 17(2): 395–405, 2016.

13. Dorsa Sadigh, S Shankar Sastry, and Sanjit A. Seshia. Verifying robustness of human-aware autonomous cars. *IFAC-PapersOnLine*, 51(34): 131–138, 2019.

14. Bhargava Reddy, Ye-Hoon Kim, Sojung Yun, Chanwon Seo, and Junik Jang. Real-time driver drowsiness detection for embedded system using model compression of deep neural networks. In *IEEE Conference on Computer Vision and Pattern Recognition Workshops*, pages 121–128, 2017.

15. Theocharis Kyriacou, Guido Bugmann, and Stanislao Lauria. Vision-based urban navigation procedures for verbally instructed robots. *Robotics and Autonomous Systems*, 51(1): 69–80, 2005.

16. SAE International Committee. Taxonomy and definitions for terms related to on-road motor vehicle automated driving systems. Technical Report J3016–201401, SAE International, 2014. http://doi.org/ 10.4271/J3016_201806.

17. Jonathan Horgan, Ciaran Hughes, John McDonald, and Senthil Yogamani. Vision-based driver assistance systems: Survey, taxonomy and advances. In *International Conference on Intelligent Transportation Systems (ITSC)*, pages 2032–2039. IEEE, 2015.

18. Zhentao Hu, Tianxiang Chen, Quanbo Ge, and Hebin Wang. Observable degree analysis for multi-sensor fusion system. *Sensors*, 18(12): 4197, 2018.

19. Shan Luo, Joao Bimbo, Ravinder Dahiya, and Hongbin Liu. Robotic tactile perception of object properties: A review. *Mechatronics*, 48: 5467, 2017.
20. Guilherme N. DeSouza and Avinash C. Kak. Vision for mobile robot navigation: A survey. *IEEE Transactions on Pattern Analysis and Machine Intelligence*, 24(2): 237–267, 2002.
21. Oliver Pink, Jan Becker, and Soren Kammel. Automated driving on public roads: Experiences in real traffic. *IT-Information Technology*, 57(4): 223–230, 2015.
22. Thomas Braunl. *Embedded Robotics: Mobile Robot Design and Applications with Embedded Systems*. Springer Science & Business Media, 2008.
23. Gregory Dudek and Michael Jenkin. *Computational Principles of Mobile Robotics*. Cambridge University Press, 2010.
24. Uwe Handmann, Thomas Kalinke, Christos Tzomakas, Martin Werner, and Werner von Seelen. Computer vision for driver assistance systems. In *Enhanced and Synthetic Vision 1998*, volume 3364, pages 136–148. International Society for Optics and Photonics, 1998.
25. Yann LeCun, Yoshua Bengio, and Geoffrey Hinton. Deep learning. *Nature*, 521(7553): 436, 2015.
26. Jiquan Ngiam, Aditya Khosla, Mingyu Kim, Juhan Nam, Honglak Lee, and Andrew Y. Ng. Multimodal deep learning. In *International Conference on Machine Learning (ICML-11)*, pages 689–696, 2011.
27. Li Deng and Dong Yu. Deep learning: Methods and applications. *Foundations and Trends® in Signal Processing*, 7(3–4): 197–387, 2014.
28. Alex Krizhevsky, Ilya Sutskever, and Geoffrey E. Hinton. Imagenet classification with deep convolutional neural networks. In *Advances in Neural Information Processing Systems*, pages 1097–1105, 2012.
29. Josh Patterson and Adam Gibson. *Deep Learning: A Practitioner's Approach*. O'Reilly Media, Inc., 2017.
30. Ian Goodfellow, Yoshua Bengio, and Aaron Courville. *Deep Learning*. MIT Press, 2016.
31. Gao Huang, Zhuang Liu, Laurens Van Der Maaten, and Kilian Q. Weinberger. Densely connected convolutional networks. In *IEEE Conference on Computer Vision and Pattern Recognition*, pages 4700–4708, 2017.
32. Sergey Ioffe and Christian Szegedy. Batch normalization: Accelerating deep network training by reducing internal covariate shift. arXiv preprint arXiv:1502.03167, 2015.
33. Risto Miikkulainen, Jason Liang, Elliot Meyerson, Aditya Rawal, Daniel Fink, Olivier Francon, Bala Raju, Hormoz Shahrzad, Arshak Navruzyan, Nigel Duffy, et al. Evolving deep neural networks. In *Artificial Intelligence in the Age of Neural Networks and Brain Computing*, pages 293–312. Elsevier, 2019.
34. Michael Hauser, Sean Gunn, Samer Saab Jr, and Asok Ray. State-space representations of deep neural networks. *Neural Computation*, 31(3): 538–554, 2019.
35. Abhinav Valada, Gabriel L. Oliveira, Thomas Brox, and Wolfram Burgard. Deep multispectral semantic scene understanding of forested environments using multi-modal fusion. In *International Symposium on Experimental Robotics*, pages 465–477. Springer, 2016.
36. Taigo M. Bonanni, Andrea Pennisi, D.D Bloisi, Luca Iocchi, and Daniele Nardi. Human-robot collaboration for semantic labeling of the environment. In *3rd Workshop on Semantic Perception, Mapping and Exploration*, pp. 1–6, 2013.
37. Abhijit Kundu, Yin Li, Frank Dellaert, Fuxin Li, and James M. Rehg. Joint semantic segmentation and 3D reconstruction from monocular video. In *European Conference on Computer Vision*, pages 703–718. Springer, 2014.
38. Ondrej Miksik, Vibhav Vineet, Morten Lidegaard, Ram Prasaath, Matthias Nießner, Stuart Golodetz, Stephen L. Hicks, Patrick Perez, Shram Izadi, and Philip H.S. Torr. The semantic paintbrush: Interactive 3d mapping and recognition in large outdoor spaces. In *33rd Annual ACM Conference on Human Factors in Computing Systems*, pages 3317–3326. ACM, 2015.

39. Marius Cordts, Mohamed Omran, Sebastian Ramos, Timo Rehfeld, Markus Enzweiler, Rodrigo Benenson, Uwe Franke, Stefan Roth, and Bernt Schiele. The cityscapes dataset for semantic urban scene understanding. In *The IEEE Conference on Computer Vision and Pattern Recognition (CVPR)*, June 2016.

40. Gabriel J. Brostow, Julien Fauqueur, and Roberto Cipolla. Semantic object classes in video: A high-definition ground truth database. *Pattern Recognition Letters*, 30(2): 88–97, 2009.

41. Clement Farabet, Camille Couprie, Laurent Najman, and Yann LeCun. Learning hierarchical features for scene labeling. *IEEE Transactions on Pattern Analysis and Machine Intelligence*, 35(8): 1915–1929, 2013.

42. Jonathan Long, Evan Shelhamer, and Trevor Darrell. Fully convolutional networks for semantic segmentation. In *IEEE Conference on Computer Vision and Pattern Recognition*, pages 3431–3440, 2015.

43. Alex Krizhevsky, Ilya Sutskever, and Geoffrey E . Imagenet classification with deep convolutional neural networks. In F. Pereira, C. J. C. Burges, L. Bottou, and K. Q. Weinberger, editors, *Advances in Neural Information Processing Systems*, 25, pages 1097–1105. Curran Associates, Inc., 2012.

44. Karen Simonyan and Andrew Zisserman. Very deep convolutional networks for large-scale image recognition. arXiv preprint arXiv:1409.1556, 2014.

45. Christian Szegedy, Wei Liu, Yangqing Jia, Pierre Sermanet, Scott Reed, Dragomir Anguelov, Dumitru Erhan, Vincent Vanhoucke, and Andrew Rabinovich. Going deeper with convolutions. In *IEEE Conference on Computer Vision and Pattern Recognition*, pages 1–9, 2015.

46. Hyeonwoo Noh, Seunghoon Hong, and Bohyung Han. Learning deconvolution network for semantic segmentation. In *IEEE International Conference on Computer Vision*, pages 1520–1528, 2015.

47. Vijay Badrinarayanan, Alex Kendall, and Roberto Cipolla. Segnet: A deep convolutional encoder-decoder architecture for image segmentation. *IEEE Transactions on Pattern Analysis and Machine Intelligence*, 39(12): 2481–2495, 2017.

48. Jun Mao, Xiaoping Hu, Xiaofeng He, Lilian Zhang, Liao Wu, and Michael J. Milford. Learning to fuse multiscale features for visual place recognition. *IEEE Access*, 7: 5723–5735, 2019.

49. Fisher Yu and Vladlen Koltun. Multi-scale context aggregation by dilated convolutions. arXiv preprint arXiv:1511.07122, 2015.

50. Liang-Chieh Chen, Yi Yang, Jiang Wang, Wei Xu, and Alan L. Yuille. Attention to scale: Scale-aware semantic image segmentation. In *EEE Conference on Computer Vision and Pattern Recognition*, pages 3640–3649, 2016.

51. Boyu Chen, Peixia Li, Chong Sun, Dong Wang, Gang Yang, and Huchuan Lu. Multi attention module for visual tracking. *Pattern Recognition*, 87: 80–93, 2019.

52. Liuyuan Deng, Ming Yang, Hao Li, Tianyi Li, Bing Hu, and Chunxiang Wang. Restricted deformable convolution based road scene semantic segmentation using surround view cameras. arXiv preprint arXiv:1801.00708, 2018.

53. Varun Ravi Kumar, Stefan Milz, Christian Witt, Martin Simon, Karl Amende, Johannes Petzold, Senthil Yogamani, and Timo Pech. Monocular fisheye camera depth estimation using sparse lidar supervision. In *International Conference on Intelligent Transportation Systems (ITSC)*, pages 2853–2858. IEEE, 2018.

54. Marius Cordts, Timo Rehfeld, Lukas Schneider, David Pfeiffer, Markus Enzweiler, Stefan Roth, Marc Pollefeys, and Uwe Franke. The stixel world: A medium-level representation of traffic scenes. *Image and Vision Computing*, 68: 40–52, 2017.

55. Gabriel J. Brostow, Jamie Shotton, Julien Fauqueur, and Roberto Cipolla. Segmentation and recognition using structure from motion point clouds. In *European Conference on Computer Vision*, pages 44–57. Springer, 2008.

56. Andreas Geiger, Philip Lenz, and Raquel Urtasun. Are we ready for autonomous driving? The kitti vision benchmark suite. In *Conference on Computer Vision and Pattern Recognition*, pp. 3354–3361, 2012.

57. Marius Cordts, Mohamed Omran, Sebastian Ramos, Timo Rehfeld, Markus Enzweiler, Rodrigo Benenson, Uwe Franke, Stefan Roth, and Bernt Schiele. The cityscapes dataset for semantic urban scene understanding. arXiv preprint arXiv:1604.01685, 2016.

58. Heiko Hirschmuller. Accurate and efficient stereo processing by semi-global matching and mutual information. In *IEEE Conference on Computer Vision and Pattern Recognition*, volume 2, pages 807–814. IEEE, 2005.

59. Gerhard Neuhold, Tobias Ollmann, Samuel Rota Bulo, and Peter Kontschieder. The mapillary vistas dataset for semantic understanding of street scenes. In *IEEE International Conference on Computer Vision,* pages 4990–4999, 2017.

60. Xinyu Huang, Xinjing Cheng, Qichuan Geng, Binbin Cao, Dingfu Zhou, Peng Wang, Yuanqing Lin, and Ruigang Yang. The apolloscape dataset for autonomous driving. In *IEEE Conference on Computer Vision and Pattern Recognition Workshops*, pages 954–960, 2018.

61. Adrien Gaidon, Qiao Wang, Yohann Cabon, and Eleonora Vig. Virtual worlds as proxy for multi-object tracking analysis. In *IEEE Conference on Computer Vision and Pattern Recognition*, pages 4340–4349, 2016.

62. German Ros, Laura Sellart, Joanna Materzynska, David Vazquez, and Antonio M. Lopez. The synthia dataset: A large collection of synthetic images for semantic segmentation of urban scenes. In *Proceedings of the IEEE Conference on Computer Vision and Pattern Recognition*, pages 3234–3243, 2016.

63. NuTonomy. Nuscenes dataset. 2012. https://www.nuscenes.org/.

64. Magnus Wrenninge and Jonas Unger. Synscapes: A photorealistic synthetic dataset for street scene parsing. arXiv preprint arXiv:1810.08705, 2018.

65. Liang-Chieh Chen, George Papandreou, Iasonas Kokkinos, Kevin Murphy, and Alan L. Yuille. Deeplab: Semantic image segmentation with deep convolutional nets, atrous convolution, and fully connected crfs. *IEEE Transactions on Pattern Analysis and Machine Intelligence*, 40(4): 834–848, 2018.

66. Fayao Liu, Chunhua Shen, Guosheng Lin, and Ian Reid. Learning depth from single monocular images using deep convolutional neural fields. *IEEE Transactions on Pattern Analysis and Machine Intelligence*, 38(10): 2024–2039, 2016.

67. Yuanzhouhan Cao, Zifeng Wu, and Chunhua Shen. Estimating depth from monocular images as classification using deep fully convolutional residual networks. *IEEE Transactions on Circuits and Systems for Video Technology*, 28(11): 3174–3182, 2018.

68. Tinghui Zhou, Matthew Brown, Noah Snavely, and David G. Lowe. Unsupervised learning of depth and ego-motion from video. In *IEEE Conference on Computer Vision and Pattern Recognition*, pages 1851–1858, 2017.

69. Clement Godard, Oisin Mac Aodha, and Gabriel J. Brostow. Unsupervised monocular depth estimation with left-right consistency. In *IEEE Computer Vision and Pattern Recognition*, volume 2, page 7, 2017.

70. Caner Hazirbas, Lingni Ma, Csaba Domokos, and Daniel Cremers. Fusenet: Incorporating depth into semantic segmentation via fusion-based cnn architecture. In *Asian Conference on Computer Vision*, pages 213–228. Springer, 2016.

71. Lingni Ma, Jorg Stückler, Christian Kerl, and Daniel Cremers. Multi-view deep learning for consistent semantic mapping with RGB-d cameras. arXiv preprint arXiv:1703.08866, 2017.

72. Di Lin, Guangyong Chen, Daniel Cohen-Or, Pheng-Ann Heng, and Hui Huang. Cascaded feature network for semantic segmentation of rgb-d images. In *IEEE International Conference on Computer Vision (ICCV)*, pages 1320–1328. IEEE, 2017.

73. Yuanzhouhan Cao, Chunhua Shen, and Heng Tao Shen. Exploiting depth from single monocular images for object detection and semantic segmentation. *IEEE Transactions on Image Processing*, 26(2): 836–846, 2017.

74. Min Bai, Wenjie Luo, Kaustav Kundu, and Raquel Urtasun. Exploiting semantic information and deep matching for optical flow. In *European Conference on Computer Vision*, pages 154–170. Springer, 2016.

75. Bruce D. Lucas and Takeo Kanade. An iterative image registration technique with an application to stereo vision. In *Imaging Understanding Workshop*, pages 121–130. Vancouver, British Columbia, 1981.

76. Gunnar Farneback. Two-frame motion estimation based on polynomial expansion. In *Scandinavian Conference on Image Analysis*, pages 363–370. Springer, 2003.

77. Alexey Dosovitskiy, Philipp Fischer, Eddy Ilg, Philip Hausser, Caner Hazirbas, Vladimir Golkov, Patrick Van Der Smagt, Daniel Cremers, and Thomas Brox. Flownet: Learning optical flow with convolutional networks. In *IEEE International Conference on Computer Vision*, pages 2758–2766, 2015.

78. Eddy Ilg, Nikolaus Mayer, Tonmoy Saikia, Margret Keuper, Alexey Dosovitskiy, and Thomas Brox. Flownet 2.0: Evolution of optical flow estimation with deep networks. In *IEEE Conference on Computer Vision and Pattern Recognition*, pages 2462–2470, 2017.

79. Zhe Ren, Junchi Yan, Bingbing Ni, Bin Liu, Xiaokang Yang, and Hongyuan Zha. Unsupervised deep learning for optical flow estimation. In *Thirty-First AAAI Conference on Artificial Intelligence*, pp. 1495–1501, 2017.

80. Junhwa Hur and Stefan Roth. Joint optical flow and temporally consistent semantic segmentation. In *European Conference on Computer Vision*, pages 163–177. Springer, 2016.

81. Karen Simonyan and Andrew Zisserman. Very deep convolutional networks for large-scale image recognition. arXiv preprint arXiv:1409.1556, 2014.

82. Mennatullah Siam, Heba Mahgoub, Mohamed Zahran, Senthil Yogamani, Martin Jagersand, and Ahmad El-Sallab. Modnet: Moving object detection network with motion and appearance for autonomous driving. arXiv preprint arXiv:1709.04821, 2017.

83. Suyog Dutt Jain, Bo Xiong, and Kristen Grauman. Fusionseg: Learning to combine motion and appearance for fully automatic segmention of generic objects in videos. arXiv preprint arXiv:1701.05384, 2017.

84. Karen Simonyan and Andrew Zisserman. Two-stream convolutional networks for action recognition in videos. In *Advances in Neural Information Processing Systems*, pages 568–576, 2014.

85. Mennatullah Siam and Mohammed Elhelw. Enhanced target tracking in uav imagery with pn learning and structural constraints. In *Proceedings of the IEEE International Conference on Computer Vision Workshops*, pages 586–593, 2013.

86. Ahmed Salaheldin, Sara Maher, and Mohamed Helw. Robust real-time tracking with diverse ensembles and random projections. In *Proceedings of the IEEE International Conference on Computer Vision Workshops*, pages 112–120, 2013.

7 Aerial Imagery Registration Using Deep Learning for UAV Geolocalization

Ahmed Nassar, and Mohamed ElHelw

CONTENTS

7.1   INTRODUCTION

The proliferation of drones, also known as unmanned aerial vehicles (UAVs), has shifted from being used for military operations to being available for the public market. This conception came through the advancement and availability of mature hardware components such as Bluetooth, Wi-Fi, radio communication, accelerometers, gyroscopes, high-performance processors with low consumption, and efficient battery technologies. This rise can be attributed to the increasing smartphone demand, which led to the race of manufacturing better hardware components, which brought benefits to other sectors such as UAVs [1]. The smartphone market has led to the affordability of these elements since it reduced prices and the size of the hardware. Now UAVs can benefit from all the different types of components and modules in many various applications, such as photography, mapping, agriculture, surveillance, rescue operations, and the recently famous delivery systems. The UAV applications mentioned previously and the ones available in the current market are currently being controlled by radio communications through a controller, a software that plans the route, or by using the on-board camera.

Conventionally, UAV navigation has relied on inertial sensors, on altitude meters, and sometimes on a global positioning system within their solution to feed commands to their motor controller. The latter modality uses the Global Navigation Satellite System (GNSS) composed of different regional systems such as GPS (which is owned by the United States), GLONASS, Galileo, Beidou, and others located at the Medium Earth Orbit with the purpose of providing autonomous geo-spatial positioning [2]. This positioning is provided in latitude, longitude, and altitude data. Both terms, "GPS" and "GNSS," are commonly used interchangeably in UAV navigation literature. The signal transmitted from the GPS satellites could be affected by the distance and the ambient/background noise emanating in numerous places with higher vulnerability to interference and/or blockade in urban areas with buildings causing loss of line of sight (LOS) to navigational satellites. However, some areas are GPS denied, meaning that the GPS signal is not available or weak, and this necessitates finding alternative approaches to UAV guidance.

A plethora of work has been proposed for UAV localization by using inertial measurement unit (IMU) readings and on-board camera to acquire features that are stored in a feature database. Scaramuzza et al. [3], Chowdhary et al. [4], and Rady et al. [5] offer examples of these methods. During UAV flight, information in the pre-acquired feature database is used to approximate UAV location based on matching currently extracted features with those in the database. Similar work [6] uses online mapping services and feature extraction to be able to localize a UAV as well as terrain classification. Mantelli et al. [7] used abBRIEF for global localization and estimated the UAV pose in 4 degrees of freedom.

Currently, the abundance of satellite and aerial high-resolution imagery [8] covering most of the globe has facilitated the emergence of new applications such as autonomous vision-only UAV navigation. In this case, UAV images are compared with aerial/satellite imagery to infer the drone location as shown in Figure 7.1. However, processing these images is computationally demanding and even more involved if data labeling is needed to process and extract actionable information. The ability to correlate two images of the same location under different viewing angles, illumination conditions, occlusions, object arrangements, and orientations is crucial to enable robust autonomous UAV navigation. In traditional computer vision, image entropy and edges are used as features and compared for similarity as presented by Fan et al. [9] and Sim et al. [10]. Other works use local feature detectors [11], such as Scale-Invariant Feature Transform (SIFT) [12] or Speeded-Up Robust Features (SURF) [13], to produce features that are then matched, providing affine transformations as investigated by Koch et al. [14] and Shukla [15]. Using Simultaneous Localization And Mapping (SLAM), UAV navigation is achieved by acquiring and correlating distinct features of indoor and/or outdoor regions in order to map the

FIGURE 7.1   Registration of two images from different domains.

UAV to a specific location. This method is more suitable to constrained environments and involves mainly micro UAVs and quadcopters. Conte et al. [16] and Yol et al. [17] use template matching with cross-correlation where the aerial image is used as a template to match another georeferenced image.

Recently, deep learning has been extensively used for aerial image analysis. To this end, semantic segmentation based on fully convolutional networks is applied on satellite imagery in order to classify each pixel within the image to a common class such as surfaces (pavements, hard surfaces), buildings, vegetation, trees, cars, clutter/background, and roads, among others [18, 19, 20, 21]. In previous work [22], the comparison of multiple images is achieved by using a Siamese Network [23]. In this case, a Siamese Network [24] was trained with two types of images: a satellite image and a perspective-transformed panoramic image. Subsequently, features from both images were extracted and the distance between the features of each image was calculated to find out if they are similar.

7.2   AN INTEGRATED FRAMEWORK FOR UAV LOCALIZATION

7.2.1   Framework Architecture

A framework is proposed with the aim of geolocalizing a UAV while finding out its orientation and height using only its on-board camera [25]. While designing the framework, the process is conceptualized based on how a person from a bird's eye or aerial view would be able to describe his/her location. Several components are used to emulate this process using traditional computer vision and recent state-of-the-art deep learning methods. To demonstrate the framework, Figure 7.3 is used to explain every module and component, and they will be described in further detail in the upcoming sections.

The primary goal is to localize the frames from the UAV with the earth observation images. The UAV sequence frames will be denoted as $S = (S_{(1)}, S_{(2)}, \ldots S_n)$ and

FIGURE 7.2 Left panel is a S_n, with its corresponding S_n ROI from Bing Maps in Famagusta, Cyprus on the right panel.

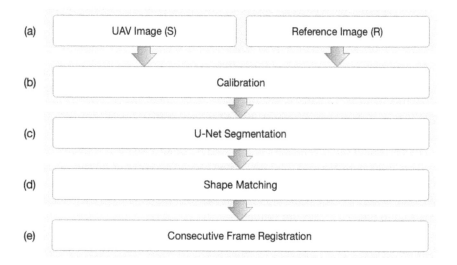

FIGURE 7.3 The pipeline of the UAV localization framework.

their labels will be represented by $L = (L_{(1)}, L_{(2)}, ...L_n)$. In this literature, the Earth Observation (EO) image will be referred to as the reference map, its notation will be R and its label is M. The extents of r on the map and its pixel GPS coordinates are known. Therefore, registering S_n onto r will provide back the GPS coordinates of the S_n. Initially, S_n and its GPS location coordinates on r are assumed to be known, since the person or system deploying the UAV knows its initial location.

Two main inputs are fed to the localization framework. The first corresponds to prerecorded top-down view footage S acquired from 1) a UAV flying over a specific region and 2) a simulated overflight generated by using Google Earth. The second, is a high-resolution EO imagery map obtained from free public online services such as Google Maps, Bing Maps, or existing datasets such as ISPRS's Potsdam, which is used for r as demonstrated in Figure 7.2. Consequently, the different components of the framework deal with those two inputs, and the techniques used will be evaluated on how they deal with these inputs to localize the aerial vehicle.

7.2.2 CALIBRATION

Calibration is an integral part of the framework and is called upon at $S_{(1)}$ and every five iterations ($S_{(n+5)}$) frames. It is responsible for finding the initial location, the difference in scale, and orientation between the S_n and its corresponding field of view in r, and to autocorrect any drift that might occur. To accomplish this task, SIFT is used to make sure that the localization estimation is checked and updated every fixed period. That is, under the condition of being the first frame, or the third sequence frame, S and r are passed to the SIFT registration, which gives out a homography that is applied to r and also updates the drone location as shown in Figure 7.4. SIFT is robust; however, it is time consuming and thus it is not used frequently to reduce computational overhead.

As stated before, knowing the initial GPS coordinates, a region of interest r is cropped from r around the coordinate, given that $r = R(i, w_{s+b})$ where i is the coordinate at the center and the width is determined by w_{s+b}, which is equal to the width of S plus a margin b. This is done for the purpose of limiting the search for the features extracted from S_n in r. As seen in Figure 7.5, r covers a wider region than the one in S. Furthermore, to get the pixel location i from the coordinate, this conversion is calculated by Equations 7.1 and 7.2, which maps the GPS bounds of R to its pixel size, which returns for each pixel its actual GPS location. This process results in reducing the search space and eliminates the possibility for the matching to go astray.

$$\text{pix}_x = \frac{(\text{width}_{max} - \text{width}_{min})(\text{lon} - \text{lon}_w)}{(\text{lon}_e - \text{lon}_w)} \tag{7.1}$$

$$\text{pix}_y = \frac{(\text{height}_{max} - \text{height}_{min})(\text{lat} - \text{lat}_n)}{(\text{lat}_s - \text{lat}_n)} \tag{7.2}$$

$$\text{lat} = \frac{\text{lat}_n + (\text{lat}_n - \text{lat}_s)(\text{pix}_h - \text{height}_{min})}{(\text{height}_{max} - \text{height}_{min})} \tag{7.3}$$

$$\text{lon} = \frac{\text{lon}_w + (\text{lon}_e - \text{lon}_w)(\text{pix}_w - \text{width}_{min})}{(\text{width}_{max} - \text{width}_{min})} \tag{7.4}$$

Where pix_x: Pixel's location in x axis, pix_y: Pixel's location in y axis, height_{min}: r height minimum (0), height_{max}: r height maximum, width_{min}: r width minimum (0), width_{max}: r width maximum, lon: current longitude/coordinate, lat: current latitude coordinate, lat_n: North latitude bound of r, lat_s: South latitude bound o r, lon_e: East latitude bound of r, and lon_w: West latitude bound of r.

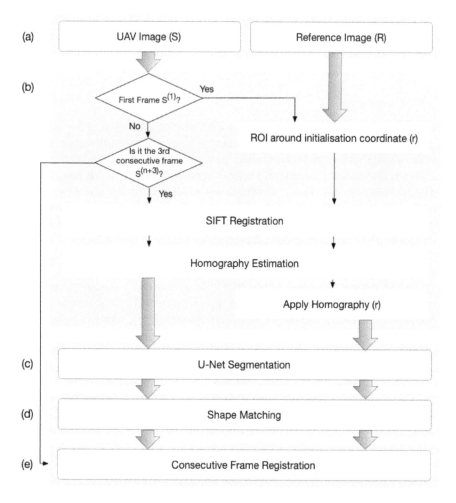

FIGURE 7.4   Calibration using SIFT registration, where location estimation is checked and updated every fixed period.

Using the search space created, features are extracted from both r and S_n. Afterward, the features are matched and used to estimate a homography matrix that in turn is used to extract transformation components. Forthwith, the rotation and scale components are maintained and are incremented upon or decremented depending on the next frame or transformation. This helps calculate the orientation of the UAV using the rotation and altitude by using scale in relation to r.

7.2.3   REGISTRATION AND DETECTION USING SEMANTIC SEGMENTATION

Using down-looking images, drone localization is improved by using available key visual landmarks such as roads and buildings. Semantic segmentation is utilized to help extract specific shape types that are subsequently matched for enhanced results.

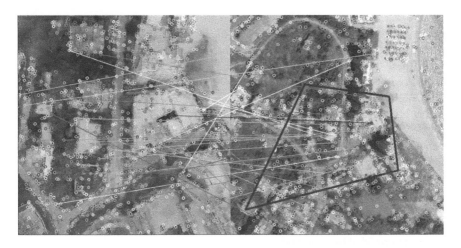

FIGURE 7.5 Registration using SIFT for calibration. The circles represent feature key points extracted, and the lines represent the matching of them. The polygon represents the boundaries of the S_n (left) in r (right) after the transformations.

To be able to accomplish this as seen in Figure 7.6, S_n and r are fed into a U-Net network and the segmented outputs are subsequently matched.

The Segmentation Network: The task to segment input images (r, S) provides classification of input image pixels and is accomplished using U-Net network whose architecture is yet another encoding and decoding model that follows the network architecture of Ronneberger et al. [26]. The segmentation network is able to classify image regions into buildings, roads, and vegetation. Due to the size of images and their number, each class was trained separately, resulting in a model for each class in each dataset.

Data Pre-processing: The images that are used to train the network, presented in Figure 7.3.2, each have a large size exceeding 2000×2000 pixels per image. Therefore, the images are split into smaller patches of size 224×224 pixels, similar to the sizes used by VGG [27] or ResNet [28]. Simple normalization is applied on each patch by dividing the pixel value by its channel's highest value. After that, the patches are stitched to form the full image. Consequently, the same is applied to the ground-truth images (M,L).

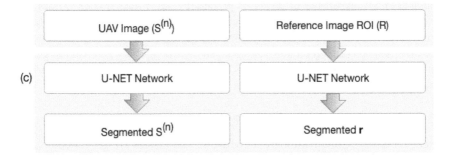

FIGURE 7.6 Two U-Net networks take in S_n and r and provide segmented shapes.

Moreover, data augmentation is applied to these patches before training. The patches are rotated randomly, which helps the model to distinguish different objects with different orientations. They are also flipped horizontally and vertically. On the whole, these techniques create more data to be used in training, which increases the model accuracy.

Optimization Algorithm: The optimization algorithm chosen was the Adaptive Moment Estimator (ADAM) [29]. It's known to be an update to RMSProp [30] and is widely acknowledged to be superior to it. By keeping the exponentially decaying average of all past gradients, ADAM is able to calculate an adaptive learning rate for all network parameters. In early experiments, ADAM converged much faster than stochastic gradient descent and NADAM [31].

Loss Function: The loss function used was Dice's coefficient, which is also known as the Sørensen index [32]. Basically, Dice's coefficient finds the similarity of objects in image segmentation by finding the overlap between a ground-truth and the output provided by a method.

$$\text{Dicescore} = \frac{2\text{TP}}{2\text{TP} + \text{FP} + \text{FN}} \tag{7.5}$$

Where
    TP: True Positive
    FP: False Positive, and
    FN: False Negative.

As shown in Equation 7.2.3, the Dice score is calculated by finding the positives and penalizing the false positives found as well as positives that the method could not find. Dice is similar to the Jaccard index [33], which is another commonly used loss function for image segmentation. However, Dice is currently being favored due to being more intuitive than Jaccard, which only counts true positives once in both the numerator and denominator.

Regularization and Normalization: To avoid overfitting, regularization using dropout layers was added after every convolutional layer with a value of 0.05. After every dropout layer, a batch normalization layer was also added to normalize the activation of each batch.

Network Initialization Using Transfer Learning: Training from scratch is computationally expensive in terms of processing and time. Transfer learning is especially relied upon to reduce the training time and in case of small datasets. In fact, it is difficult to have a dataset that will be capable of catering to truly deep networks. Models trained on large datasets, such as ImageNet [27], contain generic weights that can be used in different tasks. The pre-trained model is trained using a random image generator that also creates the ground-truth for those images.

7.2.4 REGISTRATION BY SHAPE MATCHING

After processing S_n and r through the network, two images are generated containing only buildings and roads as seen in Figure 7.7. The two images are then each split into two layers: buildings and roads. Using computer vision techniques such as

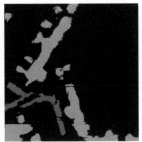

FIGURE 7.7 Buildings segmented using U-Net (the left panel is an EO image, in the middle panel is the result of segmentation, and the right panel is an EO image overlaid with segmentation).

contouring, the boundaries of different objects (buildings or roads) are determined. Subsequently, region features such as area, shape, and location of each object are computed. Consequently, the contour of every segment in each image is selected and compared to objects in the other image for a match; this process is illustrated in steps in Figure 7.8.

All matches are given a score based on contour area, spatial distance, radius, Hausdorff distance [34], centroid, and orientation using Hu moment [35] as shown in Figure 7.9. The Hausdorff distance measures the distance between two sets of points created from shapes. The Hu moment is used to calculate the orientation and centroid of the shape. The Hausdorff distance and Hu moments are used as features in order to find the similarity between the shapes.

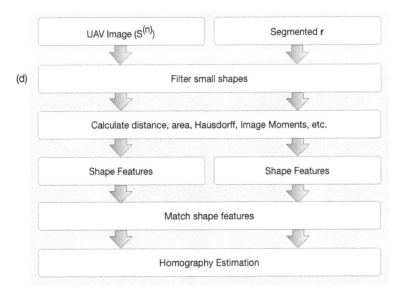

FIGURE 7.8 The segmented images are processed to find their properties such as area, distance, location, image moments, etc., which for each shape or object acts as a feature. These features are matched to create a homography.

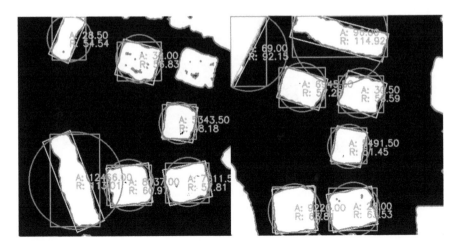

FIGURE 7.9   The left panel shows segmented S_n images, and the right panel shows segmented r images. The contours are computed around a segmented image along with the area, and radius is calculated.

Afterwards, the matches with the highest scores are chosen as presented in Figure 7.10. However, matches are eliminated if the score is close and/or within a short distance, since this might be an unreliable match between 2 objects that are of the same size and within the same distance. For example, a grid of buildings that look the same would confuse the matches. To conclude, as done with the calibration step, using a threshold the top matches that are found to be reliable are used to calculate another homography. This homography, which is estimated from semantic segmentation, is applied to calculate the current location of the UAV accurately.

FIGURE 7.10   Matches of segmented buildings.

7.2.5 DETECTION BY SEMANTIC SEGMENTATION

An additional feature that semantic segmentation adds to this work is that it provides contextual knowledge of the UAV surroundings. Knowing if the UAV is flying above a building or road is very beneficial in surveillance, delivery, and search and rescue missions. For instance, building and road detection is achieved by using the classified image with a box drawn around each contour; having defined the pixel values of each class, each contour can be identified.

7.2.6 REGISTRATION OF SEQUENCE FRAMES

Seeking a real-time motion estimation technique for S, the same method performed in the calibration component (Section 7.2.2). Differently, the image registration process is being done between S_n and S_{n+f} with f being the frame rate. The frame rate is used to skip in-between frames since it will be exhausting to register all consecutive frames with insignificant changes. In fact, registering two consecutive images from a sequence of the same domain can exploit the fact that given smooth platform motion, the difference between the frames is insignificant, as seen in Figure 7.12. Therefore, ORB features were an appropriate choice for this task since they are efficient alternatives to SURF and SIFT. It is important to note that ORB cannot deal with scale invariance effectively, and thus SIFT is used in the intermittent calibration step (check Section 7.2.2) for updating UAV scale and height. As shown in Figure 7.11, the UAV location is updated depending on the homography estimated from consecutive frames.

Likewise, ORB produces a homography that is used to apply a translation that is added to the current pixel location to calculate the final UAV coordinate using Equations. 7.3 and 7.4. It is important to note that in case of insufficient matches, the calibration step, discussed in Section 7.2.2, is called upon for enhanced reliability.

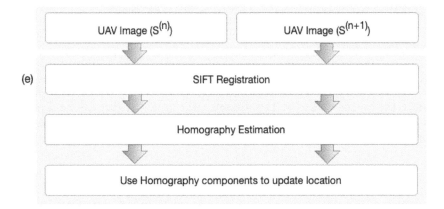

FIGURE 7.11 Two consecutive images from the UAV footage are taken and a homography is estimated from them to update the current UAV location.

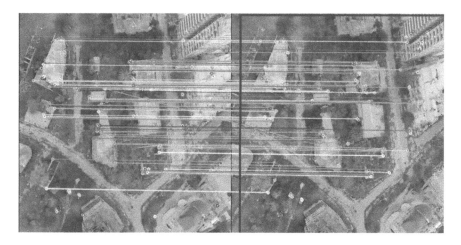

FIGURE 7.12   Registration of *S* using ORB.

7.3   RESULTS AND ANALYSIS

7.3.1   PLATFORM

To be able to carry out the experiments, many libraries and configurations were tried before settling on the current setup. Naturally, the operating system chosen was Ubuntu*, a Debian-based Linux operating system that is available for free and is supported on many laptops, servers, and embedded development kits such as NVIDIA Jetson Tx1/Tx2†. As deep learning is involved in this work, the GPU performance was harnessed by using CUDA 8‡ and cuDNN v5§. The code was mostly written in C++ to be able to cross-compile the code on multiple platforms and obtain fast execution [36]. The Python programming language was also used to integrate with deep learning libraries through Python wrappers. To train the network, IRISA-UBS Lab Cluster¶ was used. The lab has the following specifications: Intel Xeon CPU E5-2687W (3.10 GHz × 20 cores) and NVIDIA 900 GTX TitanX.

7.3.2   DATASETS

Due to lack of established datasets to benchmark UAV path navigation using cameras, datasets had to be collected and ground-truth manually created to be able to produce performance metrics. The datasets used in this project can be split into two categories: aerial datasets and satellite imagery datasets. Multiple sources are used to cater to different views (rooftops, roads, color). A dataset was also created

---

* https://www.ubuntu.com
† https://www.nvidia.com/en-us/autonomous-machines/embedded-systems-dev-kits-modules/
‡ https://www.geforce.com/hardware/technology/cuda
§ https://developer.nvidia.com/cudnn
¶ http://cluster-irisa.univ-ubs.fr/wiki/doku.php/accueil

using images obtained from online mapping services for different locations and their ground-truth created by hand using OpenStreetMap (OSM)*.

ISPRS WG II/4 Potsdam: This dataset [20] is provided by the ISPRS Working Group II/4 for the sole purpose of detecting objects and aiding in 3D reconstruction in urban scenes. The dataset covers two cities: Vaihingen (Germany), a small village with small buildings, and Potsdam (Germany), a dense city with big blocks of large buildings and narrow streets. In this work, we focus on the Potsdam dataset because it contains RGB information and the images reflect a state-of-the-art airborne image dataset in a very high resolution. Figure 7.13 shows samples of the dataset.

The datasets come with ground-truth or label images to classify six classes: Impervious Surfaces, Buildings, Low Vegetation, Trees, Cars, and Clutter/ Background. The Potsdam dataset is made of 38 patches of the same size trimmed from a larger mosaic and a Digital Surface Model (DSM). The resolution of both data types is 5 cm. The images come in different channel compositions: IRRG (3 channels IR-R-G), RGB (3 channels R-G-B), and RGBIR (4 channels R-G-B-IR). This dataset has proven very beneficial throughout this work due to semantically segmented labels and high resolution.

Potsdam Extended: In this work, Google Satellite Images (GSI) are used and thus images from GSI with the same extents and zoom level for 12 of the original ISPRS dataset locations were obtained. Subsequently, the ground-truth from ISPRS is used and updated if there is a difference between recent GSI and ISPRS. Along

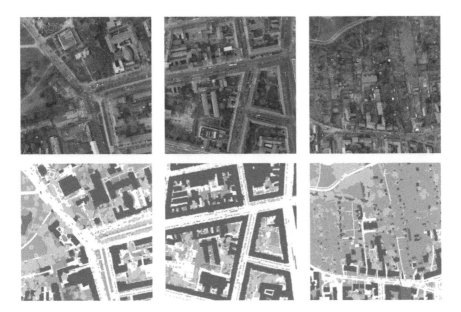

FIGURE 7.13 A sample of the tiles available from the ISPRS Potsdam 2D Semantic Labeling Challenge.

---

* https://www.openstreetmap.org/

with the corresponding ground-truth from ISPRS, several tiles were generated from areas outside the ISPRS dataset and ground-truth produced using OSM as shown in Figure 7.13. The difference between both ISPRS datasets and Google Satellite Image can be seen in Figures 7.14 and 7.15 (Table 7.1).

UAV Videos: One of the major difficulties that faced this work was the unavailability of multiple data sources for one location, especially the top-down view video from a UAV. To address this problem, previously acquired UAV footage from the Internet was used.

The first video used is shot in Famagusta (Cyprus), acquired with top-down view and was available on YouTube by the name of "Above Varosha: one of the most famous ghost cities" [37]. The UAV motion in the video was smooth as it was gliding seamlessly with no vibrations or sudden changes in direction or camera orientation (Figure 7.16). Figure 7.17 shows frame samples from the video. For every 32 frames, ground-truth position was manually computed as GPS coordinates by comparison with the other GSI map. The excerpt of the video that was good enough to be used was 43 secs in length.

Having high-resolution aerial imagery from ISPRS of Potsdam inspired the idea of simulating a UAV footage of this area. This was implemented by using Google Earth 3D and navigating a camera with top-down view. The corresponding video footage was captured while gliding over a region in Potsdam. The captured video length is 2 minutes.

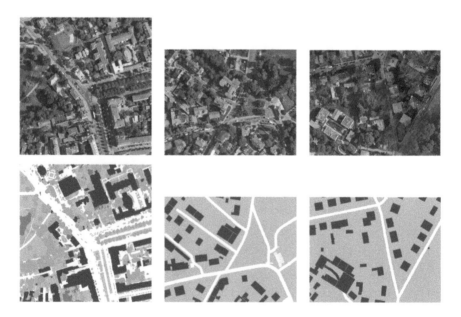

FIGURE 7.14 A sample of the Potsdam Extended dataset. On the top row, the left panel is the corresponding Google Satellite Image copying ISPRS's tile and using its ground-truth. The middle and right panels are tiles produced from other areas in Potsdam. The bottom row is OSM as the ground-truth.

FIGURE 7.15 A comparison between and ISPRS Potsdam (left panel) and a Google Satellite image (right panel).

TABLE 7.1
Comparison of the Different Datasets

Dataset name	Resolution	Area (Sq. km)	Classes
Potsdam	5	2.16	6
Potsdam Extended	9.1	1.24	3
Famagusta	19	1.13	3

FIGURE 7.16 A sample of frames acquired from the Famagusta UAV video.

FIGURE 7.17 A sample of frames acquired from the Potsdam UAV video.

7.3.3 EARTH OBSERVATION IMAGE REGISTRATION

The calibration component is responsible for registering two images from two sources together (S, r) to obtain the transformation describing UAV motion. The registration component that is trying to register consecutive images S from the on-board camera has a much easier job, since the images are from the same source with minor differences. It is important to note that this experiment is carried out without the corrective assistance of the semantic segmentation registration. Furthermore, the experiment is evaluated on a ground-truth manually created.

The Calibration Component Experiment: The calibration step starts first on $S_{(1)}$ where the GPS location is known, being the source, and goes through the registration process with r cropped from r. After every five frames processed by the framework, this component is set to action. The Famagusta Sequence is 1374 frames in total. After every 32 frames (video captured at 32 fps), a frame is processed, since it is inefficient and time consuming to process every frame. Subsequently, the calibration step is called nine times. On the other hand, the Potsdam Sequence is composed from 10,800 frames, and so the calibration component is called every 60 frames (video captured at 60 fps). As explained before, since ground-truth does not exist, the registration had to be evaluated using the following algorithm:

1. Manually, another path is created from the center points of the UAV sequence frames, which provides our ground-truth.
2. Features are extracted from the S_n and r.
3. If 70% of the matches are found have a distance of 0.2 or less, then a homography is created.
4. Apply homography to the reference map ROI, which results in a warped image.
5. A GPS coordinate is estimated for the center pixel of the warped frame using the method explained in Section 7.2.2 and Equation 7.2.2.
6. The distance between the estimated GPS coordinate and the ground-truth is calculated by using Haversine Distance Formula, as illustrated by Figure 7.18.

The UAV Sequence Experiment: As explained in Section 7.2.6, the registration of the S is done using ORB features due to their efficiency. Carrying over from the calibration step, the image registration process done between S_n and S_{n+f} yields a translation that is added to the location estimated from the calibration step. Therefore, the registration of S is called for the same number of frames. In fact, since the UAV is not likely to travel that far in one second, S_n and S_{n+f} are likely to cover nearly identical scene areas.

In our experiment, these steps are taken to ensure that a registration is satisfactory before taking it into account:

1. A path is created from the center points of the UAV sequence frames to set up the ground-truth.
2. Features are then extracted from both consecutive frames passed from the sequence.

FIGURE 7.18 The left panel is S_n, and the right panel is the estimated r. The actual UAV GPS coordinate is {52.404758, 13.053362}, while the estimated GPS coordinate is {52.404737, 13.053338}.

3. A higher tolerance is set here in comparison to Section 7.3.3 since it is an undemanding process; therefore a homography is created if only 80% of the matches found have a distance of 0.15 or less.
4. The translation component provided from the homography is added to the previous GPS component, to estimate the current location.
5. The estimated locations create an estimated flight path of the UAV.

Local Feature Registration Results: The output of the pipeline excluding the semantic segmentation pipeline performs relatively well overall. As shown in Table 7.2, the estimated GPS coordinates are deviating 8.3 m from the actual flight path. As expected, the SIFT calibration step helps the system to "catch up" when it is lagging due to differences in scale (altitude) or photogrammetric differences. Without the SIFT calibration step, longer videos or flights continue to drift until system failure or until the system is no longer capable of comparing or estimating the location.

Qualitatively as seen in Figures 7.19 and 7.20 using this method alone, the deviation to the eye does not seem significant at that height (300 m). Also, it is apparent in

TABLE 7.2

Deviation in Meters Comparing the Two Approaches

Dataset	Without local features	With local features
Famagusta	26 m	10.4 m
Potsdam	52.1 m[†]	8.3 m

[†] Only 60% of the Potsdam Sequence was used to produce this result before localization drifted way and was not capable of finding accurate position.

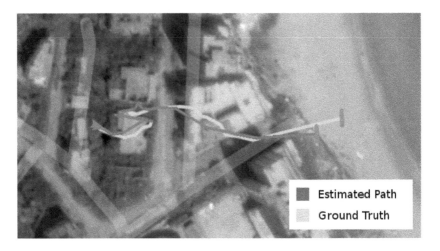

FIGURE 7.19   The actual path of the UAV and the estimated one visualized on a map using their GPS coordinates for the Famagusta Sequence.

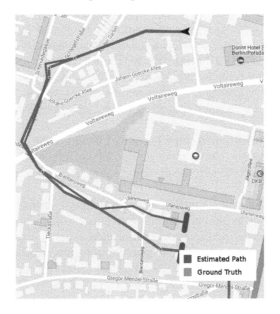

FIGURE 7.20   The actual path of the UAV and the estimated one visualized on a map using their GPS coordinates for the Potsdam Sequence. The intersecting rounded line represents where the experiment ended due to accumulated drift.

several cases that the pipeline performs well if the buildings are exactly perpendicular from a top-down view and their facades are not visible.

7.3.4   SEMANTIC SEGMENTATION

As explained in Section 7.2.3, the semantic segmentation component was devised for two reasons. The first is to be able to replicate the methodology humans would use to navigate by comparing two aerial or satellite images together trying to find matching structures to correct position. The second reason is to avail more information to the UAV, hence making it aware of the types of objects it is flying over, which could be beneficial in route planning or decision making. This component builds upon the previous steps to increase the accuracy of localization and reduce the deviation or drift of the UAV from its planned trajectory. In this section, the setup of the experiment will be presented with all the different components and choices made for the network including the training process. Afterward, the metrics chosen for evaluation are presented along with a discussion of the results and the outcome of the experiments.

The Semantic Segmentation Experiment: As a result of hardware constraints, many models were pre-trained for this experiment. Each region (Famagusta, Potsdam) had its separate model, and each class (buildings, roads, etc.) was trained separately using U-Net (check Section 7.2.3). The pre-trained model was trained with a random image generator [38] that was used to initially start training the different models. The data was split into 224×224 pixel patches, and then the patches were augmented using vertical, horizontal, and random rotation and fed to the U-Net network.

Training: Many configurations and different runs have been performed for each model with different hyper-parameters. The training is stopped instantly when the average F1-score on the validation dataset stopped providing better scores. The average number of epochs trained for each model was 6 and each epoch took almost 3.5 hours to train.

Metrics: To evaluate the semantic segmentation method, the average F1-score is used as an error metric on each trained class. Precision is the outcome of how many positive predictions were predicted in fact positive. Recall is the measure of how many positive predictions are predicted correctly. The F1-score is the average or the harmonic mean of both precision and recall.

$$F1 = 2 \frac{\text{precision} \times \text{recall}}{\text{precision} + \text{recall}} \qquad (7.6)$$

$$\text{Precision} = \frac{TP}{TP + FP} \qquad (7.7)$$

$$\text{Recall} = \frac{TP}{TP + FN} \qquad (7.8)$$

Semantic Segmentation Results: The experiment was run over two regions, Potsdam and Famagusta. The performance of this step was evaluated experimentally using two ground-truth values since each city has its UAV image sequence and its satellite or aerial reference map as presented in Section 7.3.2. The performance of the semantic segmentation in our pipeline was evaluated in terms of localization.

To evaluate the model fairly, the ISPRS Potsdam 2D Semantic Labeling Challenge was used as a benchmark to compare how our model performs against other work.

However, this work was compared to work that solely uses RGB images in their training and doesn't utilize DSM (Digital Surface Models) (Table 7.3).

Experiment I was trained purely on ISPRS Potsdam dataset using 19 images for training and 5 for validation. The purpose of this experiment was to find out how well U-Net works in comparison to the other work based on benchmark scores. The experiment was run only for the building and roads classes, which are the most important landmarks for this region since it is an urban area. In general, the Experiment I model performed better across the 2 classes and resulted in an average score of 84.7%. Unfortunately, the building class score lagged behind, but this experiment was the basis on which the other experiments were built (Table 7.4).

After testing the model generated from Experiment I on the UAV sequence with source images from GSI, the results were not satisfactory. So **Experiment II** was trained on the Potsdam Extended dataset (Section 1.3.2) using Experiment I model for pre-training. The first convolutional layer was frozen while training this experiment since the first layer contains the basic shapes and edges. The scores were satisfactory when tested and the predictions provided sharp edges with hollow shapes as shown in Figure 7.21.

TABLE 7.3

Comparison to Other Work Using ISPRS Potsdam 2D Semantic Labeling Challenge (RGB Images Only)*

Method	Average	Building	Roads
CASIA2[a] [39]	95.5%	97%	93.3%
Experiment I	**84.7%**	85.1%	**84.3%**
Kaiser et al. [40]	83.85%	**91.3%**	76.4%

* The first score is highlighted in bold.

[a] This entry is added for a full picture. It is the highest score available on the ISPRS Potsdam 2D Challenge; however, this method uses DSM data.

TABLE 7.4

Semantic Segmentation Results Using Our Models*

Method	Average	Building	Roads	Vegetation
Experiment II	84.43%	87.98%	**85.2%**	**80.1%**
Experiment III	**85.7%**	**88%**	83.4%	–
Experiment IV	_76%_	76.3%	74.7%	77.4%

* The first score is highlighted in bold, the second is underlined, and the third is in italics.

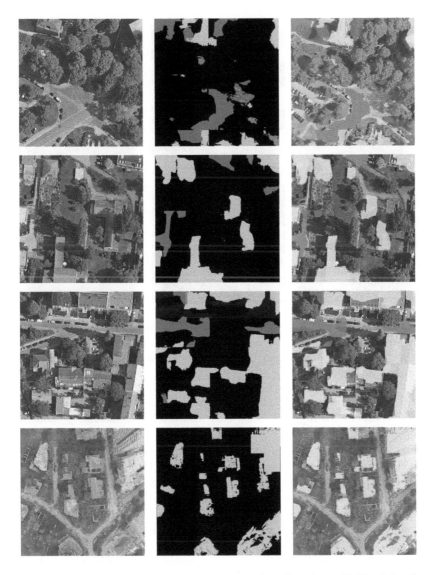

FIGURE 7.21 A sample of the predictions made using Experiment II. The left column shows the target images, the middle column shows the predictions, and the right column is the combination of the previous columns. These target images are Potsdam's (first 3 rows) and Famagusta's (last row) S. Buildings are labeled cyan, and roads are labeled red.

Experiment III was carried out to validate if freezing Experiment II's first convolutional layers was a good idea. Therefore, in this experiment the first convolutional layer was unfrozen. The Experiment III model provided a fuller shape but with inaccurate edges in comparison to Experiment II as shown in Figure 7.22. However, this model provided the highest score. To demonstrate the procedure of only training from the online map services using OSM as a ground-truth, **Experiment IV** was

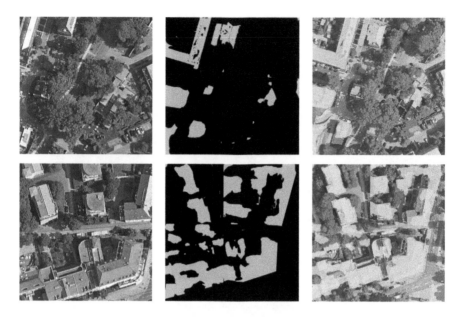

FIGURE 7.22   A sample of the predictions made using Experiment III. The left column is the target image, the middle column is the prediction, and the right column is the prediction overlaying the target image. These target images are Potsdam's *S*. Buildings are labeled in cyan.

arranged. As expected, the results were behind Experiments II and III by nearly 10%. This proves that high-resolution data with pixel-wise accurate ground-truths definitely had an effect (Table 7.5).

In general, our task was not to come up with the state-of-the-art segmentation method, but to choose the best segmentation method for the proposed framework. Although Experiment III provided the highest score, practically Experiment II was the model used due to its sharper shapes, whereas hollow shapes were remedied using morphological operations. Qualitatively, the buildings that are close to each other are challenging to segment. There is also a discrepancy in segmenting buildings from sidewalks, which are sometimes considered part of the building, resulting in added difficulty for the matching process during the registration step. Another

TABLE 7.5

The Specifications That Trained Our Network, after Trying Many Variations in Terms of Filter Size or Optimizers

Method	Filter size	Batch size	LR	Optimizer	Epochs
Experiment **I**	3×3	224	0.001	ADAM	6
Experiment **II**	3×3	224	0.001	ADAM	6
Experiment **III**	3×3	224	0.001	ADAM	7
Experiment **IV**	3×3	224	0.001	ADAM	8

point to consider when using OSM as a ground-truth is that OSM labels treat trees and vegetation as one class. This inconsistency or misclassification between vegetation and tree pixels was accounted for by merging. However, for our application the difference between trees and vegetation is not that vital. The result of the segmentation step usually contains inaccuracies that have to be remedied using morphological techniques such as dilation and erosion, which are applied to fill empty hollows in the shapes and remove small regions.

7.3.5 SEMANTIC SEGMENTATION REGISTRATION USING SHAPE MATCHING RESULTS

Similarly to the evaluation provided in Section 7.3.3(c) for the local feature-based registration, the same is done for the registration using shape matching. This method takes S_n and its corresponding $S_{(r)}$ and matches the equivalent shapes segmented to register them. Once again, the path estimated by our framework is compared to the ground-truth of GPS coordinates manually labeled. Practically, the shape-matching step reduces drift errors and significantly improves localization.

As seen in Table 7.6 the deviation from the ground-truth is reduced to a 5.1-meter error for Famagusta, and considerably for Potsdam to 3.6 meters as shown in Figure 7.23 and illustrated in Figure 7.25. This result could be explained by the low quality of Famagusta's r. As for Potsdam, the quality of both r and S contributed to better localization as shown in Figure 7.24 and illustrated in Figure 7.26. However,

TABLE 7.6
Deviation in Meters with the Addition of Semantic Segmentation to the Pipeline

Dataset	Without local features	With local features	With semantic segmentation
Famagusta	26 m	10.4 m	5.1 m
Potsdam	52.1 m	6.3 m	3.6 m

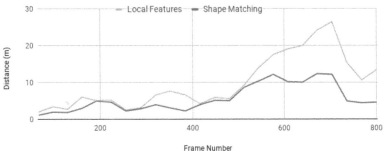

FIGURE 7.23 This graph shows the deviation from the actual GPS coordinate in meters for every time the calibration function is called for the Famagusta Sequence.

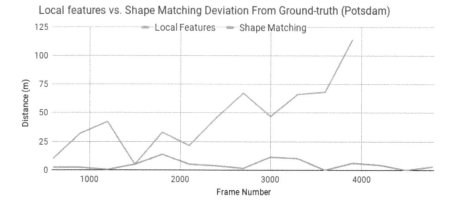

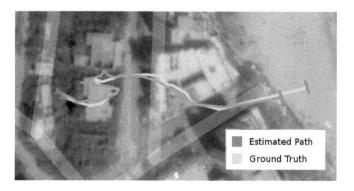

FIGURE 7.24 This graph shows the deviation from the actual GPS coordinate in meters for every time the calibration function is called for the Potsdam Sequence.

FIGURE 7.25 The ground-truth path of the UAV and the estimated path visualized on a map using their GPS coordinates using registration by shape matching in Famagusta.

it has been observed that in Potsdam, building blocks are tightly spaced, so when segmented they create huge blobs that were very difficult to match at a low altitude. Better localization would be achieved in less dense areas or with finer segmentation.

In this section, the feasibility of our framework was investigated. The requirements of the system were first laid out and the framework was introduced. The process of acquiring datasets, their specifications, and the creation of their ground-truths were presented along with justification for picking specific datasets. Afterward, the main registration components were broken down and evaluated focusing on their effectiveness within the framework for localizing the UAV separately.

7.4   CONCLUSION

The essence of this work is to demonstrate the capability of using computer vision to geolocalize a UAV without using any inertial sensors or GPS but using only an on-board camera. The proposed method is then demonstrated in its entirety and the process from start to end is explained in an orderly fashion. Each component is broken down into

FIGURE 7.26   The ground-truth path of the UAV and the estimated path visualized on a map using their GPS coordinates using registration by shape matching in Potsdam.

its subcomponents and its purpose is clarified and evaluated. The proposed method is advantageous over other methods that use only local feature detectors to match the UAV image to an offline reference map, even when searching within a window as the experiments have shown. We believe that segmentation and matching shapes helps avoid matching the wrong features in areas that look similar or have an urban pattern. However, a major disadvantage of the method is areas with no recognizable landmarks or features, but this is a common disadvantage of all methods that depend on vision solely. Future work is to experiment with larger datasets that cover multiple cities with different visual features, which would increase the robustness of the proposed method. Another element is to use a trained neural network to compare the shapes of the segmented objects and replace the demanding dictionary search and heuristics employed by the current method.

BIBLIOGRAPHY

1. K. D. Atherton, "Senate hearing: Drones are "basically flying smart-phones"." https://www.popsci.com/technology/article/2013–03/ how-drone-smartphone, March 2013.
2. B. Hofmann-Wellenhof, H. Lichtenegger, and E. Wasle, *GNSS-Global Navigation Satellite Systems: GPS, GLONASS, Galileo, and More*. Springer Science & Business Media, 2007.

3. D. Scaramuzza, M. C. Achtelik, L. Doitsidis, F. Friedrich, E. Kosmatopoulos, A. Martinelli, M. W. Achtelik, M. Chli, S. Chatzichristofis, L. Kneip, et al., "Vision-controlled micro flying robots: From system design to autonomous navigation and mapping in gps-denied environments," *IEEE Robotics & Automation Magazine*, vol. 21, no. 3, pp. 26–40, 2014.

4. G. Chowdhary, E. N. Johnson, D. Magree, A. Wu, and A. Shein, "Gps-denied indoor and outdoor monocular vision aided navigation and control of unmanned aircraft," *Journal of Field Robotics*, vol. 30, no. 3, pp. 415–438, 2013.

5. S. Rady, A. Kandil, and E. Badreddin, "A hybrid localization approach for uav in GPS denied areas," in *IEEE/SICE International Symposium on System Integration*, pp. 1269–1274, IEEE, 2011.

6. A. Masselli, R. Hanten, and A. Zell, "Localization of unmanned aerial vehicles using terrain classification from aerial images," in *Intelligent Autonomous Systems*, 13, pp. 831–842, Springer, 2016.

7. M. Mantelli, D. Pittol, R. Neuland, A. Ribacki, R. Maffei, V. Jorge, E. Prestes, and M. Kolberg, "A novel measurement model based on abbrief for global localization of a UAV over satellite images," *Robotics and Autonomous Systems*, vol. 112, pp. 304–319, 2019.

8. V. Sowmya, K. Soman, and M. Hassaballah, "Hyperspectral image: Fundamentals and advances," in *Recent Advances in Computer Vision*, pp. 401–424, Springer, 2019.

9. B. Fan, Y. Du, L. Zhu, and Y. Tang, "The registration of UAV down-looking aerial images to satellite images with image entropy and edges," in *Intelligent Robotics and Applications*, pp. 609–617, 2010.

10. D.-G. Sim, R.-H. Park, R.-C. Kim, S. U. Lee, and I.-C. Kim, "Integrated position estimation using aerial image sequences," *IEEE Transactions on Pattern Analysis and Machine Intelligence*, vol. 24, no. 1, pp. 1–18, 2002.

11. M. Hassaballah and A. I. Awad, "Detection and description of image features: An introduction," in *Image Feature Detectors and Descriptors*, pp. 1–8, Springer, 2016.

12. D. G. Lowe, "Distinctive image features from scale-invariant keypoints," *International Journal of Computer Vision,* vol. 60, no. 2, pp. 91–110, 2004.

13. H. Bay, A. Ess, T. Tuytelaars, and L. Van Gool, "Speeded-up robust features (SURF)," *Computer Vision and Image Understanding*, vol. 110, no. 3, pp. 346–359, 2008.

14. T. Koch, P. d'Angelo, F. Kurz, F. Fraundorfer, P. Reinartz, and M. Korner, "The tumdlr multimodal earth observation evaluation benchmark," in *IEEE Conference on Computer Vision and Pattern Recognition (CVPR) Workshops*, June 2016.

15. P. Shukla, S. Goel, P. Singh, and B. Lohani, "Automatic geolocation of targets tracked by aerial imaging platforms using satellite imagery," *International Archives of Photogrammetry, Remote Sensing and Spatial Information Sciences*, vol. 40, no. 1, p. 381, 2014.

16. G. Conte and P. Doherty, "Vision-based unmanned aerial vehicle navigation using georeferenced information," *EURASIP Journal on Advances in Signal Processing*, vol. 2009, p. 10, 2009.

17. A. Yol, B. Delabarre, A. Dame, J.-E. Dartois, and E. Marchand, "Vision-based absolute localization for unmanned aerial vehicles," in *2014 IEEE/RSJ International Conference on Intelligent Robots and Systems*, pp. 3429–3434, IEEE, 2014.

18. N. Audebert, B. Le Saux, and S. Lefvre, "Semantic segmentation of earth observation data using multimodal and multi-scale deep networks," in *Asian Computer Vision Conference*, pp. 180–196, Springer, Cham, November 2016.

19. D. Marmanis, J. D. Wegner, S. Galliani, K. Schindler, M. Datcu, and U. Stilla, "Semantic segmentation of aerial images with an ensemble of CNNs," *ISPRS Annals of the Photogrammetry, Remote Sensing and Spatial Information Sciences*, vol. 3, p. 473, 2016.

20. F. Rottensteiner, G. Sohn, J. Jung, M. Gerke, C. Baillard, S. Benitez, and U. Breitkopf, "The ISPRS benchmark on urban object classification and 3d building reconstruction," *ISPRS Annals of the Photogrammetry, Remote Sensing and Spatial Information Sciences*, vol. 1, no. 3, pp. 293–298, 2012.

21. V. Mnih, K. Kavukcuoglu, D. Silver, A. Graves, I. Antonoglou, D. Wierstra, and M. Riedmiller, "Playing atari with deep reinforcement learning," arXiv preprint arXiv:1312.5602, 2013.
22. S. Lefvre, D. Tuia, J. D. Wegner, T. Produit, and A. S. Nassar, "Toward seamless multiview scene analysis from satellite to street level," *Proceedings of the IEEE*, vol. 105, pp. 1884–1899, October 2017.
23. S. Chopra, R. Hadsell, and Y. LeCun, "Learning a similarity metric discriminatively, with application to face verification," in *IEEE International Conference on Computer Vision and Pattern Recognition*, vol. 1, pp. 539–546, IEEE, 2005.
24. J. Bromley, I. Guyon, Y. LeCun, E. Sackinger, and R. Shah, "Signature verification using a "siamese" time delay neural network," in *Advances in Neural Information Processing Systems*, pp. 737–744, 1994.
25. A. Nassar, K. Amer, R. ElHakim, and M. ElHelw, "A deep CNN-based framework for enhanced aerial imagery registration with applications to uav geolocalization," in *IEEE Conference on Computer Vision and Pattern Recognition Workshops*, pp. 1513–1523, 2018.
26. O. Ronneberger, P. Fischer, and T. Brox, "U-net: Convolutional networks for biomedical image segmentation," arXiv preprint arXm:1505.0Ä597, 2015.
27. K. Simonyan and A. Zisserman, "Very deep convolutional networks for large-scale image recognition," arXiv preprint arXm:1Ä09.1556, 2014.
28. K. He, X. Zhang, S. Ren, and J. Sun, "Deep residual learning for image recognition," in *IEEE Conference on Computer Vision and Pattern Recognition*, pp. 770–778, 2016.
29. D. Kingma and J. Ba, "Adam: A method for stochastic optimization," arXiv preprint arXiv:1412.6980, 2014.
30. T. Tieleman and G. Hinton, "Lecture 6.5-rmsprop: Divide the gradient by a running average of its recent magnitude," *COURSERA: Neural Networks for Machine Learning*, vol. 4, no. 2, pp. 26–31, 2012.
31. T. Dozat, "Incorporating nesterov momentum into adam," in *International Conference on Learning Representations*, 2016.
32. T. Sørensen, "A method of establishing groups of equal amplitude in plant sociology based on similarity of species and its application to analyses of the vegetation on danish commons," *Biol. Skr.*, vol. 5, pp. 1–34, 1948.
33. P. Jaccard, "Le coefficient generique et le coefficient de communaute dans la flore marocaine." Mémoires de la Société Vaudoise des Sciences Naturelles, 2: 385–403, 1926.
34. D. P. Huttenlocher, G. A. Klanderman, and W. J. Rucklidge, "Comparing images using the hausdorff distance," *IEEE Transactions on Pattern Analysis and Machine Intelligence*, vol. 15, no. 9, pp. 850–863, 1993.
35. M.-K. Hu, "Visual pattern recognition by moment invariants," *IRE Transactions on Information Theory*, vol. 8, no. 2, pp. 179–187, 1962.
36. R. Hundt, "Loop recognition in C++/Java/Go/Scala," *Proceedings of Scala Days*, p. 38, 2011.
37. A. FPOHOB, "Above varosha: One of the most famous ghost cities." https://youtu.be/AYKC0dsrh4U, February 2016.
38. ZFTurbo, "ZF UNET 224 pretrained model." https://github.com/ZFTurbo/ZF_UNET_224_Pretrained_Model, 2017.
39. Y. Liu, B. Fan, and C. Pan, "Method description for potsdam: 2D labelling challenge." http://www2.isprs.org/commissions/comm3/wg4/ 2d-sem-label-potsdam.html.
40. P. Kaiser, J. D. Wegner, A. Lucchi, M. Jaggi, T. Hofmann, and K. Schindler, "Learning aerial image segmentation from online maps," *IEEE Transactions on Geoscience and Remote Sensing*, vol. 55, no. 11, pp. 6054–6068, 2017.

8 Applications of Deep Learning in Robot Vision

Javier Ruiz-del-Solar and Patricio Loncomilla

CONTENTS

8.1 INTRODUCTION

Deep learning is a hot topic in the pattern recognition, machine learning, and computer vision research communities. This fact can be seen clearly in the large number of reviews, surveys, special issues, and workshops that are being organized, and the special sessions in conferences that address this topic (e.g., [1–9]). Indeed, the explosive growth of computational power and training datasets, as well as technical improvements in the training of neural networks, has allowed a paradigm shift in pattern recognition, from using hand-crafted features (e.g., Histograms of Oriented Gradients, Scale Invariant Feature Transform, and Local Binary Patterns) [10] together with statistical classifiers to the use of data-driven representations, in which features and classifiers are learned together. Thus, the success of this new paradigm is that "it requires very little engineering by hand" [4], because most of the parameters are learned from the data, using general-purpose learning procedures. Moreover, the existence of public repositories with source code and parameters of trained deep neural networks (DNNs), as well as that of specific deep learning frameworks/tools such as *Tensorflow* [11], has promoted the use of deep learning methods.

The use of the deep learning paradigm has facilitated addressing several computer vision problems in a more successful way than with traditional approaches. In fact, in several computer vision benchmarks, such as the ones addressing image classification, object detection and recognition, semantic segmentation, and action recognition, just to name a few, most of the competitive methods are now based on the use of deep neural networks. In addition, most recent presentations at the flagship conferences in this area use methods that incorporate deep learning.

Deep learning has already attracted the attention of the robot vision community. However, given that new methods and algorithms are usually developed within the computer vision community and then transferred to the robot vision one, the question is whether or not new deep learning solutions to computer vision and recognition problems can be directly transferred to robotics applications. This transfer is not straightforward considering the multiple requirements of current solutions based on deep neural networks in terms of memory and computational resources, which in many cases include the use of GPUs. Furthermore, following [12], it must be considered that robot vision applications have different requirements from standard computer vision applications, such as real-time operation with limited on-board computational resources, and the constraining observational conditions derived from the robot geometry, limited camera resolution, and sensor/object relative pose.

Currently, there are several reviews and surveys related to deep learning [1–7]. However, they are not focused on robotic vision applications, and they do not consider their specific requirements. In this context, the main motivation of this chapter is to address the use of this new family of algorithms in robotics applications. This chapter is intended to be a guide for developers of robot vision systems. Therefore, the focus is on the practical aspects of the use of deep neural networks rather than on theoretical issues. It is also important to note that as convolutional neural networks (CNNs) [7–13] are the most commonly used deep neural networks in vision applications, the analysis provided in this chapter will be focused on them.

This chapter is organized as follows: In Section 2, requirements for robot vision applications are discussed. In Section 3, relevant papers related to robot vision are described and analyzed. In Section 4, current and future challenges of deep learning in robotic applications are discussed. Finally, in Section 5, conclusions related to current robot vision systems and its usability are drawn.

8.2   REQUIREMENTS IN ROBOT VISION

The aim of robotics is to build machines that can behave, i.e., execute tasks, autonomously while interacting with their environment. For reaching autonomy, a robot must be able both to perceive the environment using sensors, and to process the sensorial data for obtaining information relevant to the task to be solved. Robot vision encompasses algorithms that use data acquired from cameras for performing several tasks, such as object detection and categorization, object grasping and manipulation, representation and classification of scenes, and spatiotemporal vision, among others. These capabilities are needed for performing complex tasks needed in applications like pick-and-place robotics, 3D mapping, automated harvesting, service robotics, and self-driving cars, just to name a few.

Robotics imposes significant restrictions in the algorithms that can be used: (i) most autonomous robots use batteries as power sources, so hardware capabilities must be adapted for increasing available operation time; (ii) robots must operate in real time and be responsive; and (iii) several processes alongside robot vision must be executed in parallel. These conditions are needed for a robot to behave and to complete tasks autonomously in the real world, as these tasks commonly involve time constraints for being solved successfully. Then, the computational workload is a crucial component that must be considered when designing solutions for robot vision tasks.

In the following sections, recent advances based on deep learning related to robot vision and future challenges are described. As stated before, deep neural networks require a considerable amount of computational power, which currently limits their usability in robotics applications. However, the development of new processing platforms for massive parallel computing with low power consumption will enable the adoption of a wide variety of deep learning algorithms in robotic applications.

8.3 ROBOT VISION METHODS

In this section, recently published studies presented at flagship journals, robotics conferences, and symposia are described and analyzed. The works have been categorized in four application areas: *Object Detection and Categorization*, *Object Grasping and Manipulation*, *Scene Representation and Classification*, and *Spatiotemporal Vision*. The most representative works and the observed tendencies related to the use of deep learning for addressing robot vision problems in each area are described in the next sections. It is important to note that many of the described works have not been validated in real robots operating under real-world conditions, but only in databases corresponding to robot applications. Therefore, in these cases it is not clear whether the proposed methods can work in real time using a real robot platform.

8.3.1 Object Detection and Categorization

Object detection and categorization is a fundamental ability in robotics. It enables a robot to execute tasks that require interaction with object instances in the real world. Deep learning is already being used for general-purpose object detection and categorization, as well as for pedestrian detection, and for detecting objects in robotics soccer. State-of-the-art methods used for object detection and categorization are based on generating object proposals, and then classifying them using a deep neural network (DNN), enabling systems to detect thousands of different object categories. As will be shown, one of the main challenges for the application of DNNs for object detection and characterization in robotics is real-time operation. It must be stressed that obtaining the required object proposals for feeding the DNNs is not real time in the general case, and that, on the other hand, general-purpose object detection and categorization methods based on DNNs are not able to run in real time in most robotics platforms. These challenges are addressed by using task-dependent methods for generating few, fast, and high-quality proposals for a limited number of possible object categories. These methods are based on using other information sources for segmenting the objects (depth information, motion, color, etc.), and/or by using object-specific, non-general-purpose, weak detectors for generating the required proposals. Also, smaller DNN architectures can be used when dealing with a limited number of object categories.

It is worth mentioning that most of the reported studies do not indicate the frame rate needed for full object detection/categorization, or they show frame rates that are far from being real time. In generic object detection methods, computation of proposals using methods like Selective Search or EdgeBoxes takes most of the time [14]. Systems like Faster R-CNN [15] that compute proposals using CNNs are able to obtain higher frame rates, but require the use of high-end GPUs to work (see Figure 8.1). Also, methods derived from YOLO (You Only Look Once) [16] offer a better runtime, but they both impose limits on the number of objects that can be detected, and have a slightly lower performance. The use of task-specific knowledge-based detectors on depth [17–20], motion [21], color segmentation [22], or weak object detectors [23] can be useful for generating fewer, faster proposals, which is the key for achieving high frame rates on CPU-based systems. Methods based on fully convolutional networks (FCNs) cannot achieve high frame rates on robotic platforms with low processing capabilities (no GPU available) because they process images with larger resolutions than normal CNNs. Thus, FCNs cannot be used trivially for real-time robotics on these kinds of platforms.

First, the use of complementary information sources for segmenting the objects will be analyzed. Robotic platforms usually have range sensors, such as Kinects/ Asus or LIDAR (Light Detection and Ranging) sensors, that are able to extract depth information that can be used for boosting the object segmentation. Methods that use RGBD data for detecting objects include those presented in [17–19, 24, 25], while methods that use LIDAR data include those of [20, 26]. These methods are able to generate proposals by using tabletop segmentation or edge/gradient information.

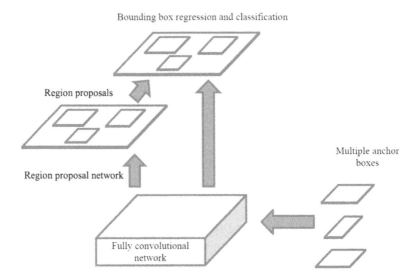

FIGURE 8.1 Faster-RCNN is an object detector that uses a region proposal network and a classification network that share the first convolutional layers [15].

Also, they generate colorized images from depth for use in a standard CNN pipeline for object recognition. These studies use CNNs and custom depth-based proposals, and then their speed is limited by the CNN model. For instance, in [27], a system is described that runs at 405 fps on a Titan X GPU, 67 fps on a Tegra X1 GPU, and 62 fps on a Core i7 6700 K CPU [27]. It must be noted that while Titan X GPU and Core i7 CPU processors are designed for desktop computers, Tegra X1 is a mobile GPU processor for embedded systems aimed at low power consumption, and so it can be used in robotic platforms.

Second, two applications that use specific detectors for generating the object proposals will be presented: pedestrian detection, and detection of objects in robotics soccer. Detecting pedestrians in real time with high reliability is an important ability in many robotics applications, especially for autonomous driving. Large differences in illumination and variable, cluttered backgrounds are hard issues to be addressed. Methods that use CNN-based methods for pedestrian detection include [23, 28–34]. Person detectors such as LDCF [23] or ACF [35] are used for generating pedestrian proposals in some methods. As indicated in [23], the system based on AlexNet [36] requires only 3 msec for processing a region, and proposal computation runs at 2 fps when using LDCF, and at 21.7 fps when using ACF in an NVIDIA GTX980 GPU, with images of size 640×480. The lightweight pedestrian detector is able to run at 2.4 fps in an NVIDIA Jetson TK1, and beats classical methods by a large margin [23]. A second approach is to use an FCN for detecting parts of persons [37]. This method is tested on an NVIDIA Titan X GPU, and it is able to run at 4 fps when processing images of size 300×300. KITTY pedestrians is a benchmark used by state-of-the-art methods. Current leading methods in the KITTY benchmark whose algorithms are described (many of the best-performing methods do not describe their algorithms), are shown in Table 8.1. Note that, as the objective of the benchmark is to evaluate accuracy, most of the methods cannot run in real time (~30 fps). This is an aspect that needs further research.

TABLE 8.1
Selected Methods for Pedestrian Detection from KITTY Benchmark*

Method	Moderate	Easy	Hard	Runtime	Computing environment
RRC [30] (code available)	75.33%	84.14%	70.39%	3.6 s	GPU @ 2.5 Ghz (Python + C/C++)
MS-CNN [31] (code available)	73.62%	83.70%	68.28%	0.4 s	GPU @ 2.5 Ghz (C/C++)
SubCNN [32]	71.34%	83.17%	66.36%	2 s	GPU @ 3.5 Ghz (Python + C/C++)
IVA [33] (code available)	70.63%	83.03%	64.68%	0.4 s	GPU @ 2.5 Ghz (C/C++)
SDP+RPN [34]	70.20%	79.98%	64.84%	0.4 s	GPU @ 2.5 Ghz (Python + C/C++)

* Note that methods not describing their algorithms are not included.

Robotics soccer is an interesting area because it requires real-time vision algorithms that are able to work with very limited computational resources. Deep learning techniques developed in robot soccer can therefore be transferred to other resource-constrained platforms, such as smartphones and Unmanned Aerial Vehicles (UAVs). Current applications for robotics soccer include soccer player detection [22, 38], ball detection [39], and visual navigation [40]. State-of-the-art soccer player detection [38] uses a model based on SqueezeNet using only two convolutional layers, an extended fire layer and a fully connected layer, while processing regions of size 28×28. This method requires only 1 msec on a NAO robot for processing a region, using only an Intel Atom CPU. Another application is ball detection, which is addressed in [39]. This method uses an architecture based on three convolutional layers and two fully connected layers. However, its runtime inside a Nao robot is not reported. Finally, in [40] a system based on image processing and DNNs that is able to detect players, balls, and the orientation of the robot inside the playing field is proposed. The system is able to work in real time inside Nao robots, without using any color information. Currently, CNNs have been used for solving very specific tasks in robotics soccer successfully, and it is expected that new algorithms (and hardware) will enable the use of deep learning in more global, high-level tasks.

8.3.2 OBJECT GRASPING AND MANIPULATION

The ability to grasp objects is of paramount importance for autonomous robots. Classical algorithms for grasping and manipulating objects require detecting both the gripper and the object in the image, and using graph-based or Jacobian-based algorithms for computing the movements for the gripper. Deep learning has generated a paradigm change since tasks like grasping, manipulation, and visual servoing can be solved directly, without the need of hand-crafting intermediate processes like detecting features, detecting objects, detecting the gripper, or performing graph-based optimizations. As problems are solved as a whole from raw data, the performance limit imposed by inaccuracies of intermediate processes can be surpassed, and, therefore, novel approaches for solving tasks related to grasping, manipulation, and visual servoing are being developed. Also, the new algorithms are able to be generalized to novel objects and situations. The performance of this new generation of methods is currently limited by the size of the training datasets and the computing capabilities, which becomes relevant when using hardware-limited robotic platforms.

Grasping objects directly, without the need of detecting specific objects, is a difficult problem because it must generalize over different kinds of objects, including those not available when training the system. This problem has recently been addressed successfully by using deep learning. Studies that are based on selecting grasp locations include [41–43] and [44], which used self-supervision during 700 hours of trial and error in a Baxter robot. That work has been extended for learning eye-hand coordination in [45], which is the state of the art. The latter system is trained without supervision by using over 800,000 grasp attempts collected over the course of two months, using between 6 and 14 manipulators at any given time. The proposed network processes two inputs: an image (size 512×512), and a motor command. The system computes the probability that a given motor command will produce a successful grasp, conditioned on the image, which enables the system to control the robot. The network is composed by seven convolutional layers, whose output is concatenated with the motor command. The concatenation layer is further processed by an additional eight convolutional layers and two fully connected layers. The system achieves effective real-time control; the robot can grasp novel objects successfully and correct mistakes by continuous servoing. It also lowers the failure rate from 35% (hand designed) to 20%. However, learned eye-hand coordination depends on the particular robot used for training the system. The computing hardware used in this work is not fully described. It must be noted that these kinds of solutions are able to run successfully in real time using GPUs.

Systems able to infer manipulation trajectories for object-task pairs using DNNs have been developed recently. This task is difficult because generalization over different objects and manipulators is needed. Proposed solutions include [46, 47]. These systems are able to learn from manipulation demonstrations and generalize over new objects. In Robobarista [46], the input of the system is composed of a point cloud, a trajectory, and a natural language space, which is mapped to a joint space. Once trained, the trajectory generating the best similarity to a new point cloud and natural language instruction is selected. Computational hardware (GPU) and runtime are not reported in this work. In [47], the proposed system is able to learn new tasks by using image-based reinforcement learning. That system is composed by an encoder and a decoder. The encoder is formed by three convolutional layers and two fully connected layers. Computational hardware and runtime are not reported in this work.

Visual servoing systems have also benefited from the use of deep learning techniques. In Saxena et al. [48], a system that is able to perform image-based visual servoing by feeding raw data into convolutional networks is presented. The system does not require the 3D geometry of the scene or the intrinsic parameters from the camera. The CNN takes a pair of monocular images as input, representing the current and desired pose, and processes them by using four convolutional layers and a fully connected layer. The system is trained for computing the transformation in 3D space (position and orientation) needed for moving the camera from its current pose to the desired one. The estimated relative motion is used for generating linear and angular speed commands for moving the camera to the desired pose. The system is trained on synthetic data and tested both in simulated environments and by using a real quadrotor. The system takes 20 msec for processing each frame when using an NVIDIA Titan X GPU, and 65 msec when using a Quadro M2000 GPU. A wifi connection was used for transferring data between the drone and a host PC.

Multimodal data delivers valuable information for object grasping. In [49], a haptic object classifier that is fed by visual and haptic data is proposed. The outputs of a visual CNN and a haptic CNN are fused by a fusion layer, followed by a loss function for haptic adjectives. The visual CNN is based on the (5a) layer of GoogleNet, while the haptic CNN is based on three one-dimensional convolutional layers followed by a fully connected layer. Examples of haptic adjectives are "absorbent," "bumpy," and "compressible." The trained system is able to estimate haptic adjectives of objects by using only visual data and it is shown to improve over systems based on classical features. Computing hardware and runtime of the method are not reported.

8.3.3 SCENE REPRESENTATION AND CLASSIFICATION

Scene representation is an important topic since it is required by mobile robots for performing a variety of tasks that depend on semantic and/or geometric information from the current scene. In particular, visual navigation is an important ability that is needed by autonomous robots for performing more general tasks that involve moving autonomously. Classical approaches for representing scenes from monocular images are based on extraction of hand-crafted features, while visual navigation requires the robot to have a metric or topological map of its environment, as well as an estimation of the depth of the obstacles/objects in the current frame. Deep learning has enabled solving a variety of scene-related tasks directly, ranging from encoding scene information to performing navigation from raw monocular data, without the need of hand-crafting sub-processes like feature extraction, map estimation, or localization. These novel task-oriented algorithms are able to solve difficult problems successfully that were not directly affordable before. In the following paragraphs the use of deep learning in applications such as place recognition, semantic segmentation, visual navigation, and 3D geometry estimation is analyzed.

Place recognition is a task that has been solved successfully by using deep learning-based approaches. Studies related to this problem are [50, 55]. These methods are able to recognize places under severe appearance changes caused by weather, illumination, and changes in viewpoint. Also, in [52], data from LIDAR is used alongside a CNN for building a tridimensional semantic map. The CNN used is based on AlexNet and requires 30 msec to run on an NVIDIA Quadro K4000 GPU. There are several datasets for testing algorithms, such as Sun [56] and Places [57, 58], and also several variants derived from them. As there is not a dominant benchmark for place recognition, different studies use different datasets, sometimes customized, and thus most methodologies cannot be compared directly. Also, the best-performing methods in ILSVRC, scene classification challenge [59] and Places challenges [60], do not describe their algorithms in detail, but achieve an impressive 0.0901 top-5 classification error. Computing hardware and runtime related to those challenges [59, 60] is not reported. Nevertheless, the related datasets used in scene classification, and the best-performing methods in each case, are shown in Table 8.2.

Semantic segmentation consists of classifying each pixel in the image, enabling to solve tasks that require a pixel-wise level of precision. Studies related to semantic

TABLE 8.2
Datasets Used for Scene Classification

Dataset	Results
Places [58]	50.0% [58] (CaffeNet), 14.56 msec @ Pascal Titan X
	85.16% [54] (AlexNet for scene classification, 8 conv for semantic segmentation)
Sun [56, 61]	38% [56] (non deep)
	66.2% [58] (AlexNet), 14.56 msec @ Pascal Titan X
Sun-RGBD [54]	41.3% [54] (input 81×81, 5 conv + 3 fc classification, 8 conv semantic segmentation)

segmentation of scenes are [26, 54, 62–70]. Also, multimodal information is used in [52, 71, 72]. These studies are useful for tasks that require pixel-wise precision like road segmentation and other tasks in autonomous driving. However, runtimes of methods based on semantic segmentation are usually not reported. PASCAL VOC and Cityscapes are two benchmarks used for semantic segmentation. While images from PASCAL VOC are general-purpose, Cityscapes is aimed at autonomous driving. Best-performing methods in both databases are reported in Tables 8.3 and 8.4. It can be noted that DeepLabv3 [64] and PSPNet [65] (see Figure 8.2) achieve good average precision at both benchmarks, and also code from [65] is available. Also, the increasing performance in average precision has a deep impact on autonomous driving, as this application requires confident segmentation for working because of the risks involved on it.

Visual navigation using deep learning techniques is an active research topic since new methodologies are able to solve navigation directly, without the need for detecting objects or roads. Examples of studies related to visual navigation are [74–78]. These methods are able to perform navigation directly over raw images captured by RGB cameras. For instance, the forest trail follower in [74] (see Figure 8.3) has an architecture composed of seven convolutional layers and one fully connected layer. It is able to run at 15 fps on an Odroid-U3 platform, enabling it to be used in a real UAV.

TABLE 8.3

PASCAL VOC Segmentation Benchmark*

Name	AP	Runtime (msec/frame)
DeepLabv3-JFT [64]	86.9	n/a
DIS [69]	86.8	140 msec (GPU model not reported)
CASIA_IVA_SDN [70]	86.6	n/a
IDW-CNN [73]	86.3	n/a
PSPNet [65] (code available)	85.3	n/a

* Only entries with reported methods are considered. Repeated entries are not considered.

TABLE 8.4

Cityscapes Benchmark*

Name	IoU class	Runtime (ms/frame)
Deeplabv3 [64]	81.3	n/a
PSPNet [65]	81.2	n/a
ResNet-38 [66]	80.6	6 msec (minibatch size 10) @ GTX 980 GPU
TuSimple_Coarse [67] (code available)	80.1	n/a
SAC-multiple [68]	78.2	n/a

* Only entries with reported methods are considered. Repeated entries are not considered.

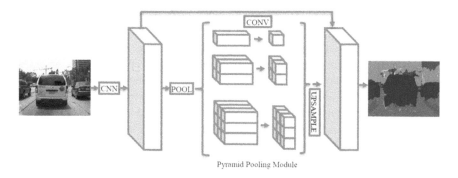

Pyramid Pooling Module

FIGURE 8.2 Pyramid Scene Parsing Network (PSPNet) is a scene-parsing algorithm based on the use of convolutional layers, followed by upsampling and concatenation layers [65].

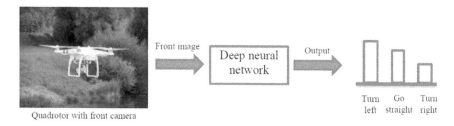

FIGURE 8.3 A deep neural network is used to decide one of three possible actions (turn left, go straight, turn right) from a single image. This system is used as guide in an autonomous quadrotor [74].

In [79], a visual navigation system for Nao robots based on deep reinforcement learning is proposed. The system consists of two neural networks—an actor and a critic—that are trained by using Deep Deterministic Policy Gradients. Both networks are conformed by convolutional, fully connected, and Long Short-Term Memory (LSTM) layers. Once trained, the system is able to run in real time in the Nao internal CPU.

The use of CNN methodologies for estimating 3D geometry is another active research topic in robotics. Studies related to 3D geometry estimation from RGB images include [80–82]. Also, depth information is added in [71, 72]. These methods provide a new approach to dealing with 3D geometry, and they are shown to overcome the classical structure-from-motion approaches, as they are able to infer depth even from only a single image.

A functional area is a zone in the real world that can be manipulated by a robot. Functional areas are classified by the kind of manipulation action the robot can realize on them. In [83], a system for localizing and recognizing functional areas by using deep learning is presented. The system enables an autonomous robot to have a functional understanding of an arbitrary indoor scene. An ontology considering 11 possible end categories is set. Functional areas are detected by using selective-search region proposals, and then by applying a VGG-based

CNN trained on the functionality categories. The system is able to generalize onto novel scenes. Neither hardware nor runtime are reported. However, its architecture is similar to RCNN [14], which requires 18 seconds for processing a frame on a K20x GPU.

8.3.4 SPATIOTEMPORAL VISION

The processing of video and other spatiotemporal sources of information is of paramount importance in robotics. Classical methods related to spatiotemporal vision require engineering subtasks such as motion detection, optical flow, local feature extraction, or image segmentation. Deep learning has the potential to change the current paradigm of spatiotemporal vision from hand-engineering complex subtasks to solving tasks as a whole. The increment in both video datasets and computational power will enable the analysis of spatiotemporal data at near real time, but only in the future. Then, spatiotemporal vision will be a goal of future robot vision as well as computer vision research.

The use of short video sequences for object recognition is an alternative for improving current image-based techniques. In [84], a convolutional LSTM is proposed for processing short video sequences containing about four frames, for robotic perception. The system is shown to improve on the baseline (smoothing CNN results from individual image frames), and it is able to run at 0.87 fps when using a GPU.

Human action recognition is an area that can benefit greatly from using video sequences instead of isolated images. Reports of research dealing with recognition of human actions include [85–93], in which driver activity anticipation is performed. In various studies, information from multiple frames is integrated by using several CNNs for each frame, by using LSTM or Dynamic Time Warping. The C3D method [90] is able to run at 313 fps on a K40 GPU, processing videos with a resolution of 320×240 pixels. Thus, those kinds of methods have the potential for running in real time. However, current state-of-the-art methods are not real time, or their runtimes are not described. The best-performing methods on the UCF-101 dataset are reported in Table 8.5. Note that both the accuracy and runtime of C3D [90] are lower than those reported by Wu et al. [92], and so a tradeoff between accuracy

TABLE 8.5
Current Results on the UCF-101 Video Dataset for Action Recognition [91]

Name	Accuracy	Runtime (msec/frame)
C3D (1 net) + linear SVM [90]	82.3%	3.2 msec/frame @ Tesla K40 GPU
VGG-3D + C3D [90]	86.7%	n/a
Ng et al. [92]	88.6%	n/a
Wu et al. [93]	92.6%	1,730 msec/frame @ Tesla K80 GPU
Guo et al. [91]	93.3%	n/a
Lev et al. [86]	94.0%	n/a

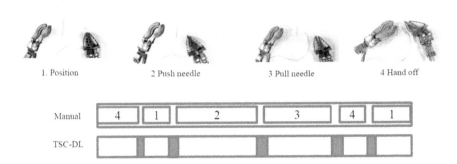

FIGURE 8.4 TSC-DL is a system that is able to segment a suturing video sequence into different stages (position needle, push needle, pull needle, and hand-off [94]).

and runtime is present. For selecting a method, it must be considered if real time is needed in the specific task to be solved.

Transfer of video-based computer vision techniques to medical robotics is an active area of research, especially in surgery-related applications. The use of deep learning in this application area has already begun. For instance, in [94], a system that is able to segment video and kinematic data for performing unsupervised trajectory segmentation of multimodal surgical demonstrations is implemented (see Figure 8.4). The system is able to segment video-kinematic descriptions of surgical demonstrations successfully, corresponding to stages such as "position," "push needle," "pull needle," and "hand-off." The system is based on a switching linear dynamic system, which considers both continuous and discrete states composed of kinematic and visual features. Visual features are extracted from the fifth convolutional layer of a VGG CNN and then reduced to 100 dimensions by using PCA. Clustering on the state space is applied by the system and the system is able to learn transitions between clusters, enabling trajectory segmentation. The system uses a video source with 10 frames per second; however, the frame rate of the full method is not reported.

8.4 FUTURE CHALLENGES

As already mentioned, the increase in the use of deep learning in robotics will depend on the possibility that DNN methods will be able to adapt to the requirements of robotics applications, such as real-time operation with limited on-board computational resources, and the ability to deal with constrained observational conditions derived from the robot geometry, limited camera resolution, and sensor/object relative pose.

The current requirements of large computational power and memory needed for most CNN models used in vision applications are relevant barriers to their adoption in resource-constrained robotic applications such as UAVs, humanoid biped robots, or autonomous vehicles. Efforts are already being made to develop CNN architectures that can be compressed and quantized for use in resource-constrained platforms [95, 96]. For instance, in [38] a CNN-based robot detector is presented that is able to run in real time, in Nao robots while playing soccer. In addition, companies such as Intel, NVIDIA, and Samsung, just to name a few, are developing CNN chips that will enable real-time vision applications [4]. For instance, mobile GPU processors like NVIDIA Tegra K1 enable efficient implementation of deep learning algorithms with low power consumption, which is relevant for mobile robotics [27]. It is expected that these methodologies will consolidate in the next few years and will then be available to the developers of robot vision applications.

The ability of deep learning models to achieve spatial invariance respect to the input data in a computationally efficient manner is another important issue. In order to address this, the spatial transformer module (STM) was recently introduced [97]. It is a self-contained module that can be incorporated into DNNs, and that is able to perform explicit spatial transformation of the features. The most important characteristic of STMs is that its parameters (i.e., the transformations of the spatial transformation) can be learned together with the other parameters of the network using the backpropagation of the loss. The further development of the STM concept will allow addressing the required invariance to the observational conditions derived from the robot geometry, limited camera resolution, and sensor/object relative pose. STMs are able to deal with rotations, translations, and scaling [97], and STMs are already being extended to deal with 3D transformations [98]. Further development of STM-inspired techniques is expected in the next few years.

Unsupervised learning is another relevant area of future development for deep learning-based vision systems. Biological learning is largely unsupervised; animals and humans discover the structure of the world by exploring and observing it. Therefore, one would expect that similar mechanisms could be used in robotics and other computer-based systems. Until now, the success of purely supervised learning, which is based largely on the availability of massive labeled data, has overshadowed the application of unsupervised learning in the development of vision systems. However, when considering the increasing availability of non-labeled digital data, and the increasing number of vision applications that need to be addressed, one would expect further development of unsupervised learning strategies for DNN models [4, 7]. In the case of robotics applications, a natural way of addressing this issue is by using deep learning and reinforcement learning together. This strategy has already been applied for the learning of games (Atari, Go, etc.) [99, 100], for solving simple tasks like pushing a box [47], and for solving more complex tasks such as navigating in simulated environments [77, 78], but it has not been used for robots learning in the wild. The visual system of most animals is active, i.e., animals decide where to look, based on a combination of internal and external stimuli. In a robotics system having a similar active vision approach, reinforcement learning can be used for deciding where to look according to the results of the interaction of the robot with the environment.

A related relevant issue is that of open-world learning [101], i.e., to learn to detect new classes incrementally, or to learn to distinguish among subclasses incrementally after the main one has been learned. If this can be done without supervision, new classifiers can be built based on those that already exist, greatly reducing the effort required to learn new object classes. Note that humans are continuously inventing new objects, fashion changes, etc., and therefore robot vision systems will need to be updated continuously, adding new classes and/or updating existing ones [101]. Some recent work has addressed these issues, based mainly on the joint use of deep learning and transfer learning methods [102, 103].

Classification using a very large number of classes is another important challenge to address. AlexNet, like most CNN networks, was developed for solving problems that contain ~1,000 classes (e.g., ILSVRC 2012 challenge). It uses fully connected units and a final softmax unit for classification. However, when working on problems with a very large number of classes (10,000+) in which dense data sampling is available, nearest neighbors could work better than other classifiers [104]. Also, the use of a hierarchy-aware cost function (based on WordNet) could enable providing more detailed information when the classes have redundancy [104]. However, in current object recognition benchmarks (like ILSVRC 2016), the number of classes has not increased, and no hierarchical information has been used. The ability of learning systems to deal with a high number of classes is an important challenge that needs to be addressed for performing open-world learning.

Another important area of research is the combination of new methods based on deep learning with classical vision methods based on geometry, which has been very successful in robotics (e.g., visual SLAM). This topic has been addressed at recent workshops [105] and conferences [106]. There are several ways in which these important and complementary paradigms can be combined. On the one hand, geometry-based methods such as Structure from Motion and Visual SLAM can be used for the training of deep learning-based vision systems; geometry-based methods can extract and model the structure of the environment, and this structure can be used for assisting the learning process of a DNN. On the other hand, DNNs can also be used for learning to compute the visual odometry automatically, and eventually to learn the parameters of a visual SLAM system. In a very recent study, STMs were extended to 3D spatial transformations [98], allowing end-to-end learning of DNNs for computing visual odometry. It is expected that STM-based techniques will allow end-to-end training for optical flow, depth estimation, place recognition with geometric invariance, small-scale bundle adjustment, etc. [98].

A final relevant research topic is the direct analysis of video sequences in robotics applications. The analysis of video sequences by using current techniques based on deep learning is very demanding of computational resources. It is expected that the increase of computing capabilities will enable the analysis of video sequences in real time in the near future. Also, the availability of labeled video datasets for learning tasks such as action recognition will enable the improvement of robot perception tasks, which are currently addressed by using independent CNNs for individual video frames.

8.5 CONCLUSIONS

In this chapter, several approaches based on deep learning that are related to robot vision applications are described and analyzed. They are classified into four categories of methods: (i) object detection and categorization, (ii) object grasping and manipulation, (iii) scene representation and classification, and (iv) spatiotemporal vision. (i) Object detection and categorization has been transferred successfully into robot vision. While the best-performing object detectors are mainly variants of Faster-RCNN, methods derived from YOLO offer a better runtime. (ii) Object grasping, manipulation, and visual servoing are currently being improved by the adoption of deep learning. While real-time processing is needed for hand-eye coordination and visual servoing to work, it is not needed for open-loop object grasping detection. Also, these methods are highly dependent on the robot used for collecting data. (iii) Methods for representing, recognizing, or classifying scenes as a whole (not pixel-wise) feed images directly into a single CNN, and then near-real-time processing can be achieved. However, pixel-wise methods based on FCNs require a GPU for running in near real time. (iv) Spatiotemporal vision applications need real-time processing capability to be useful in robotics. However, current research using deep learning is experimental, and the frame rates for most of the methods are not reported. Platforms with low computing capability (only CPU) are not able to run most of the methods at a useful frame rate. The adoption of general deep learning methods in robot vision applications requires addressing several problems: (i) Deep neural networks require large amounts of data for being trained. Obtaining large amounts of data for specific robotic applications is a challenge to address each time a specific task needs to be solved. (ii) Deep learning is able to deal with difficult problems. However, it cannot generalize to conditions that are not represented by the available data. (iii) The need of low power consumption is a critical condition to be considered. However, the availability of specialized hardware for computing algorithms based on CNNs with low power consumption would make it possible to broaden the use of deep neural networks in robotics.

ACKNOWLEDGMENTS

This work was partially funded by FONDECYT grant 1161500 and CONICYT PIA grant AFB18004.

BIBLIOGRAPHY

1. Yanming Guo, Yu Liu, Ard Oerlemans, Songyang Lao, Song Wu, and Michael S. Lew. Deep learning for visual understanding: A review. *Neurocomputing*, 187, 27–48, 2016.
2. Soren Goyal and PaulBenjamin. Object recognition using deep neural networks: A survey. http://arxiv.org/abs/1412.3684, 2014.
3. Jiuxiang Gu, Zhenhua Wang, Jason Kuen, Lianyang Ma, Amir Shahroudy, Bing Shuai, Ting Liu, Xingxing Wang, Gang Wang, Jianfei Cai, and Tsuhan Chen. Recent advances in convolutional neural networks. *Pattern Recognition*, 77, 354–377, 2018.
4. Yann LeCun, Yoshua Bengio, and Geoffrey Hinton. Deep learning. *Nature*, 521, 436–444, 28 May 2015, doi:10.1038/nature14539,
5. Li Deng. A tutorial survey of architectures, algorithms, and applications for deep learning. *APSIPA Transactions on Signal and Information Processing*, 3, 1–29, 2014.
6. Jürgen Schmidhuber. Deep learning in neural networks: An overview. *Neural Networks*, 61, 85–117, 2015.
7. Suraj Srinivas, Ravi Kiran Sarvadevabhatla, Konda Reddy Mopuri, Nikita Prabhu, Srinivas S. S. Kruthiventi, and R. Venkatesh Babu. A taxonomy of deep convolutional neural nets for computer vision. *Frontiers in Robotics and AI*, 2, 2016.

8. Xiaowei Zhou, Emanuele Rodola, Jonathan Masci, Pierre Vandergheynst, Sanja Fidler, and Kostas Daniilidis. Workshop geometry meets deep learning, ECCV 2016, 2016.

9. Yi Li, Yezhou Yang, Michael James, Danil Prokhorov. Deep Learning for Autonomous Robots, Workshop at RSS 2016, 2016.

10. Awad, Ali Ismail, and Mahmoud Hassaballah. Image feature detectors and descriptors. *Studies in Computational Intelligence.* Springer International Publishing, Cham, 2016.

11. Martín Abadi, Paul Barham, Jianmin Chen, Zhifeng Chen, Andy Davis, Jeffrey Dean, et al. Tensorflow: A system for large-scale machine learning. 12th Symposium on Operating Systems Design and Implementation (OSDI 16), pp. 265–283, 2016.

12. Patricio Loncomilla, Javier Ruiz-del-Solar, and Luz Martínez, Object recognition using local invariant features for robotic applications: A survey, *Pattern Recognition,* 60, 499–514, 2016.

13. Y. LeCun, B. Boser, J. Denker, D. Henderson, R. Howard, W. Hubbard, and L. Jackel. Backpropagation applied to handwritten zip code recognition. *Neural Computation,* 1(4), 541–551, 1989.

14. R. Girshick, J. Donahue, T. Darrell, and J. Malik. Rich feature hierarchies for accurate object detection and semantic segmentation. *IEEE Conference on Computer Vision and Pattern Recognition (CVPR),* 580–587, 2014.

15. Shaoqing Ren, Kaiming He, Ross Girshick, and Jian Sun. Faster R-CNN: Towards real-time object detection with region proposal networks. *IEEE Transactions on Pattern Analysis and Machine Intelligence,* 39, 6, 1137–1149, 2017.

16. Joseph Redmon and Ali Farhadi. YOLO9000: Better, faster, stronger. *IEEE Conference on Computer Vision and Pattern Recognition (CVPR),* 6517–6525, 2017.

17. Max Schwarz, Hannes Schulz, and Sven Behnke. RGB-D object recognition and pose estimation based on Pre-trained convolutional neural network features. *IEEE International Conference on Robotics and Automation (ICRA),* 1329–1335, 2005.

18. Saurabh Gupta, Ross Girshick, Pablo Arbeláez, and Jitendra Malik. Learning rich features from RGB-D images for object detection and segmentation. *European Conference on Computer Vision (ECCV),* 345–360, 2014.

19. Andreas Eitel, Jost Tobias Springenberg, Luciano Spinello, Martin Riedmiller, and Wolfram Burgard. Multimodal deep learning for robust RGB-D object recognition. *IEEE/RSJ International Conference on Intelligent Robots and Systems (IROS),* 681–687, 2015.

20. Joel Schlosser, Christopher K. Chow, and Zsolt Kira. Fusing LIDAR and images for pedestrian detection using convolutional neural networks. *IEEE International Conference on Robotics and Automation (ICRA),* 2198–2205, 2016.

21. G. Pasquale, C. Ciliberto, F. Odone, L. Rosasco, and L. Natale. Teaching iCub to recognize objects using deep convolutional neural networks. *4th International Conference on Machine Learning for Interactive Systems (MLIS'15),* 43, 21–25, 2015.

22. Dario Albani, Ali Youssef, Vincenzo Suriani, Daniele Nardi, and Domenico Daniele Bloisi. A deep learning approach for object recognition with NAO soccer robots. *RoboCup International Symposium,* 392–403, 2016.

23. Denis Tomè, Federico Monti, Luca Baroffio, Luca Bondi, Marco Tagliasacchi, and Stefano Tubaro. Deep convolutional neural networks for pedestrian detection. *Signal Processing: Image Communication,* 47, C, 482–489, 2016.

24. Hasan F. M. Zaki, Faisal Shafait, and Ajmal Mian. Convolutional hypercube pyramid for accurate RGB-D object category and instance recognition. *IEEE International Conference on Robotics and Automation (ICRA), 2016,* 1685–1692, 2016.

25. Judy Hoffman, Saurabh Gupta, Jian Leong, Sergio Guadarrama, and Trevor Darrell. Cross-modal adaptation for RGB-D detection. *IEEE International Conference on Robotics and Automation (ICRA),* 5032–5039, 2016.

26. Bo Li, Tianlei Zhang, and Tian Xia. Vehicle detection from 3D lidar using fully convolutional network. *Proceedings of Robotics: Science and System Proceedings*, 2016.
27. Nvidia: GPU-Based deep learning inference: A performance and power analysis. Whitepaper, 2015.
28. Jan Hosang, Mohamed Omran, Rodrigo Benenson, and Bernt Schiele. Taking a deeper look at pedestrians. *IEEE Conference on Computer Vision and Pattern Recognition (CVPR)*, 4073–4082, 2015.
29. Rudy Bunel, Franck Davoine, and Philippe Xu. Detection of pedestrians at far distance. *IEEE International Conference on Robotics and Automation (ICRA)*, 2326–2331, 2016.
30. Jimmy Ren, Xiaohao Chen, Jianbo Liu, Wenxiu Sun, Jiahao Pang, Qiong Yan, Yu-Wing Tai, and Li Xu. Accurate single stage detector using recurrent rolling convolution. *IEEE Conference on Computer Vision and Pattern Recognition (CVPR)*, 752–760, 2017.
31. Zhaowei Cai, Quanfu Fan, Rogerio S. Feris, and Nuno Vasconcelos. A unified multi-scale deep convolutional neural network for fast object detection. *European Conference on Computer Vision (ECCV)*, 354–370, 2016.
32. Yu Xiang, Wongun Choi, Yuanqing Lin, and Silvio Savarese. Subcategory-aware convolutional neural networks for object proposals and detection. *IEEE Winter Conference on Applications of Computer Vision (WACV)*, 924–933, 2017.
33. Yousong Zhu, Jinqiao Wang, Chaoyang Zhao, Haiyun Guo, and Hanqing Lu. Scale-adaptive deconvolutional regression network for pedestrian detection. *Asian Conference on Computer Vision*, 416–430, 2016.
34. Fan Yang, Wongun Choi, and Yuanqing Lin. Exploit all the layers: Fast and accurate CNN object detector with scale dependent pooling and cascaded rejection classifiers. *IEEE International Conference on Computer Vision and Pattern Recognition (CVPR)*, 2129–2137, 2016.
35. Piotr Dollar, Ron Appel, Serge Belongie, and Pietro Perona. Fast feature pyramids for object detection. *IEEE Transactions on Pattern Analysis and Machine Intelligence*, 36, 8, 1532–1545, 2014.
36. Alex Krizhevsky, Ilya Sutskever, and Geoffrey E. Hinton. ImageNet classification with deep convolutional neural networks. *Advances in Neural Information Processing Systems*, 25, 1097–1105, 2012.
37. Gabriel L. Oliveira, Abhinav Valada, Claas Bollen, Wolfram Burgard, and Thomas Brox. Deep learning for human part discovery in images. *IEEE International Conference on Robotics and Automation (ICRA)*, 1634–1641, 2016.
38. Nicolás Cruz, Kenzo Lobos-Tsunekawa, and Javier Ruiz-del-Solar. Using convolutional neural networks in robots with limited computational resources: Detecting NAO robots while playing soccer. RoboCup 2017: Robot World Cup XXI, 19–30, 2017.
39. Daniel Speck, Pablo Barros, Cornelius Weber, and Stefan Wermter. Ball localization for robocup soccer using convolutional neural networks. *RoboCup International Symposium*, 19–30, 2016.
40. Francisco Leiva, Nicolas Cruz, Ignacio Bugueño, and Javier Ruiz-del-Solar. Playing soccer without colors in the SPL: A convolutional neural network approach. *RoboCup Symposium*, 2018 (in press).
41. Ian Lenz, Honglak Lee, and Ashutosh Saxena. Deep learning for detecting robotic grasps. *The International Journal of Robotics Research*, 34, 4–5, 705–724, 2015.
42. Joseph Redmon and Anelia Angelova. Real-time grasp detection using convolutional neural networks. *IEEE International Conference on Robotics and Automation (ICRA)*, 1316–1322, 2015.

43. Lerrel Pinto and Abhinav Gupta. Supersizing self-supervision: Learning to grasp from 50K tries and 700 robot hours. *IEEE International Conference on Robotics and Automation (ICRA)*, 3406–3413, 2016.

44. Di Guo, Tao Kong, Fuchun Sun, and Huaping Liu. Object discovery and grasp detection with a shared convolutional neural network. *IEEE International Conference on Robotics and Automation (ICRA)*, 2038–2043, 2016.

45. Sergey Levine, Peter Pastor, Alex Krizhevsky, and Deirdre Quillen. Learning hand-eye coordination for robotic grasping with deep learning and large-scale data collection. *The International Journal of Robotics Research*, 37, 4–5, 421–436, 2018.

46. Jaeyong Sung, Seok Hyun Jin, and Ashutosh Saxena. Robobarista: Object part based transfer of manipulation trajectories from crowd-sourcing in 3D pointclouds. *Robotics Research*, 3, 701–720, 2018.

47. Chelsea Finn, Xin Yu Tan, Yan Duan, Trevor Darrell, Sergey Levine, and Pieter Abbeel. Deep spatial autoencoders for visuomotor learning. *IEEE International Conference on Robotics and Automation (ICRA)*, 512–519, 2016.

48. Aseem Saxena, Harit Pandya, Gourav Kumar, Ayush Gaud, and K. Madhava Krishna. Exploring convolutional networks for end-to-end visual servoing. *IEEE International Conference on Robotics and Automation (ICRA)*, 3817–3823, 2017.

49. Yang Gao, Lisa Anne Hendricks, Katherine J. Kuchenbecker, and Trevor Darrell. Deep learning for tactile understanding from visual and haptic data. *IEEE International Conference on Robotics and Automation (ICRA)*, 536–543, 2016.

50. Manuel Lopez-Antequera, Ruben Gomez-Ojeda, Nicolai Petkov, and Javier Gonzalez-Jimenez. Appearance-invariant place recognition by discriminatively training a convolutional neural network. *Pattern Recognition Letters*, 92, 89–95, 2017.

51. Yi Hou, Hong Zhang, and Shilin Zhou. Convolutional neural network-based image representation for visual loop closure detection. *IEEE International Conference on Information and Automation (ICRA)*, 2238–2245, 2015.

52. Niko Sunderhauf, Feras Dayoub, Sean McMahon, Ben Talbot, Ruth Schulz, Peter Corke, Gordon Wyeth, Ben Upcroft, and Michael Milford. Place categorization and semantic mapping on a mobile robot. *IEEE International Conference on Robotics and Automation (ICRA)*, 5729–5736, 2006.

53. Niko Sunderhauf, Sareh Shirazi, Adam Jacobson, Feras Dayoub, Edward Pepperell, Ben Upcroft, and Michael Milford. Place recognition with convnet landmarks: Viewpoint-robust, condition-robust, training-free. *Robotics: Science and System Proceedings*, 2015.

54. Yiyi Liao, Sarath Kodagoda, Yue Wang, Lei Shi, and Yong Liu. Understand Scene Categories by objects: A semantic regularized scene classifier using convolutional neural networks. *IEEE International Conference on Robotics and Automation (ICRA)*, 2318–2325, 2016.

55. Peter Uršic, Rok Mandeljc, Aleš Leonardis, and Matej Kristan. Part-based room categorization for household service robots. *IEEE International Conference on Robotics and Automation (ICRA)*, 2287–2294, 2016.

56. Jianxiong Xiao, James Hays, Krista A. Ehinger, Aude Oliva, and Antonio Torralba. Sun database: Large-scale scene recognition from abbey to zoo. *IEEE Conference on Computer Vision and Pattern recognition (CVPR)*, 3485–3492, 2010.

57. Bolei Zhou, Agata Lapedriza, Aditya Khosla, Aude Oliva, and Antonio Torralba. Places: A 10 million image database for scene recognition. *IEEE Transactions on Pattern Analysis and Machine Intelligence*, 40, 6, 1452–1464, 2018.

58. Bolei Zhou, Agata Lapedriza, Jianxiong Xiao, Antonio Torralba, and Aude Oliva. Learning deep features for scene recognition using places database. *Advances in Neural Information Processing Systems* 27 (NIPS), 487–495, 2014.

59. ILSVRC 2016 scene classification, 2016. http://image-net.org/challenges/LSVRC/2016/results.
60. Places Challenge 2016. http://places2.csail.mit.edu/results2016.html.
61. Jianxiong Xiao, James Hays, Krista A. Ehinger, Aude Oliva, and Antonio Torralba. SUN database: Large-scale scene recognition from abbey to zoo. *IEEE Conference on Computer Vision and Pattern recognition (CVPR)*, 3485–3492, 2010.
62. Caio Cesar Teodoro Mendes, Vincent Fremont, and Denis Fernando Wolf. Exploiting fully convolutional neural networks for fast road detection. *IEEE International Conference on Robotics and Automation (ICRA)*, 3174–3179, 2016.
63. Shuran Song, Samuel P. Lichtenberg, and Jianxiong Xiao. Sun rgb-d: A rgb-d scene understanding benchmark suite. *IEEE Conference on Computer Vision and Pattern Recognition (CVPR)*, 567–576, 2015.
64. Liang-Chieh Chen, George Papandreou, Florian Schroff, and Hartwig Adam. Rethinking atrous convolution for semantic image segmentation. Technical Report. arXiv preprint arXiv:1706.05587, 2017.
65. Hengshuang Zhao, Jianping Shi, Xiaojuan Qi, Xiaogang Wang, and Jiaya Jia. Pyramid scene parsing network. *IEEE Conference on Computer Vision and Pattern Recognition (CVPR)*, 6230–6239, 2017.
66. Zifeng Wu, Chunhua Shen, and Anton van den Hengel. Wider or deeper: Revisiting the resnet model for visual recognition. *Pattern Recognition*, 90, 119–133, 2019.
67. Panqu Wang, Pengfei Chen, Ye Yuan, Ding Liu, Zehua Huang, Xiaodi Hou, and Garrison Cottrell. Understanding convolution for semantic segmentation. *IEEE Winter Conference on Applications of Computer Vision (WACV)*, 1451–1460, 2018.
68. Rui Zhang, Sheng Tang, Yongdong Zhang, Jintao Li, and Shuicheng Yan. Scale-adaptive convolutions for scene parsing. *IEEE Conference on Computer Vision and Pattern Recognition (CVPR)*, 2031–2039, 2017.
69. Ping Luo, Guangrun Wang, Liang Lin, and Xiaogang Wang. Deep dual learning for semantic image segmentation. *IEEE Conference on Computer Vision and Pattern Recognition (CVPR)*, 2718–2726, 2017.
70. Jun Fu, Jing Liu, Yuhang Wang, and Hanqing Lu. Stacked deconvolutional network for semantic segmentation. *IEEE Transactions on Image Processing*, 99, 2017.
71. Cesar Cadena, Anthony Dick, and Ian D. Reid. Multi-modal auto-encoders as joint estimators for robotics scene understanding. *Proceedings of Robotics: Science and System Proceedings*, 2016.
72. Farzad Husain, Hannes Schulz, Babette Dellen, Carme Torras, and Sven Behnke. Combining semantic and geometric features for object class segmentation of indoor scenes. *IEEE Robotics and Automation Letters*, 2 (1), 49–55, 2016.
73. Guangrun Wang, Ping Luo, Liang Lin, and Xiaogang Wang. Learning object interactions and descriptions for semantic image segmentation. *IEEE Conference on Computer Vision and Pattern Recognition (CVPR)*, 5859–5867, 2017.
74. Alessandro Giusti, Jérôme Guzzi, Dan C. Ciresan, Fang-Lin He, Juan P. Rodríguez, Flavio Fontana, Matthias Faessler, Christian Forster, Jürgen Schmidhuber, Gianni Di Caro, Davide Scaramuzza, and Luca M. Gambardella. A machine learning approach to visual perception of forest trails for mobile robots. *IEEE Robotics and Automation Letters*, 1, 2, 661–667, 2016.
75. Dan C. Ciresan, Ueli Meier, and Jürgen Schmidhuber. Multi-column deep neural networks for image classification. *IEEE Conference on Computer Vision and Pattern Recognition (CVPR)*, 3642–3649, 2012.
76. Lei Tai and Ming Liu. A robot exploration strategy based on q-learning network. *IEEE International Conference on Real-time Computing and Robotics (RCAR)*, 57–62, 2016.
77. Yuke Zhu, Roozbeh Mottaghi, Eric Kolve, Joseph J. Lim, Abhinav Gupta, Li Fei-Fei, and Ali Farhadi. Target-driven visual navigation in indoor scenes using deep

reinforcement learning. *IEEE International Conference on Robotics and Automation (ICRA)*, 3357–3364, 2017.

78. Piotr Mirowski, Razvan Pascanu, Fabio Viola, Hubert Soyer, Andrew J. Ballard, Andrea Banino, Misha Denil, Ross Goroshin, Laurent Sifre, Koray Kavukcuoglu, Dharshan Kumaran, and Raia Hadsell. Learning to navigate in complex environments. *International Conference on Learning Representations (ICLR)*, 2017.

79. Kenzo Lobos-Tsunekawa, Francisco Leiva, and Javier Ruiz-del-Solar. Visual navigation for biped humanoid robots using deep reinforcement learning. *IEEE Robotics and Automation Letters*, 3, 4, 3247–3254, 2018.

80. Gabriele Costante, Michele Mancini, Paolo Valigi, and Thomas A. Ciarfuglia. Exploring representation learning with CNNs for frame to frame ego-motion estimation. *IEEE Robotics and Automation Letters*, 1, 1, 18–25, 2016.

81. Shichao Yang, Daniel Maturana, and Sebastian Scherer. Real-time 3D scene layout from a single image using convolutional neural networks. *IEEE International Conference on Robotics and Automation (ICRA)*, 2183–2189, 2016.

82. Alex Kendall and Roberto Cipolla. Modelling uncertainty in deep learning for camera relocalization. *IEEE International Conference on Robotics and Automation (ICRA)*, 4762–4769, 2016.

83. Chengxi Ye, Yezhou Yang, Cornelia Fermuller, and Yiannis Aloimonos. What can i do around here? Deep functional scene understanding for cognitive robots. *IEEE International Conference on Robotics and Automation (ICRA)*, 4604–4611, 2017.

84. Ivan Bogun, Anelia Angelova, and Navdeep Jaitly. Object recognition from short videos for robotic perception. arXiv:1509.01602v1 [cs.CV], 4 September 2015.

85. Jimmy Ren, Xiaohao Chen, Jianbo Liu, Wenxiu Sun, Jiahao Pang, Qiong Yan, Yu-Wing Tai, and Li Xu. Accurate single stage detector using recurrent rolling convolution. *IEEE Conference on Computer Vision and Pattern Recognition (CVPR)*, 752–760, 2017.

86. Guy Lev, Gil Sadeh, Benjamin Klein, and Lior Wolf. RNN fisher vectors for action recognition and image annotation. *European Conference on Computer Vision (ECCV)*, 833–850, 2016.

87. Farzad Husain, Babette Dellen, and Carme Torra. Action recognition based on efficient deep feature learning in the spatio-temporal domain. *IEEE Robotics and Automation Letters*, 1, 2, 984–991, 2016.

88. Xiaochuan Yin and Qijun Chen. Deep metric learning autoencoder for nonlinear temporal alignment of human motion. *IEEE International Conference on Robotics and Automation (ICRA)*, 2160–2166, 2016.

89. Ashesh Jain, Avi Singh, Hema S Koppula, Shane Soh, and Ashutosh Saxena. Recurrent neural networks for driver activity anticipation via sensory-fusion architecture. *IEEE International Conference on Robotics and Automation (ICRA)*, 3118–3125, 2016.

90. Du Tran, Lubomir Bourdev, Rob Fergus, Lorenzo Torresani, and Manohar Paluri. Learning spatiotemporal features with 3D convolutional networks. *IEEE International Conference on Computer Vision (ICCV)*, 4489–4497, 2015.

91. Huiwen Guo, Xinyu Wu, and Wei Fengab. Multi-stream deep networks for human action classification with sequential tensor decomposition. *Signal Processing*, 140, 198–206, 2017.

92. Joe Yue-Hei Ng, Matthew Hausknecht, Sudheendra Vijayanarasimhan, Oriol Vinyals, Rajat Monga, and George Toderici. Beyond short snippets: Deep networks for video classification. *IEEE Conference on Computer Vision and Pattern Recognition (CVPR)*, 4694–4702, 2015.

93. Zuxuan Wu, Yu-Gang Jiang, Xi Wang, Hao Ye, and Xiangyang Xue. Multi-stream multi-class fusion of deep networks for video classification. *ACM Multimedia*, 791–800, 2016.

94. Adithyavairavan Murali, Animesh Garg, Sanjay Krishnan, Florian T. Pokorny, Pieter Abbeel, Trevor Darrell, and Ken Goldberg. TSC-DL: Unsupervised trajectory segmentation of multi-modal surgical demonstrations with deep learning. *IEEE International Conference on Robotics and Automation (ICRA)*, 4150–4157, 2016.

95. Song Han, Huizi Mao, and William J. Dally. Deep compression: Compressing deep neural networks with pruning, trained quantization and huffman coding. *International Conference on Learning Representations (ICLR'16)*, 2016.

96. Song Han, Xingyu Liu, Huizi Mao, Jing Pu, Ardavan Pedram, Mark A. Horowitz, and William J. Dally. EIE: Efficient inference engine on compressed deep neural network. *International Conference on Computer Architecture (ISCA)*, 243–254, 2016.

97. Max Jaderberg, Karen Simonyan, Andrew Zisserman, and Koray Kavukcuoglu. Spatial transformer networks. *Advances in Neural Information Processing Systems*, 28, 2017–2025, 2015.

98. Ankur Handa, Michael Bloesch, Viorica Patraucean, Simon Stent, John McCormac, and Andrew Davison. gvnn: Neural network library for geometric computer vision. *Computer Vision – ECCV 2016 Workshops*, 67–82, 2016.

99. Volodymyr Mnih, Koray Kavukcuoglu, David Silver, Andrei A. Rusu, Joel Veness, Marc G. Bellemare, Alex Graves, Martin Riedmiller, Andreas K. Fidjeland, Georg Ostrovski, Stig Petersen, Charles Beattie, Amir Sadik, Ioannis Antonoglou, Helen King, Dharshan Kumaran, Daan Wierstra, Shane Legg, and Demis Hassabis. Human-level control through deep reinforcement learning. *Nature*, 518, 529–533, 2015.

100. David Silver, Aja Huang, Chris J. Maddison, Arthur Guez, Laurent Sifre, George van den Driessche, Julian Schrittwieser, Ioannis Antonoglou, Veda Panneershelvam, Marc Lanctot, Sander Dieleman, Dominik Grewe, John Nham, Nal Kalchbrenner, Ilya Sutskever, Timothy Lillicrap, Madeleine Leach, Koray Kavukcuoglu, Thore Graepel, and Demis Hassabis. Mastering the game of go with deep neural networks and tree search. *Nature*, 529, 7587, 484–489, 2016.

101. Rodrigo Verschae and Javier Ruiz-del-Solar. Object detection: Current and future directions. *Frontiers in Robotics and AI*, 29, 2, 2015.

102. Yoshua Bengio. Deep learning of representations for unsupervised and transfer learning. *International Conference on Unsupervised and Transfer Learning Workshop*, 27, 17–37, 2012.

103. Marc-Andr Carbonneau, Veronika Cheplygina, Eric Granger, and Ghyslain Gagnon. Multiple instance learning. *Pattern Recognition*, 77, 329–353, 2018.

104. Jia Deng, Alexander C. Berg, Kai Li, and Li Fei-Fei. What does classifying more than 10,000 image categories tell us? *European Conference on Computer Vision: Part V (ECCV'10)*, 71–84, 2010.

105. Stefan Leutenegger, Thomas Whelan, Richard A. Newcombe, and Andrew J. Davison. Workshop the future of real-time SLAM: Sensors, processors, representations, and algorithms, *ICCV*, 2015.

106. Ken Goldberg. Deep grasping: Can large datasets and reinforcement learning bridge the dexterity gap? Keynote Talk at ICRA, 2016.

9 Deep Convolutional Neural Networks

Foundations and Applications in Medical Imaging

Mahmoud Khaled Abd-Ellah, Ali Ismail Awad,
Ashraf A. M. Khalaf, and Hesham F. A. Hamed

CONTENTS

9.1　INTRODUCTION

Machine learning has become a primary focus and one of the most mainstream subjects among research groups. The main uses of machine learning include feature extraction, feature reduction, feature classification, forecasting, clustering, regression, and ensembles with different algorithms. Classic algorithms and techniques are optimized to provide effective self-learning [1]. Since machine learning is applied in a wide range of studies, many techniques have been provided, such as Bayesian network, clustering, decision tree learning, and deep learning as shown in Figure 9.1. Meanwhile, a few analysts in the machine learning field have attempted to learn models that combine the learning of features. These models normally consist of different layers of nonlinearity. This led to the generation of the first deep learning models. Early models such as deep belief networks [2], stacked autoencoders (SAEs) [3], and restricted Boltzmann machines [4] indicated guarantees on small datasets. This was known as the "unsupervised pretraining" phase. It was believed that these "pretrained" models would be a good initialization for supervised tasks, such as classification.

Deep learning first appeared in 2006 as a new subfield of machine learning research. In Mosavi and Varkonyi-Koczy [5], it was named the hierarchical learning algorithm, and it provided research fields related to pattern recognition. Deep learning of key factors mainly depends on supervised or unsupervised learning and nonlinear processing in multiple layers [6]. With supervised learning, the class target label is available; however, the absence of the class target label means an unsupervised system. During the ImageNet large-scale visual recognition challenge (ILSVRC) competition in 2012 [7], different scaled algorithms were applied to a large dataset, which required, in addition to other tasks, the grouping of an image into one of a thousand classes. Surprisingly, a convolutional neural network (CNN) significantly reduced the error rate on

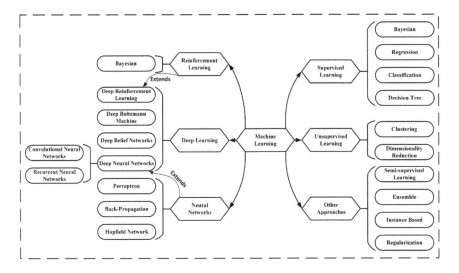

FIGURE 9.1 Machine learning is the overall category that includes supervised learning, unsupervised learning, reinforcement learning, neural networks, deep learning, and other approaches.

that task, outperforming the traditional methodologies. Moreover, this could be accomplished by applying end-to-end supervised training without unsupervised pretraining. Throughout the following few years, "ImageNet uses deep neural networks in classification" [8] has been the most powerful and poignant paper in the computer vision field.

Convolutional neural networks, a specific type of deep learning model, have since been broadly embraced by the computer vision community. Specifically, the model prepared by Alex Krizhevsky, famously called "AlexNet", has been utilized and adapted for different vision tasks. The output of a layer is provided as input to the next layer in the algorithm and is called nonlinear processing in multiple layers. A hierarchy is used between layers to consider and organize the importance of data. As an example, applying a deep learning algorithm can perform grayscale image coloring automatically. Recurrent neural networks (RNNs) can add sound to a silent video [1]. Consequently, this chapter principally discusses CNNs because these models are more applicable to the computer vision field. This chapter can also serve as a guide for specialists starting in deep learning/computer vision.

Computer vision problems such as image characterization and object recognition have customarily employed handcrafted features [9] such as the scale-invariant feature transform (SIFT) [10] and the histogram of oriented gradients (HOG) [11]. Specifically, representations dependent on the bag-of-visual-words descriptor [12] have achieved success in image classification. These works were normally followed by learning algorithms, such as support vector machines (SVMs). Accordingly, the implementation of these algorithms essentially depends on the features utilized, implying that advancements in computer vision depend on building better arrangements of features [13].

This chapter provides a comprehensive foundation for deep convolutional neural networks. The basic components of CNNs are summarized. Then, we present some later developments in various aspects of CNNs, including the convolutional layer, pooling layer, activation function, fully connected layers (FCLs), regularization, optimization, normalization, and network depth, and we present the advances in each phase. Next, some network operations are discussed, specifically fine-tuning, reducing overfitting, and using CNN activations as features. Then, some commonly used networks, such as AlexNet, GoogleNet, VGG-19, and ResNet, are presented. The theoretical knowledge of CNN types along with the three most common architectures and learning algorithms are provided. Moreover, these networks are called region-based CNNs (R-CNNs), fully convolutional networks (FCNs), and hybrid learning networks.

CNNs are widely being applied in different domains because of their outstanding performance, such as in image classification, object detection, speech recognition, face detection, vehicle recognition, facial expression recognition, segmentation of MR brain images, and many more. This chapter also introduces various processing applications of deep convolutional neural networks in medical imaging, such as feature extraction, tumor detection, and tumor segmentation. This study will serve as a reference and resource for researchers who are interested in this field.

This chapter is arranged as follows. It first presents the general foundations of CNNs (Section 9.2). Section 9.3 is devoted to reporting efficient learning approaches. The network operations and a brief introduction to the current programming tools available for executing these algorithms are described in Section 9.4. An overview of the commonly used CNN is documented in Section 9.5. Then, a discussion of different CNN types for specific issues is presented in Section 9.6. Section 9.7 is dedicated to discussing deep learning applications in medical imaging. Section 9.8 describes the network design of the proposed system and provides simulation results and evaluation criteria. Finally, conclusions are outlined in Section 9.9.

9.2 CONVOLUTIONAL NEURAL NETWORKS (CNNS)

The possibility of a CNN is not new. This model appeared to function admirably for handwritten digit recognition [14] in early 1998. However, because of the inability of these models to scale to substantially larger images, they gradually lost support. This loss of support was to a great extent because of memory and equipment limitations and the inaccessibility of large amounts of training data. With the expansion in computational power due to the wide accessibility of GPUs and the emergence of extensive datasets such as ImageNet [7], it became possible to train larger and more complex models. This ability first appeared with the mainstream AlexNet model, which was discussed previously and was the start of using the deep neural network in computer vision.

9.2.1 CNN LAYERS

In this section, we will discuss the essential building layers of CNNs. This discussion assumes that the reader knows about traditional neural networks (TNNs), which

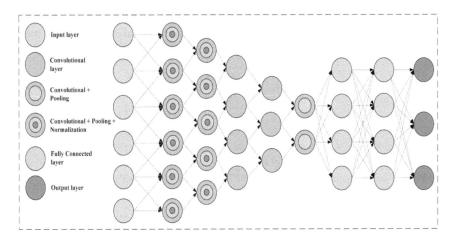

FIGURE 9.2   Representation of CNN network weights. A ReLU nonlinearity is added after every convolutional layer.

we will call "fully connected layers (FCLs)". Figure 9.2 presents a portrayal of the weights in the AlexNet network. Although convolutional layers are used in the first five layers, the last three layers are fully connected.

9.2.1.1   Convolutional Layers

Utilizing TNNs for image classification is unfeasible for the following reason: the number of input nodes will be very large. Consider a 200×200 image, for which we would have 40,000 nodes. If the hidden layer has 20,000 nodes, the size of the input weights matrix would be $40,000 \times 20,000 = 800$ million. This number is only for one layer, and as we increase the number of layers, this number significantly increases more quickly. Moreover, image vectorization ignores the image spatial structure. How can we develop a network that overcomes these disadvantages?

Two-dimensional convolutions are used instead of matrix multiplications. Additionally, it normally considers the 2D structure of the image. Meanwhile, convolutions can also be thought of as a TNN with two requirements: local connectivity and weight sharing [15]. There is additional proof from visual neuroscience that comparable calculations occur inside the human brain. Hubel and Wiesel [16] discovered two types of cells in the essential visual cortex: the basic cells and the complex cells. The basic cells reacted primarily to arranged edges and gratings, which are reminiscent of Gabor features, an exceptional class of convolutional channels. The complex cells were also sensitive to these edges and gratings. CNNs are profound learning networks that have been enhanced in recent decades. CNNs can automatically learn features from inputs, particularly images, similar to the case in this exploration [17]. The convolutional layers are utilized to convolve the image with weights to acquire a feature map. The weights of the portions associate the feature map to the previous layer. In the example shown in Figure 9.2, the numbers $11 \times 11 \times 3$, along with 96 underneath C1, show that the first layers have 96 filters, every one a spatial measurement of 11×11, with one for each of the 3 RGB channels.

Therefore, the subsequent component initiation after convolution has 96 channels, and we are now able to apply another arrangement of multichannel features for these features as well. The 2D convolution can be defined as in Equation 9.1 [17].

$$y(i,j) = x(i,j) * w(i,j) = \sum_{m}\sum_{n} x(m,n)w(i-m, j-n). \qquad (9.1)$$

where x is the input and w is the weight that is provided from the previous layer. The convolution matrix size is $m \times n$, and the convolutional layer output is y. The input of the CNN and the convolutional kernel dimensions should be the same. A multidimensional kernel should be used if the CNN is a multidimensional array [17].

9.2.1.2 Max-Pooling Layer

The Neocognitron display inspired the consideration of basic cells as convolutions. Proceeding in a similar way, the complex cells can be considered a max-pooling task. This task can be thought of as a maximum filter, where every $n \times n$ area is supplanted with its maximum value [18]. The max-pooling task requires the following:

1. Selection of the most astounding activation in a nearby region, in this way providing a low level of spatial invariance. This is comparable to the task of complex cells.
2. Size reduction of the following layer by a factor of n^2. Fewer parameters should be learned in the following layers when the activation size is reduced.

Different types of pooling include stochastic pooling [19], average pooling, and winner-takes-all pooling [20]. However, these types of pooling are not as commonly utilized as max-pooling. Other pooling types occasionally cannot extract good features; for example, average pooling provides an average value by taking all features into account, which is a very generalized computation. Therefore, poor accuracy will be obtained with average pooling when the systems do not require all features from the convolutional layer. Another example, winner-takes-all pooling, is similar to max-pooling; however, it keeps the feature dimension fixed, and no downsampling occurs. However, some classification tasks have used different pooling or mixed pooling and outperform max-pooling. Thus, the selection of pooling type depends on the database type.

9.2.1.3 Nonlinearity and ReLU

Nonlinearities are necessary between layers to ensure that the network is more expressive than a linear one. This makes the network more powerful and includes the capability to learn complicated and complex features from data. The nonlinearity enables the network to generate nonlinear mappings from inputs to output; furthermore, it can cut-off or adjust the developed output. A nonlinear layer is employed to reduce the generated output or saturate the output [21]. Thus, the convolutional layers are followed by a nonlinear layer. Customary feedforward neural networks utilize the sigmoid or tanh nonlinearities, as shown in Equations 9.2 and 9.3, respectively. Therefore, the new convolutional systems utilize a nonlinear activation function. The most commonly applied activation function for deep learning applications in the literature is rectified linear unit (ReLU) because of the following reasons:

1. It <u>exhibits</u> faster learning, which is characterized as in Equation 9.4 [17].
2. ReLU provides better performance and generalization compared to other activation functions used in deep learning.
3. Sigmoid and tanh functions cause a problem called the vanishing gradient in the backpropagation, which is solved with ReLU [21].

CNNs with this nonlinearity have been found to train quicker than others.

$$\sigma(x) = \frac{1}{1 + e^{-x}} \tag{9.2}$$

$$\tanh(x) = \frac{e^x - e^{-x}}{e^x + e^{-x}} \tag{9.3}$$

$$\mathrm{Re}\,\mathrm{LU}(x) = \max(0, x) \tag{9.4}$$

Recently, Maas et al. [22] presented another type of nonlinearity, called the leaky-ReLU. This nonlinearity is characterized as shown in Equation 9.5 [18].

$$\mathrm{leaky} - \mathrm{ReLU}(x) = \max(0, x) + \alpha \min(0, x) \tag{9.5}$$

where α is a predecided parameter. He et al. [23] enhanced this by recommending that the α parameter be additionally learned, prompting a considerably richer model.

9.2.1.4 Fully Connected Layers (FCLs)

The TNN was created by matrix multiplications rotated with nonlinearities such as sigmoid. These layers are currently called fully connected layers, where every unit in the previous layer is linked to each unit in the following layer. This is in contrast to convolutional layers, where there is only a local connection. In current networks, utilizing numerous FCLs is mainly avoided because it utilizes a great number of parameters [24]. One or more FCLs can be used, as shown in Figure 9.2.

9.3 OPTIMIZATION APPROACHES

A highly expressive network is of little use without a good learning algorithm to effectively learn the network's parameters. The generic CNN era endeavored to produce such effective algorithms. The tasks in computer vision show that a powerful model can be learned by a very simple supervised training procedure. Minimizing the loss function is the general approach for learning. The sigmoid cross-entropy or softmax loss functions are used in classification tasks, while those including regression utilize the Euclidean error function. In Figrue 9.2, the FC8 layer output classifies one of the thousand classes of the dataset.

9.3.1 Gradient-Based Optimization

The backpropagation algorithm is generally used to train neural networks [25], which speeds up the gradient descent (GD) computation by the chain rule. However, for large datasets with (at least thousands) images, utilizing GD is unfeasible. In such cases, the stochastic gradient descent (SGD) is regularly utilized, where one calculates gradients for individual images instead of the whole dataset. It has been discovered that training with SGD is superior to training with GD. The main disadvantage of SGD is that it converges slowly [26]. To overcome this disadvantage, SGD is commonly utilized with a minibatch, where the minibatch generally contains a few data points (≈ 100) instead of all data points.

Momentum [27] has a place with a group of techniques that aim to accelerate the SGD convergence. This technique is mainly used in deep network training and is regularly considered a basic component. Augmentations such as Adam [28], Nesterov's accelerated GD [29], Adagrad [30], and Adadelta [31] are known to work similarly well. For a point-by-point discussion on how these techniques function, more details can be found in [32].

9.3.2 Dropout Property

This occurs when the model network has more terms than necessary or uses more complicated techniques with numerous training parameters, especially with a small amount of training data or limited data, to combat this effective regularization technique. It has been observed that traditional regularizers such as $\ell 1$ or $\ell 2$ regularization on the weights of the neural network are inadequate in this view.

Another nice property of CNNs is named "dropout", which is a simple regularization technique to solve the overfitting problem. Dropout is a strong regularization method [33], and it has been shown to enhance generalization for extensive neural networks. In dropout, we arbitrarily drop neurons in training with a probability p. Therefore, only an irregular subset of neurons in SGD iterations are trained. Applying dropout properties is valuable only in the training stage. At test time, we utilize all neurons, and every neuron activation is multiplied by p to represent the scaling. Hinton et al. [33] demonstrated that this strategy was equivalent to training a large range of neural networks with shared parameters and then utilizing their geometry, intending to obtain a single expectation. Numerous augmentations to dropout, such as DropConnect [34] and Fast Dropout [35], have appeared to work better in specific cases. Maxout [36] is a nonlinearity that enhances the execution of a network that utilizes dropout.

9.3.3 Batch Normalization

Batch normalization (BN) [37] is another helpful regularizer that enhances the generalization and that also definitely accelerates convergence. This system handles the issue of inward covariate shift, where the circulation of each layer's data sources changes persistently amid the training phase. This is the result of variable weights for each layer, which results in a changing output activation distribution. This ability requires careful preparation and generally slows training. To overcome this issue, the layer output activations are normalized by BN to guarantee that its value is inside a small interval. In particular, minibatch normalization is performed with BN, which applies the average of the mean-variance statistics. Extra move and scale parameters are additionally learned to frustrate the normalization effect if necessary. Recently, BN has been observed to be a fundamental part of extra deep network training.

9.3.4 NETWORK DEPTH

The universal approximation hypothesis [38] states that one hidden layer is adequate to display any continuous function. However, it has been reported [6] that such systems require an exponentially extensive number of neurons compared to using many hidden layers. Although the inspiration for making deep neural networks was clear, for quite a while researchers did not have a method that could proficiently train deep neural networks (greater than 3 layers). Training much deeper networks became possible for researchers after the introduction of greedy layerwise pretraining [2], which made deep learning networks the mainstream in machine learning.

Recently, many deep neural networks have appeared; for example, AlexNet [8] has 7 layers. Recent models such as VGGnet [39] and GoogleNet [40] have 19 and 22 layers, respectively, and appeared to perform much better than AlexNet. The main disadvantage of the much deeper networks is that such networks are very hard to train. Srivastava et al. [41] used deep neural networks with 100 layers to present highway systems. He et al. [24] utilized an adaptation of this method with 152 layers to obtain cutting-edge results on the ILSVRC 2015 database.

9.4 OPERATIONAL PROPERTIES

9.4.1 USING PRETRAINED CNNS

The main advantage of the AlexNet network is that it can be used for different tasks for which it was not originally intended. It is very easy to download a pretrained model and then adjust the model to suit the current application. We explain two such approaches to utilize networks in this way [18].

9.4.1.1 Fine-Tuning

This process refers to modifying a pretrained network that is used for image classification for an alternate task that is performed by utilizing the trained weights as an initial condition and starting SGD again for the other task. The learning rate will be much lower than that for the original network. When the new task is the same as the original task, we can make the earlier layers fixed, and only the later semantic layers should be relearned. However, if the new task is completely different, we should either retrain all layers or train the network from scratch. The number of layers to retrain also depends on the quantity of information available for learning the new task. The larger the dataset is, the higher is the number of layers that can be retrained. The reader is referred to Yosinski et al. [42] for more information.

9.4.1.2 CNN Activations as Features

As mentioned previously, the later layers in AlexNet appear to learn visually semantic characteristics. These representations are essential in making 1,000-class classifications. Because these demonstrate a large number of classes, the FC7 layer can be used as a global feature identifier. The provided features are better than the traditional features, such as SIFT or HOG, for different computer vision tasks. Donahue et al. [43] first presented the use of CNNs as features and performed tests to decide their proportionality for different errands. Fully connected layer activations are additionally utilized for image recovery [44]. Indeed, these are used as generic features and can be utilized in many tasks [45].

9.4.2 DEEP LEARNING DIFFICULTIES

9.4.2.1 Overfitting Reduction

Training a deep learning model with a limited number of training images is a challenging task due to the number of learnable parameters. Various strategies have been developed by researchers, including the following:

1. Applying a nonlinear activation function such as ReLU [46].
2. Using a good initialization and increasing the momentum slowly [47].
3. Applying dropout to randomly deactivate connections on each iteration [48].
4. Performing normalization by using batch normalization layers [37].
5. Using denoising autoencoder layers to reconstruct the corrupted inputs [49].

9.5 COMMONLY USED DEEP NEURAL NETWORKS

Deep learning uses considerable amounts of computational energy and acquires advanced results in multiple difficult tasks. Among the various deep learning strategies, methods based on convolutional neural networks (CNNs) are more appropriate for tasks related to images. The main goal of this section is to provide an overview of the commonly used CNN methods, such as AlexNet, VGG-19, deep residual networks, and GoogleNet, as reported in Table 9.1.

TABLE 9.1

Comparison among Well-Known Convolutional Neural Networks Using ImageNet Dataset

Year	CNN	No. of layers	No. of parameters	Error rate
2012	AlexNet [8]	7	60 million	15.3%
2014	GoogleNet [40]	22	4 million	6.67%
2014	VGGNet [39]	19	138 million	7.3%
2015	ResNet [24]	152	50 million	3.6%

9.5.1 ALEXNET

AlexNet has been discussed previously. Figure 9.2 shows max-pooling grouped with layers 1, 2, and 5, while the last two fully connected layers are combined with drop-out because they contain the most parameters. Local response normalization has been provided to layers 1 and 2, which does not affect the network performance as in [50]. The ILSVRC 2012 database, which has 1.2 million images for 1,000 classes, is used to train the network. Two GPUs are used to train the network for one month. With more powerful GPUs, the network can be trained in a few days [50]. Learning algorithms for hyperparameters such as dropout, weight decay, momentum, and learning rate were manually tuned. Blob-like features and Gabor-like oriented edges are learned in the prior layers. The highest-order features, such as shapes, are learned by the next layers. Semantic attributes such as wheels or eyes are learned by the last layers.

9.5.2 GOOGLENET

For GoogleNet [40], the best method at ILSVRC-2014; the classification performance is significantly improved with very deep networks. The number of parameters has been increased due to the large number of layers, and a number of design tricks are used, such as using a 1×1 convolutional layer after the traditional convolutional layer, which reduces the number of parameters and increases the expressive power of the CNN. It is believed that having at least one 1×1 convolutional layer is much the same as having a multilayer perceptron network preparing the convolutional layer output that goes before it. Another trick used by the authors is to involve the internal layers of the network in the calculation of the target function rather than the final softmax layer (as in AlexNet).

9.5.3 VGG-19

The VGG-19 [39] network design is one of the highest-performing deep CNN (DCCN) networks. The VGG design provides a fascinating feature in that it divides the large convolutional filters into small ones. The small filters are carefully selected with a number of parameters approximately the same as the parameters in the larger convolutional filters, which they should replace. The net impact of this design choice is productivity and regularization-like impact on parameters because of the smaller size of the included filters.

9.5.4 DEEP RESIDUAL NETWORKS (RESNET)

Deep residual networks (ResNet) [24] increase the number of trained layers to the highest depth with 152–200 layers. The network used residual connections between layers that were used to train network layers to adapt the residual map instead of adapting to the underlying map. In other words, the layers did not learn a mapping of $h(x)$ directly, but it will learn $f(x) = h(x)-x$ and then add x to be $f(x)+x$. This straightforward method was observed to be helpful for learning networks with several layers, providing sophisticated accuracy in many tasks.

TABLE 9.2

Software List for Convolutional Neural Networks

Package	Interfaces	AutoDiff
Theano	Python	Yes
MatConvNet	MATLAB	No
Chainer	Python	No
Caffe	Python, MATLAB, C++, command line	No
Torch	Torch, C	Yes
TensorFlow	C++, Python	Yes
CNTK	C++, command line	No

9.5.5 TOOLS AND SOFTWARE

In this section, we will discuss different software programs that are available for deep neural network learning. Table 9.2 includes the commonly used packages. Caffe is the most commonly utilized software among computer vision applications. It is composed with C++ and is frequently utilized through the Python and command line interfaces. The structure of this system makes it simple to utilize pretrained neural systems for different tasks. In general, Theano, Torch, and TensorFlow appear to be more prominent. These structures are intended to make it simple to alter low-level subtle elements of neural system models. Theano and TensorFlow are utilized in Python, while Torch is utilized in Lua. Likewise, a couple of wrapper libraries exist for Theano, for example, Lasagne [51] and Keras [52], which provide normally utilized functionalities and some help for pretrained models.

In this specific situation, automatic differentiation (AutoDiff) [53] is an essential feature to have in such a structure. When determining new layers and regularizers for neural systems, one also needs to indicate what the gradient through that layer would resemble. AutoDiff calculates the derivative of complex functions automatically in a closed shape using derivative definitions of primitives.

9.6 CONVOLUTIONAL NEURAL NETWORK TYPES

9.6.1 REGION-BASED CONVOLUTIONAL NEURAL NETWORKS (R-CNN)

Most CNN networks are trained to classify images using images that include one object. During testing, most of the images include multiple objects, but the CNN network can recognize a single object. The main problem with CNN network design is the need to cover more tasks than image classification, such as localization and detection. The major aim of object detection is to detect different objects in each image of various sizes. Meanwhile, a group of object-like regions is detected in a specific image, named region proposals, which have gained much attention [54]. Region proposal methods work at a low level and produce hundreds of objects such as image patches at multiple levels. To appoint a network classification to cover the task of localizing objects, the network must search for image regions with different scales simultaneously.

Girshick et al. [55] used region proposals to solve the problem of object localization. The method obtained the best results on the PASCAL VOC 2012 detection dataset. It is named region-based CNN (R-CNN) because it uses regions of the images followed by a CNN. Different works have used the R-CNN method to extract features in small regions to solve many target applications in computer vision.

9.6.2 FULLY CONVOLUTIONAL NETWORKS (FCNs)

The achievements of convolutional neural networks in object detection [55] and classification tasks [8, 40] have motivated specialists to utilize deep neural networks for more difficult tasks, such as scene parsing and semantic segmentation. Scene parsing and semantic segmentation assign a label for each pixel in the image to which it belongs (e.g., table, sofa, road, and so forth), which is called per-pixel classification. The region features are extracted to a support vector machine (SVM) to classify them into classes.

Long et al. [56] provided a fully convolutional network (FCN) structure to learn per-pixel tasks, such as semantic segmentation, in a comprehensive manner. FCN layers perform a location-invariant activity. The FCN network does not have any inner product or densely connected layers, as in CNN networks that are used for classification. The classification CNN networks have a restriction on input size because the inner product layers have a limitation on the input size. However, FCN does not include any of the inner product layers; it can basically work with any input size. Long et al. [56] introduced a deconvolution layer that obtains the spatial resolution back to the upsampling operation from the output of the sub-sampled image. They performed the upsampling operation at intermediate layers, which are concatenated to acquire features at the pixel level.

9.6.3 HYBRID LEARNING NETWORKS

Multitask learning is basically a machine learning model, wherein the goal is to prepare the learning network to perform well on numerous tasks. Multitask learning systems will generally explore shared portrayals that exist among the functions to obtain a superior generalization execution over their counterparts created for one task. In CNNs, multitask learning is acknowledged utilizing distinctive methodologies. One class of methodologies uses a multitask loss function with hyperparameters normally controlling the losses. For instance, Girshik et al. [57] utilize a multitask loss to train the network for bounding-box and classification regression functions along these lines, enhancing execution for object recognition.

9.7 DEEP LEARNING APPLICATIONS IN MEDICAL IMAGING

Over the past decades, we have seen the significance of medical imaging, e.g., magnetic resonance (MR), mammography, computed tomography (CT), X-ray, ultrasound, positron emission tomography (PET), and so forth, for the early treatment, diagnosis, and detection of diseases [58]. In medical centers, the interpretation of medical images has generally been performed by human specialists, for example, radiologists and doctors. Because of vast varieties in pathology and the potential exhaustion of human specialists, the use of computer-aided diagnosis systems has increased. As medical imaging technologies improve, medical image analysis will improve with the support of machine learning methods. Feature representation or meaningful feature extraction is important for machine learning prosperity to achieve the target tasks. Specialists, including radiologists and physicians, use their knowledge about the object domains in designing task-related or meaningful features, making it difficult for non-specialists to exploit automated learning systems for their own examinations [59].

However, deep learning has mitigated these obstacles by retaining the feature extraction step in a learning step [60]. Deep learning requires only the input dataset. Thus, feature extraction is performed internally in deep learning networks, which enables non-specialists in machine learning to adequately utilize deep learning for their own applications. Deep learning strategies are very successful due to the following reasons:

1. Developments in learning algorithms.
2. Advancements in graphics processing units (GPUs) and central processing units (CPUs).
3. Dataset availability.

Generally, deep learning strategies are very successful when the available number of image samples in the training stage is very large. For instance, 1 million images were given in the ILSVRC competition. However, the medical field has a limited number of samples in the dataset, less than 1,000 samples. This increases the challenge of developing a deep learning model with a low number of training samples without experiencing overfitting. Various strategies have been developed by researchers [59], such as the following:

1. Reducing the dimension of the input by using 2D or 3D patches rather than the full images.
2. Using data augmentation to train the network from scratch.
3. Using a pretrained model as a feature extractor and retraining it with the target task.
4. Using pretrained model parameters and then fine-tuning them with the related medical images.
5. Transforming the fully connected layer weights of a pretrained model into convolutional kernels.

There are several medical imaging applications that depend on the type of medical images. Medical imaging applications can be categorized based on the human parts (such as brain tumor, heart, organ/body part, cell, pulmonary nodules, lymph nodes, interstitial lung disease, cerebral microbleeds, and noticeably sclerosis lesion), medical image type (MRI, CT, X-ray, ultrasound, and PET), and application type (detection, segmentation, and classification). In this section, we present the deep learning methods used for feature representation, structure detection, and segmentation.

9.7.1 Deep Learning for Feature Extraction

The methods for medical image processing mainly characterize the local anatomical characteristics based on representations of the morphological features designed by experts. Moreover, the represented features of an image for a specific problem are hardly ever reusable for other types. For instance, the methods of brain MR image segmentation designed for 1.5-Tesla T1-weighted images are not viable for 7.0-Tesla T1-weighted or T2-weighted images, or for different imaging modalities or distinct organs [61, 62]. Supervised learning methods require large amounts of training data that are manually labeled by experts to find the most essential and relevant features for specific tasks, while the extracted features may deviate due to the complexity of the anatomical structures. Therefore, a change in image features makes the training process start again.

Wu et al. [61, 62] presented a general framework for feature extraction that detects the essential characteristics for the accurate segmentation and detection of brain regions and anatomical structures. Furthermore, this framework can be used for various types of medical images. A stacked autoencoder (SAE) is used with a sparsity constraint to learn extracted features hierarchically from layer to layer, which contains hierarchical encoding and decoding systems. A nonlinear deterministic mapping is used to map the input to an activation vector y. This procedure is repeated by using the vector y as input for the next layer until the high-level feature presentations are obtained. The decoding system is used to minimize the reconstruction errors between the reconstructed patch from the decoder and the input patch.

From the application mentioned above, we found that

1. Feature extraction by deep learning represents the local characteristics of the image very well.
2. Deep learning can learn the intrinsic features of new medical imaging modalities for new image analysis methods.
3. It is fully adaptive to understand the data of the image and apply to different medical imaging applications, such as prostate localization [63, 64], hippocampus segmentation [65], and brain tumor detection [17].

9.7.2 Deep Learning for Tumor Detection

Automatic detection aims to classify medical images as abnormal and normal due to the presence or absence of abnormal or suspicious regions (tumors). Computer-aided detection (CAD) is very important for medical image diagnosis. The main goal is to maximize the detection rate. Typically, the traditional pipeline of CAD is as follows:

1. Image processing techniques are applied to the input images.
2. A set of features presents the preprocessed images, such as statistical or morphological information.

3. A classifier such as support vector machine (SVM) is used to make a decision. Different detection applications are applied to different medical image modalities to detect or segment suspicious or abnormal regions, such as organ or body part detection (liver, neck, pelvis, lungs, and legs), cell detection, and brain tumor detection [66, 67].

Recently, deep learning models have improved detection performance in different medical imaging applications, such as lymph node detection [68], pulmonary nodule detection [69], interstitial lung disease classification [70], multiple sclerosis lesion detections [71], cerebral microbleed detection [72], brain tumor detection [17, 73], and cell detection [74].

9.7.3 DEEP LEARNING FOR TUMOR SEGMENTATION

Automatic segmentation is used for quantitative assessment of the brain at all ages. Removing non-brain regions such as the skull is the main preprocessing step. Kleesiek et al. [75] used 3D convolutional neural networks for skull extraction in T1-weighted MRI images. Moeskops et al. [76] presented an accurate tissue segmentation method. They used a multiscale CNN with multiple convolution sizes and multiple patch sizes, which provided very good segmentation results with an average Dice ratio from 0.82 to 0.91. The brain MRI segmentation of infants is more difficult than that of adults due to severe partial volume effects, reduced tissue contrast, ongoing white matter (WM) myelination, and increased noise in the infant's brain [77].

Zhang et al. demonstrated multimodality MR segmentation using four CNN architectures for infant brain tissues. Three convolutional layers were applied, followed by a fully connected layer and a softmax layer for tissue classification [78]. More recently, in [79], the isointense-phase brain image was segmented by multiple fully convolutional networks (MFCNs); the average Dice ratios were 0.873 for gray matter (GM), 0.852 for cerebrospinal fluid (CSF), and 0.887 for WM. Moreover, several methods for brain tumor segmentation can be found in the literature [73, 80, 81, 82, 83, 84, 85].

9.8 BRAIN TUMOR DETECTION USING DEEP CONVOLUTIONAL NEURAL NETWORK

The combination of machine learning approaches and brain magnetic resonance imaging (MRI) is crucial for the brain tumor diagnosis process by clinicians and radiologists. Although most tumor diagnosis studies have focused on segmentation and localization operations, few studies have focused on tumor detection as a time- and effort-saving operation in which normal images can be discarded early. This study introduces a new network structure for accurate brain tumor detection using a deep convolutional neural network (DCNN). The proposed structure is designed to identify the presence or absence of a brain tumor in MRI images.

9.8.1 DCNN Design

A CNN architecture is proposed for brain tumor detection using MRI images. A block diagram of the proposed architecture is shown in Figure 9.3 The sequence of processes occurs through the proposed structure as follows: brain MRI images are fed to the network's input, and then preprocessing is performed to reduce the calculation complexity and hence reduce the processing time. Because the database used contains 2D slices, the images are resized to 96×96 pixels because of the varying MRI image sizes.

Later, the network, which includes convolutional layers and fully connected layers, is used to classify the input image as a normal or abnormal (contains tumor) image. The convolutional layers are used to extract features in the feature extraction path. Moreover, it contains 7 convolutional layers with a ReLU activation function and 7 max-pooling layers. The convolutional layers use a small filter size of 5×5 pixels to provide features in the images. The max-pooling layer is used after each convolutional layer to downsample the convolutional layer output.

Furthermore, the output path contains a BN layer connected to a ReLU layer and a fully connected layer that is followed by a dropout layer. Dropout with different values was tested, and we found that a ratio of 0.8 performed with slightly better outcomes. At the final stage, another fully connected layer with the softmax function is used for the image classification process.

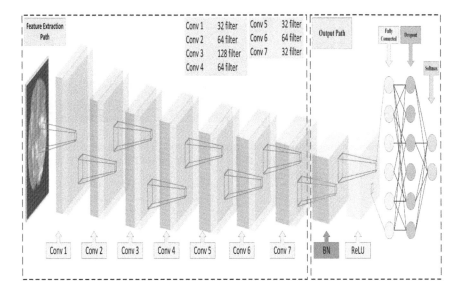

FIGURE 9.3   A block diagram of the proposed DCNN architecture for brain tumor detection.

9.8.2 Simulation Experiments and Evaluation Criteria

9.8.2.1 Environment Setup

We extracted the dataset from the Reference Image Database to Evaluate Response (RIDER) [86]. The dataset consists of 109 normal images and 240 abnormal images. We show a sample image of the dataset used in the experimental work in Figure 9.4. Table 9.3 provides the specifications of the training and testing subsets. We developed the proposed DCNN code in a Jupyter notebook using Keras and TensorFlow toolkits. The machine is equipped with an Intel Core i7 processor that works at 3.2 GHz with 24 GB of RAM and the Ubuntu desktop 64-bit experimental environment.

9.8.2.2 Evaluation Criteria

Several metrics can test classification algorithms. These metrics can be categorized as scalar metrics, such as accuracy (ACC), geometric mean (GM), sensitivity (or recall), specificity (or inverse recall), and positive prediction value (PPV) (precision), and graphical metrics, such as detection error trade-off (DET), precision-recall, and receiver operating characteristics (ROC) curves. We present most of these metrics in this section with details.

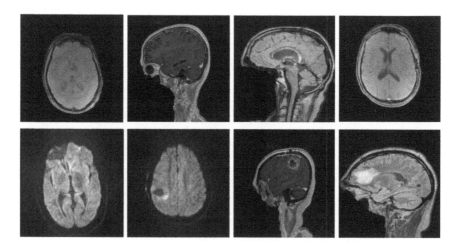

FIGURE 9.4 Sample images used for the experimental work. The image dataset is extracted from the RIDER database. The top row shows normal brain MRI images, while abnormal brain images are shown in the bottom row.

TABLE 9.3
Image Distribution in the Performance Evaluation Dataset

	Image type	No. of training sets	No. of testing sets
RIDER	Normal	45	64
	Abnormal	77	163

The accuracy (ACC) is a widely used metric for the classification performance, which is the ratio between the correctly classified samples to the total number of samples, as shown in Equation 9.6.

$$ACC = \frac{TP + TN}{TP + TN + FP + FN}. \tag{9.6}$$

where TP, TN, FP, and FN indicate the true positive, true negative, false positive, and false negative, respectively [66]

Sensitivity (SV), recall, hit rate, and true positive rate (TPR) represent the true positive samples to the total number of positive samples, as shown in Equation 9.7. Specificity (SP), inverse recall, and true negative rate (TNR) express the correctly classified negative samples to the total number of negative samples, as shown in Equation 9.8 [67].

$$SV = \frac{TP}{TP + FN}. \tag{9.7}$$

$$SP = \frac{TN}{TN + FP}. \tag{9.8}$$

The precision or positive prediction value (PPV) reflects the ratio of correctly classified positive samples to all classified positive samples, as shown in Equation 9.9. Negative predictive value (NPV) or inverse precision represents the ratio of correctly classified negative samples to the entire classified negative samples, as shown in Equation 9.10. Equation 9.11 refers to the balanced classification rate (BCR) or balanced accuracy (BA), which combines the sensitivity and specificity metrics [17].

$$Precision = PPV = \frac{TP}{TP + FP}. \tag{9.9}$$

$$NPV = \frac{TN}{TN + FN}. \tag{9.10}$$

$$BCR = BA = \frac{SV + SP}{2} = \frac{1}{2}\left(\frac{TP}{TP + FN} + \frac{TN}{TN + FP}\right). \tag{9.11}$$

The false positive rate (FPR) represents the ratio between the false positive classified samples to all negative samples, as shown in Equation 9.12. However, the false negative rate (FNR) expresses the false negative classified samples to all positive samples, as shown in Equation 9.13.

$$FPR = \frac{FP}{TN + FP}. \tag{9.12}$$

$$FNR = \frac{FN}{TP + FN}. \tag{9.13}$$

In addition to the above metrics, there are many different metrics that can be considered for evaluating various applications. The new metrics include Youden's index (YI) or Bookmaker informedness (BM), Matthews correlation coefficient (MCC), discriminant power (DP), F-measure, markedness (MK), geometric mean (GM), Jaccard, and optimization precision (OP), as shown in Equations 9.14, 9.15, 9.16, 9.17, 9.18, 9.19, 9.20, and 9.21, respectively [87].

$$YI = SV + SP - 1. \tag{9.14}$$

$$MCC = \frac{TP \times TN - FP \times FN}{\sqrt{(TP + FP)(TP + FN)(TN + FP)(TN + FN)}}. \tag{9.15}$$

$$DP = \frac{\sqrt{3}}{\pi} \left(\log\left(\frac{SV}{1 - SP} \right) + \log\left(\frac{SP}{1 - SV} \right) \right). \tag{9.16}$$

$$F\text{-measure} = \frac{2PPV \times SV}{PPV + SV}. \tag{9.17}$$

$$MK = PPV + NPV - 1. \tag{9.18}$$

$$GM = \sqrt{SV + SP}. \tag{9.19}$$

$$Jaccard = \frac{TP}{TP + FP + FN}. \tag{9.20}$$

$$OP = ACC - \frac{|SV - SP|}{SV + SP}. \tag{9.21}$$

9.8.2.3 Experimental Results

To test the proposed system, standard evaluation metrics such as accuracy, sensitivity, and specificity have been considered [17, 66]. Multiple tests were run to confirm the performance of the proposed network in fulfilling the tumor detection task. A comparison with other methods found in the literature is presented in Table 9.4. The empirical works show that the proposed DCNN structure can achieve better results in terms of accuracy, sensitivity, and specificity compared to other approaches in the literature. Although the network training time was not considered, the average testing time per MRI image was measured to be 0.24 seconds.

TABLE 9.4

Comparison of the Proposed Brain Tumor Detection Model against Some Approaches in the Literature

Methods	Performance evaluation criteria (%)		
	Sensitivity	Specificity	Accuracy
Abd-Ellah et al. [66]	83.43	25.00	66.96
Mohsen et al. [88]	97	97	96.97
Proposed DCNN	**98.43**	**97.54**	**97.79**

9.9 CONCLUSION

The major advantage of deep learning over the traditional machine learning techniques is that it can freely reveal significant features in a high-dimensional database. Recently, CNNs have led to great developments in processing text, speech, images, videos, and other applications. In this chapter, the basic understanding of CNNs and the recent enhancements of CNNs were explained and presented. CNN improvements have been discussed from various prospects, specifically the general model of CNN, layer design, loss function, activation function, optimization, regularization, normalization, fast computation, and network depth, reviewing the benefits of each phase of CNN. Additionally, the use of pretrained CNNs and deep learning difficulties have been discussed. This chapter also introduced the commonly used deep convolutional neural networks, namely, AlexNet, GoogleNet, VGG-19, and ResNet. It discussed the commonly used programs for deep CNN learning. The chapter focused on different types of CNNs, which are region-based CNNs, fully convolutional networks, and hybrid learning networks.

This chapter outlined the main deep learning applications in medical imaging processing in terms of feature extraction, tumor detection, and tumor segmentation. A deep convolutional neural network (DCNN) structure was proposed for brain tumor detection from MRI images. The proposed DCNN was evaluated using the RIDER database using 122 and 227 images for training and testing, respectively. It achieved accurate detection within a time of 0.24 seconds per image in the testing phase. This chapter promises to provide comprehensive knowledge to researchers, learners, and those who are interested in this field.

BIBLIOGRAPHY

1. Marra, F., Poggi, G., Sansone, C., Verdoliva, L.: A deep learning approach for iris sensor model identification. *Pattern Recognition Letters* 113, 4653 (2018), integrating Biometrics and Forensics.
2. Hinton, G.E., Osindero, S., Teh, Y.W.: A fast learning algorithm for deep belief Nets. *Neural Computation* 18(7), 1527–1554 (2006), https://doi.org/10.1162/neco.2006.18.7.1 527, pMID: 16764513.

3. Vincent, P., Larochelle, H., Lajoie, I., Bengio, Y., Manzagol, P.A.: Stacked denoising autoencoders: Learning useful representations in a deep network with a local denoising criterion. *Journal of Machine Learning Research* 11, 3371–3408 (Dec 2010).
4. Hinton, G.E.: Training products of experts by minimizing contrastive divergence. *Neural Computation* 14(8), 1771–1800 (2002), https://doi.org/10.1162/089976602760128018.
5. Mosavi, A., Varkonyi-Koczy, A.R.: Integration of machine learning and optimization for robot learning. In: Jabłoński, R., Szewczyk, R. (eds.) *Recent Global Research and Education: Technological Challenges*, pp. 349–355. Springer International Publishing, Cham (2017).
6. Bengio, Y.: Learning deep architectures for AI. *Foundations and Trends in Machine Learning* 2(1), 1–127 (2009), http://dx.doi.org/10.1561/2200000006.
7. Russakovsky, O., Deng, J., Su, H., Krause, J., Satheesh, S., Ma, S., Huang, Z., Karpathy, A., Khosla, A., Bernstein, M., Berg, A.C., Fei-Fei, L.: ImageNet large scale visual recognition challenge. *International Journal of Computer Vision (IJCV)* 115(3), 211–252 (2015), https://doi.org/10.1007/s11263-015-0816-y.
8. Krizhevsky, A., Sutskever, I., Hinton, G.E.: ImageNet classification with deep convolutional neural networks. In: Pereira, F., Burges, C.J.C., Bottou, L., Weinberger, K.Q. (eds.) *Advances in Neural Information Processing Systems* 25, (NIPS 2012), pp. 1097–1105. Curran Associates, Inc. (2012).
9. Hassaballah, M., Awad, A.I.: Detection and description of image features: An introduction. In: Awad, A.I., Hassaballah, M. (eds.) *Image Feature Detectors and Descriptors: Foundations and Applications, Studies in Computational Intelligence*, Vol. 630, pp. 1–8. Springer International Publishing, Cham (2016).
10. Lowe, D.G.: Distinctive image features from scale-invariant key-points. *International Journal of Computer Vision* 60(2), 91–110 (2004), https://doi.Org/10.1023/B:VISI.0000029664.99615.94.
11. Dalal, N., Triggs, B.: Histograms of oriented gradients for human detection. In: *2005 IEEE Computer Society Conference on Computer Vision and Pattern Recognition (CVPR'05)*. Vol. 1, pp. 886–893. IEEE (2005), https://doi.org/10.1109/CVPR.2005.177.
12. Yang, J., Jiang, Y.G., Hauptmann, A.G., Ngo, C.W.: Evaluating bag-of-visual-words representations in scene classification. In: *Proceedings of the International Workshop on Workshop on Multimedia Information Retrieval*. pp. 197–206. ACM, New York, NY, USA (2007), https://doi.org/10.1145/1290082.1290111.
13. Awad, A.I., Hassaballah, M.: *Image Feature Detectors and Descriptors: Foundations and Applications, Studies in Computational Intelligence*, Vol. 630. Springer International Publishing, Cham, 1st edn. (2016).
14. Lecun, Y., Bottou, L., Bengio, Y., Haffner, P.: Gradient-based learning applied to document recognition. *Proceedings of the IEEE* 86(11), 2278–2324 (1998), https://doi.org/10.1109/5-726791.
15. Bishop, C.: *Pattern Recognition and Machine Learning*. Springer-Verlag New York, 1 edn. (2006).
16. Hubel, D.H., Wiesel, T.N.: Receptive fields, binocular interaction and functional architecture in the cat's visual cortex. *The Journal of physiology* 160(1), 106–154 (1962).
17. Abd-Ellah, M.K., Awad, A.I., Khalaf, A.A.M., Hamed, H.F.A.: Two-phase multi-model automatic brain tumour diagnosis system from magnetic resonance images using convolutional neural networks. *EURASIP Journal on Image and Video Processing* 97(1), 1–10 (2018).
18. Srinivas, S., Sarvadevabhatla, R.K., Mopuri, K.R., Prabhu, N., Kruthiventi, S.S., Babu, R.V.: Chapter 2—An introduction to deep convolutional neural nets for computer vision. In: Zhou, S.K., Greenspan, H., Shen, D. (eds.) *Deep Learning for Medical Image Analysis*, pp. 25–52. Academic Press (2017).

19. Zeiler, M., Fergus, R.: Stochastic pooling for regularization of deep convolutional neural networks. In: *Proceedings of the International Conference on Learning Representation (ICLR)*. pp. 1–9 (2013).
20. Srivastava, R.K., Masci, J., Kazerounian, S., Gomez, F., Schmidhuber, J.: Compete to compute. In: Burges, C.J.C., Bottou, L., Welling, M., Ghahramani, Z., Weinberger, K.Q. (eds.) *Advances in Neural Information Processing Systems* 26, (NIPS 2013), pp. 2310–2318. Curran Associates, Inc. (2013).
21. Albawi, S., Mohammed, T.A., Al-Zawi, S.: Understanding of a convolutional neural network. In: *2017 International Conference on Engineering and Technology (ICET)*. pp. 1–6 (Aug 2017), https://doi.org/10.1109/ICEngTechnol.2017.8308186.
22. Maas, A.L.: Rectifier nonlinearities improve neural network acoustic models. In: Dasgupta, S., McAllester, D. (eds.) *Proceedings of the 30th International Conference on Machine Learning. Proceedings of Machine Learning Research*, Vol. 28. PMLR, Atlanta, Georgia, USA (17–19 June 2013).
23. He, K., Zhang, X., Ren, S., Sun, J.: Delving deep into rectifiers: Surpassing human-level performance on ImageNet classification. In: *2015 IEEE International Conference on Computer Vision (ICCV)*. pp. 1026–1034. IEEE, Santiago, Chile (2015), https://doi.org/10.1109/ICCV.2015.123.
24. He, K., Zhang, X., Ren, S., Sun, J.: Deep residual learning for image recognition. In: *2016 IEEE Conference on Computer Vision and Pattern Recognition (CVPR)*. pp. 770–778. IEEE, Las Vegas, NV, USA (June 2016), https://doi.org/10.1109/CVPR.2016.90.
25. Rumelhart, D.E., Hinton, G.E., Williams, R.J.: Learning representations by back-propagating errors. *Nature* 323, 333–336 (1986), https://doi.org/10.1038/323533a0.
26. Bottou, L.: Large-scale machine learning with stochastic gradient descent. In: Lechevallier, Y., Saporta, G. (eds.) *Proceedings of COMP-STAT'2010*. pp. 177–186. Physica-Verlag HD, Heidelberg, Paris, France (2010), https://doi.org/10.1007/978–3-7908–2604–3_16.
27. B.T.Polyak: Some methods of speeding up the convergence of iteration methods. *USSR Computational Mathematics and Mathematical Physics* 4(5), 1–17 (1964), https://doi.org/10.1016/0041–5553(64)90137–5.
28. Kingma, D.P., Ba, J.: Adam: A method for stochastic optimization. *CoRR* abs/1412.6980 (2014), http://arxiv.org/abs/1412.6980.
29. Nesterov, Y.: A method of solving a convex programming problem with convergence rate $O(1/sqr(k))$. *Soviet Mathematics Doklady* 27(2), 372–376 (1983).
30. Duchi, J., Hazan, E., Singer, Y.: Adaptive subgradient methods for online learning and stochastic optimization. *Journal of Machine Learning Research* 12, 2121–2159 (2011).
31. Zeiler, M.D.: ADADELTA: An adaptive learning rate method. *CoRR* abs/1212.5701 (2012), http://arxiv.org/abs/1212.5701.
32. Sutskever, I., Martens, J., Dahl, G., Hinton, G.: On the importance of initialization and momentum in deep learning. In: Dasgupta, S., McAllester, D. (eds.) *Proceedings of the 30th International Conference on Machine Learning. Proceedings of Machine Learning Research*, Vol. 28, pp. 1139–1147. PMLR, Atlanta, Georgia, USA (17–19 June 2013).
33. Hinton, G.E., Srivastava, N., Krizhevsky, A., Sutskever, I., Salakhutdinov, R.: Improving neural networks by preventing co-adaptation of feature detectors. *CoRR* abs/1207.0580 (2012), http://arxiv.org/abs/1207.0580.
34. Wan, L., Zeiler, M., Zhang, S., Cun, Y.L., Fergus, R.: Regularization of neural networks using DropConnect. In: Dasgupta, S., McAllester, D. (eds.) *Proceedings of the 30th International Conference on Machine Learning. Proceedings of Machine Learning Research*, Vol. 28, pp. 1058–1066. PMLR, Atlanta, Georgia, USA (17–19 June 2013).
35. Wang, S., Manning, C.: Fast dropout training. In: Dasgupta, S., McAllester, D. (eds.) *Proceedings of the 30th International Conference on Machine Learning. Proceedings of Machine Learning Research*, Vol. 28, pp. 118–126. PMLR, Atlanta, Georgia, USA (17–19 June 2013).

36. Goodfellow, I., Warde-Farley, D., Mirza, M., Courville, A., Bengio, Y.: Maxout networks. In: Dasgupta, S., McAllester, D. (eds.) *Proceedings of the 30th International Conference on Machine Learning. Proceedings of Machine Learning Research*, Vol. 28, pp. 1319–1327. PMLR, Atlanta, Georgia, USA (17–19 June 2013).

37. Ioffe, S., Szegedy, C.: Batch normalization: Accelerating deep network training by reducing internal covariate shift. In: *Proceedings of the 32nd International Conference on International Conference on Machine Learning*, Vol. 37. pp. 448–456. ICML'15, JMLR.org (2015).

38. Hornik, K.: Approximation capabilities of multilayer feedforward networks. *Neural Networks* 4(2), 251–257 (1991), https://doi.org/10.1016/0893–6080(91)90009-T.

39. Simonyan, K., Zisserman, A.: Very deep convolutional networks for large-scale image recognition. *CoRR* abs/1409.1556 (2014), http://arxiv.org/abs/1409.1556.

40. Szegedy, C., Liu, W., Jia, Y., Sermanet, P., Reed, S., Anguelov, D., Erhan, D., Vanhoucke, V., Rabinovich, A.: Going deeper with convolutions. In: *2015 IEEE Conference on Computer Vision and Pattern Recognition (CVPR)*. pp. 1–9. Boston, MA, USA (2015), https://doi.org/10.1109/CVPR.2015.7298594.

41. Srivastava, R.K., Greff, K., Schmidhuber, J.: Training very deep networks. In: *Proceedings of the 28th International Conference on Neural Information Processing Systems*, Vol. 2. pp. 2377–2385. NIPS'15, MIT Press, Montreal, Canada (2015).

42. Yosinski, J., Clune, J., Bengio, Y., Lipson, H.: How transferable are features in deep neural networks? In: Ghahramani, Z., Welling, M., Cortes, C., Lawrence, N.D., Weinberger, K.Q. (eds.) *Advances in Neural Information Processing Systems* 27, pp. 3320–3328. Curran Associates, Inc. (2014).

43. Donahue, J., Jia, Y., Vinyals, O., Hoffman, J., Zhang, N., Tzeng, E., Darrell, T.: DeCAF: A deep convolutional activation feature for generic visual recognition. In: Xing, E.P., Jebara, T. (eds.) *Proceedings of the 31st International Conference on Machine Learning. Proceedings of Machine Learning Research*, Vol. 32, pp. 647–655. PMLR, Bejing, China (22–24 June 2014).

44. Babenko, A., Slesarev, A., Chigorin, A., Lempitsky, V.: Neural codes for image retrieval. In: Fleet, D., Pajdla, T., Schiele, B., Tuytelaars, T. (eds.) *Computer Vision— ECCV 2014*. pp. 584–599. Springer International Publishing, Cham (2014), https://doi.org/10.1007/978–3-319–10590–1_38.

45. Razavian, A.S., Azizpour, H., Sullivan, J., Carlsson, S.: CNN features off-the-shelf: An astounding baseline for recognition. In: *2014 IEEE Conference on Computer Vision and Pattern Recognition Workshops*. pp. 512–519 (June 2014), https://doi.org/10.1109/CVPRW.2014.131.

46. Glorot, X., Bordes, A., Bengio, Y.: Deep sparse rectifier neural networks. In: Gordon, G., Dunson, D., Dudík, M. (eds.) *Proceedings of the Fourteenth International Conference on Artificial Intelligence and Statistics. Proceedings of Machine Learning Research*, Vol. 15, pp. 315–323. PMLR, Fort Lauderdale, FL, USA (11–13 Apr 2011).

47. Glorot, X., Bengio, Y.: Understanding the difficulty of training deep feedforward neural networks. In: Teh, Y.W., Titterington, M. (eds.) *Proceedings of the Thirteenth International Conference on Artificial Intelligence and Statistics. Proceedings of Machine Learning Research*, Vol. 9, pp. 249–256. PMLR, Chia Laguna Resort, Sardinia, Italy (13–15 May 2010).

48. Srivastava, N., Hinton, G., Krizhevsky, A., Sutskever, I., Salakhutdinov, R.: Dropout: A simple way to prevent neural networks from overfitting. *Journal of Machine Learning Research* 15, 1929–1958 (2014).

49. Vincent, P., Larochelle, H., Lajoie, I., Bengio, Y., Manzagol, P.A.: Stacked denoising autoencoders: Learning useful representations in a deep network with a local denoising criterion. *Journal of Machine Learning Research* 11, 3371–3408 (2010).

50. Chatfield, K., Simonyan, K., Vedaldi, A., Zisserman, A.: Return of the devil in the details: Delving deep into convolutional Nets. In: Valstar, M., French, A., Pridmore, T. (eds.) *Proceedings of the British Machine Vision Conference*. BMVA Press (2014), http://dx.doi.org/10.5244/C.28.6.

51. Lasagne: https://lasagne.readthedocs.io/en/latest/, Accessed: September 01, 2019.

52. Keras: https://keras.io/, Accessed: September 01, 2019.

53. Neidinger, R.: Introduction to automatic differentiation and MAT-LAB object-oriented programming. *SIAM Review* 52(3), 545–563 (2010), https://doi.org/10.1137/080743627.

54. Hosang, J., Benenson, R., Dollar, P., Schiele, B.: What makes for effective detection proposals? *IEEE Transactions on Pattern Analysis and Machine Intelligence* 38(4), 814–830 (April 2016), https://doi.org/10.1109/TPAMI.2015.2465908.

55. Girshick, R., Donahue, J., Darrell, T., Malik, J.: Rich feature hierarchies for accurate object detection and semantic segmentation. In: *2014 IEEE Conference on Computer Vision and Pattern Recognition*. pp. 580–587. IEEE, Columbus, OH, USA (June 2014), https://doi.org/10.1109/CVPR.2014.81.

56. Long, J., Shelhamer, E., Darrell, T.: Fully convolutional networks for semantic segmentation. In: *2015 IEEE Conference on Computer Vision and Pattern Recognition (CVPR)*. pp. 3431–3440. IEEE, Boston, MA, USA (June 2015), https://doi.org/10.1109/CVPR.2015.7298965.

57. Girshick, R.: Fast R-CNN. In: *2015 IEEE International Conference on Computer Vision (ICCV)*. pp. 1440–1448. IEEE, Santiago, Chile (Dec 2015), https://doi.org/10.1109/ICCV.2015.169.

58. Brody, H.: Medical imaging. *Nature* 502, S81 (10/30 2013).

59. Shen, D., Wu, G., Suk, H.I.: Deep learning in medical image analysis. *Annual review of biomedical engineering* 19, 221–248 (June 21 2017).

60. Schmidhuber, J.: Deep learning in neural networks: An overview. *Neural Networks* 61, 85–117 (2015).

61. Wu, G., Kim, M., Wang, Q., Munsell, B.C., Shen, D.: Scalable high-performance image registration framework by unsupervised deep feature representations learning. *IEEE Transactions on Biomedical Engineering* 63(7), 1505–1516 (July 2016), https://doi.org/10.1109/TBME.2015.2496253.

62. Wu, G., Kim, M., Wang, Q., Gao, Y., Liao, S., Shen, D.: Unsupervised deep feature learning for deformable registration of MR brain images. In: Mori, K., Sakuma, I., Sato, Y., Barillot, C., Navab, N. (eds.) *Medical Image Computing and Computer-Assisted Intervention—MICCAI 2013*. Lecture Notes in Computer Science, Vol. 8150, pp. 649–656. Springer Berlin Heidelberg, Berlin, Heidelberg (2013).

63. Liao, S., Gao, Y., Oto, A., Shen, D.: Representation learning: A unified deep learning framework for automatic prostate MR segmentation. In: Mori, K., Sakuma, I., Sato, Y., Barillot, C., Navab, N. (eds.) *Medical Image Computing and Computer-Assisted Intervention—MICCAI 2013*. Lecture Notes in Computer Science, Vol. 8150, pp. 254–261. Springer Berlin Heidelberg, Berlin, Heidelberg (2013).

64. Guo, Y., Gao, Y., Shen, D.: Deformable MR prostate segmentation via deep feature learning and sparse patch matching. *IEEE Transactions on Medical Imaging* 35(4), 1077–1089 (April 2016), https://doi.org/10.1109/TMI.2015.2508280.

65. Kim, M., Wu, G., Shen, D.: Unsupervised deep learning for hippocampus segmentation in 7.0 Tesla MR images. In: Wu, G., Zhang, D., Shen, D., Yan, P., Suzuki, K., Wang, F. (eds.) *Machine Learning in Medical Imaging*. Lecture Notes in Computer Science, Vol. 8184, pp. 1–8. Springer International Publishing, Cham (2013), https://doi.org/10.1007/978-3-319-02267-3_1.

66. Abd-Ellah, M.K., Awad, A.I., Khalaf, A.A.M., Hamed, H.F.A.: Classification of brain tumor MRIs using a kernel support vector machine. In: Li, H., Nykanen, P., Suomi, R., Wickramasinghe, N., Widen, G., Zhan, M. (eds.) *Building Sustainable Health Ecosystems, WIS 2016. Communications in Computer and Information Science*, Vol. 636. Springer, Cham, Tampere, Finland (2016).

67. Abd-Ellah, M.K., Awad, A.I., Khalaf, A.A.M., Hamed, H.F.A.: Design and implementation of a computer-aided diagnosis system for brain tumor classification. In: *2016 28th International Conference on Microelectronics (ICM)*. pp. 73–76. IEEE, Giza, Egypt (2016).

68. Roth, H.R., Lu, L., Liu, J., Yao, J., Seff, A., Cherry, K., Kim, L., Summers, R.M.: Improving computer-aided detection using convolutional neural networks and random view aggregation. *IEEE Transactions on Medical Imaging* 35(5), 1170–1181 (May 2016), https://doi.org/10.1109/TMI.2015.2482920.

69. Ciompi, F., de Hoop, B., van Riel, S.J., Chung, K., Scholten, E.T., Oudkerk, M., de Jong, P.A., Prokop, M., van Ginneken, B.: Automatic classification of pulmonary peri-fissural nodules in computed tomography using an ensemble of 2D views and a convolutional neural network out-of-the-box. *Medical Image Analysis* 26(1), 195–202 (September 2015), https://doi.org/10.1016/j.media.2015.08.001.

70. Gao, M., Bagci, U., Lu, L., Wu, A., Buty, M., Shin, H.C., Roth, H., Papadakis, G.Z., Depeursinge, A., Summers, R.M., Xu, Z., Mollura, D.J.: Holistic classification of CT attenuation patterns for interstitial lung diseases via deep convolutional neural networks. *Computer methods in biomechanics and biomedical engineering: Imaging & Visualization* 6(1), 1–6 (2016), https://doi.org/10.1080/21681163.2015.1124249.

71. Brosch, T., Tang, L.Y.W., Yoo, Y., Li, D.K.B., Traboulsee, A., Tam, R.: Deep 3D convolutional encoder networks with shortcuts for multi-scale feature integration applied to multiple sclerosis lesion segmentation. *IEEE Transactions on Medical Imaging* 35(5), 1229–1239 (May 2016), https://doi.org/10.1109/TMI.2016.2528821.

72. Dou, Q., Chen, H., Yu, L., Zhao, L., Qin, J., Wang, D., Mok, V.C., Shi, L., Heng, P.: Automatic detection of cerebral microbleeds from MR images via 3D convolutional neural networks. *IEEE Transactions on Medical Imaging* 35(5), 1182–1195 (May 2016), 10.1109/TMI.2016.2528129.

73. Abd-Ellah, M.K., Awad, A.I., Khalaf, A.A.M., Hamed, H.F.A.: A review on brain tumor diagnosis from MRI images: Practical implications, key achievements, and lessons learned. *Magnetic Resonance Imaging* 61, 300–318 (2019), https://doi.org/10.1016/j.mri.2019.05.028.

74. Ciresan, D.C., Giusti, A., Gambardella, L.M., Schmidhuber, J.: Mitosis detection in breast cancer histology images with deep neural networks. In: Mori, K., Sakuma, I., Sato, Y., Barillot, C., Navab, N. (eds.) *Medical Image Computing and Computer-Assisted Intervention—MICCAI 2013*. pp. 411–418. Springer International Publishing, Berlin, Heidelberg (2013), https://doi.org/10.1007/978-3-642-40763-5_51.

75. Kleesiek, J., Urban, G., Hubert, A., Schwarz, D., Maier-Hein, K., Bendszus, M., Biller, A.: Deep MRI brain extraction: A 3D convolutional neural network for skull stripping. *NeuroImage* 129, 460–469 (2016).

76. Moeskops, P., Viergever, M.A., Mendrik, A.M., de Vries, L.S., Benders, M.J.N.L., Isgum, I.: Automatic segmentation of MR brain images with a convolutional neural network. *IEEE Transactions on Medical Imaging* 35(5), 1252–1261 May 2016).

77. Weisenfeld, N.I., Warfield, S.K.: Automatic segmentation of newborn brain MRI. *NeuroImage* 47(2), 564–572 (2009).

78. Zhang, W., Li, R., Deng, H., Wang, L., Lin, W., Ji, S., Shen, D.: Deep convolutional neural networks for multi-modality isointense infant brain image segmentation. *NeuroImage* 108, 214–224 (2015).

79. Nie, D., Wang, L., Gao, Y., Shen, D.: Fully convolutional networks for multi-modality isointense infant brain image segmentation. In: *2016 IEEE 13th International Symposium on Biomedical Imaging (ISBI)*. pp. 1342–1345 (April 2016).

80. Zhao, X., Wu, Y., Song, G., Li, Z., Fan, Y., Zhang, Y.: Brain tumor segmentation using a fully convolutional neural network with conditional random fields. In: Crimi, A., Menze, B., Maier, O., Reyes, M., Winzeck, S., Handels, H. (eds.) *Brainlesion: Glioma, Multiple Sclerosis, Stroke and Traumatic Brain Injuries*. Lecture Notes in Computer Science, Vol. 10154, pp. 75–87. Springer International Publishing, Cham (2016).

81. Casamitjana, A., Puch, S., Aduriz, A., Vilaplana, V.: 3D convolutional neural networks for brain tumor segmentation: A comparison of multi-resolution architectures. In: Crimi, A., Menze, B., Maier, O., Reyes, M., Winzeck, S., Handels, H. (eds.) *Brainlesion: Glioma, Multiple Sclerosis, Stroke and Traumatic Brain Injuries*. Lecture Notes in Computer Science, Vol. 10154, pp. 150–161. Springer International Publishing, Cham (2016).

82. Pereira, S., Oliveira, A., Alves, V., Silva, C.A.: On hierarchical brain tumor segmentation in MRI using fully convolutional neural networks: A preliminary study. In: *2017 IEEE 5th Portuguese Meeting on Bioengineering (ENBENG)*. pp. 1–4 (Feb 2017).

83. Havaei, M., Davy, A., Warde-Farley, D., Biard, A., Courville, A., Bengio, Y., Pal, C., Jodoin, P.M., Larochelle, H.: Brain tumor segmentation with deep neural networks. *Medical Image Analysis* 35, 18–31 (2017), https://doi.Org/10.1016/j.media.2016.05.004.

84. Wang, G., Li, W., Ourselin, S., Vercauteren, T.: Automatic brain tumor segmentation using convolutional neural networks with test-time augmentation. In: Crimi, A., Bakas, S., Kuijf, H., Keyvan, F., Reyes, M., van Walsum, T. (eds.) *Brainlesion: Glioma, Multiple Sclerosis, Stroke and Traumatic Brain Injuries*. pp. 61–72. Springer International Publishing, Cham (2019), https://doi.org/10.1007/978-3-030-11726-9_6.

85. Abd-Ellah, M.K., Khalaf, A.A.M., Awad, A.I., Hamed, H.F.A.: TPUAR-Net: Two parallel U-Net with asymmetric residual-based deep convolutional neural network for brain tumor segmentation. In: Karray, F., Campilho, A., Yu, A. (eds.) *Image Analysis and Recognition. ICIAR 2019*. Lecture Notes in Computer Science, Vol. 11663, pp. 106–116. Springer International Publishing, Cham (2019).

86. The Cancer Imaging Archive: RIDER NEURO MRI database. https://wiki.canceri magingarchive.net/display/Public/RIDER+NEURO+MRI (2016), Accessed: February 5th, 2019.

87. Tharwat, A.: Classification assessment methods. *Applied Computing and Informatics* (2018), https://doi.org/10.1016/j.aci.2018.08.003.

88. Mohsen, H., El-Dahshan, E.S.A., El-Horbaty, E.S.M., Salem, A.B.M.: Classification using deep learning neural networks for brain tumors. *Future Computing and Informatics Journal* 3, 68–71 (2018).

10 Lossless Full-Resolution Deep Learning Convolutional Networks for Skin Lesion Boundary Segmentation

Mohammed A. Al-masni, Mugahed A. Al-antari, and Tae-Seong Kim

CONTENTS

10.1 INTRODUCTION

Skin cancer is one of the most widespread diagnosed cancers in the world. Indeed, melanoma (i.e., malignant skin tumor) usually starts when melanocyte cells begin to grow out of control [1]. Among different types of skin cancers, melanoma is the most aggressive type due to its greater capability of spreading into other organs and its higher death rate [2]. According to the annual report of the American Cancer Society (ACS), in the United States in 2018, it is estimated that about 99,550 cases were diagnosed as new cases of skin cancer (excluding basal and squamous cells skin cancers), and the estimated deaths from this disease reached up to 13,460 cases [3]. Melanoma makes up the majority among various skin cancer types, with estimated new cases and expected death cases of 91.7% and 69.2%, respectively. Moreover, it is reported that melanoma is the most fatal skin cancer, with deaths from melanoma making up 1.53% of total cancer deaths, and it represents 5.3% of all new cancer cases [3, 4]. Regarding the survival statistics from the Surveillance, Epidemiology, and End Results (SEER) Cancer Statistics Review (CSR) report [5], the survival rate of patients diagnosed early with melanoma over five years is about 91.8%. However, this rate decreases to 63% when the disease spreads into lymph nodes. Therefore, a skin lesion detected and correctly diagnosed in its earliest stage is highly curable, and such detection reduces the mortality rate. This highlights the significance of timely diagnosis and appropriate treatment of melanoma for patients' survival.

Visual inspection via the naked eye during the medical examination of skin cancers is hindered by the similarity among normal tissues and skin lesions, which may produce an incorrect diagnosis [6, 7]. In order to allow better visualization of skin lesions and improve the diagnosis of melanoma, different non-invasive imaging modalities have been developed. Dermoscopy (also known as dermatoscopy or epiluminescence microscopy) has become a gold standard imaging technology that assists dermatologists to improve the screening of skin lesions through visualizing prominent features present under the skin surface. The key idea of dermoscopy is that it acquires a magnified high-resolution image while reducing or filtering out skin surface reflections, utilizing polarized light. In clinical practice, skin lesion diagnosis is based on visual assessment of dermoscopy images. Although dermoscopy images improve diagnostic precision, examination of dermoscopy images by dermatologists via visual inspection is still tedious, time consuming, complex, subjective, and fault prone [6, 7]. Hence, automated computerized diagnostic systems for skin lesions are highly demanded to support and assist dermatologists in clinical decision-making.

Indeed, automatic segmentation of skin lesion boundaries from surrounding tissue is a key prerequisite procedure in any computerized diagnostic system of skin cancer. Accurate segmentation of the lesion boundaries plays a critical role in obtaining more prominent, specific, and representative features, which are utilized to distinguish between different skin lesion diseases [8–11]. However, automatic segmentation of skin lesions is still a challenging task due to their large variations in size, shape, texture, color, and location in the dermoscopy images. Some challenging examples of skin lesion dermoscopy images are shown in Figure 10.1. These examples show the presence of artifacts for screening: artificial segmentation such as

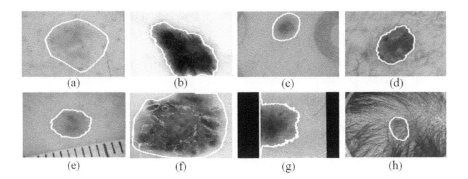

FIGURE 10.1   Examples of some challenging cases of skin lesions such as (a) low contrast, (b) irregular fuzzy boundaries, (c) color illumination, (d) blood vessels, (e) ruler mark artifact, (f) bubbles, (g) frame artifact, and (h) hair artifact. White line contours indicate the segmentation of the lesions by expert dermatologists.

ruler marks, ebony frames, air bubbles, and color illumination, and natural artifacts such as hair and blood vessels.

This chapter presents a novel deep learning method for skin lesion boundary segmentation called full-resolution convolutional networks (FrCN) [12]; this method produces very accurate segmentation. The FrCN method is a resolution-preserving model that leads to learning high-level features and improved segmentation performance. It directly learns the full-resolution attributes of each individual pixel of the input image. This is achieved by removing all the subsampling layers from the network. The evaluation of the efficiency and effectiveness of the FrCN segmentation method is presented using the well-established public International Skin Imaging Collaboration (ISIC) 2017 challenge dataset. In addition, the performance of the FrCN method is compared against well-known deep learning techniques such as U-Net, SegNet, and fully convolutional network (FCN) methods under the same experimental conditions.

10.2   RELATED WORKS

10.2.1   CONVENTIONAL SEGMENTATION METHODS

In general, conventional segmentation methods for the skin lesion boundaries can be categorized into the following five groups:

- *Thresholding methods* assign one or more appropriate threshold values to distinguish the skin lesions from the surrounding normal tissues. Pixel values greater than a particular threshold are placed as the foreground class, while the rest are placed as the background class. Improved adaptive threshold techniques such as histogram-based and Otsu threshold have been introduced to select optimal threshold value [7, 13–16].
- *Unsupervised clustering techniques* use the color space features from RGB dermoscopy images to acquire homogenous regions of the skin

lesions [17–20]. Cluster analysis methods include a connectivity model such as distance clustering or a centroid model such as the k-means clustering.

- *Region-based methods* utilize the region splitting or merging algorithms to generate groups of homogenous pixels. However, *edge-based methods* use the edge operator to provide edges between different regions [7, 21–23].
- *Active contour methods* (e.g., snakes active contours) exploit the evolution curve technique to segment skin lesions [7, 24, 25].
- *Supervised segmentation approaches* segment the skin lesion boundaries by learning multiple features and representations. Support vector machines (SVMs), decision trees (DTs), and artificial neural networks (ANN) are examples of commonly utilized classifiers [20, 26].

For further details of these traditional techniques, the latest comprehensive reviews provide more details of the skin lesion boundary segmentation algorithms [8, 27–30]. All of these methods utilize low-level features that rely only on pixel-level attributes. Hence, these classical segmentation techniques still do not deliver satisfactory performance and cannot get over the challenges of hair artifacts and low contrast.

10.2.2 INITIAL DEEP LEARNING SEGMENTATION METHODS

The great success of deep learning approaches in supervised convolutional neural networks (CNNs) for object detection and image classification have motivated researchers to develop models for the pixel-level segmentation problem. Therefore, segmentation methods employing deep learning techniques have gotten major attention in the fields of object localization, semantic segmentation, and image classification [31–37]. Deep leaning based on CNN is a powerful approach since it has the ability to extract more prominent features or representations from a whole image rather than hand-crafted features. Hence, it improves the performance of challenging segmentation tasks [38–40]. The initial CNN method for segmentation problems was introduced by Ciresan et al. in 2012 [41]. The aim was to provide full automation segmentation of the biological neuron membranes while eliminating the user intervention in the stacks of electron microscopy images. This method performed pixel-wise segmentation, which determined if each predicted pixel does or does not belong to the foreground object (i.e., binary classification). The process works first by dividing the input image into smaller regions of interest (ROIs) or patches and then passing them into a CNN classifier. The prediction of each sliding window represented the segmentation of the center pixel in the corresponding patch. At the end of this process, each pixel of the input image was segmented as membrane or non-membrane pixel. Similar work was proposed by Cernazanu-Glavan and Holban in order to segment bone pixels from chest X-ray images [42]. Due to the large size of the images, the segmentation process was performed only for some rib regions. Melinščak et al. presented a comparable CNN technique to segment retinal vessels as a binary classification task [43]. The prediction of each ROI determined the segmentation of its central pixel. This technique still needs further enhancement since the deep features extracted from each patch do not exactly reflect the attributes of the center pixel. In addition, it requires complex computation and a lot of execution time, since the network should be processed separately for each patch.

10.2.3 Recent Deep Learning Segmentation Methods

Lately, novel approaches utilizing deep CNN methods to improve the performance of challenging segmentation tasks have been under development. The first end-to-end pixel-wise semantic segmentation approach was developed by Long et al. [44, 45]. Their segmentation method was inspired from the conventional CNN recognizer, in which the authors adapted well-known contemporary classification models such as GoogLeNet, AlexNet, and Visual Geometry Group (VGG)Net into their CNN by replacing the fully connected neural network (FC-NN) layers with convolutional layers to produce spatial maps instead of prediction scores. Their model was named fully convolutional networks (FCN). FCN is robust and has the capability of yielding hierarchies of attributes. However, the subsampling operations after each convolutional layer cause the loss of the spatial resolution of the input pixels. Due to this, upsampling using deconvolution and bilinear interpolation process was applied in the last convolutional layer in order to produce a map that contains dense prediction with the same size as the original input image. FCN produces a significant enhancement of the pattern analysis, statistical modeling, and computational learning in the visual object classes from the PASCAL VOC dataset as compared to traditional techniques [46].

Another architecture, called U-Net, was developed to convert the CNN classification models to a suitable segmentation network [47]. U-Net was an extension and amendment of the FCN architecture [45], in which it performed well with fewer training images and produced more accurate segmentations on the stacks of electron microscopy images. In fact, U-Net consisted of two consecutive and symmetric paths generating a U-shaped model. The contracting path, which is typical in the FCN architecture, is responsible for extracting the relevant representations and patterns using the convolution and subsampling operations. In the expansive path, the upsampling process followed by a convolution operator was applied opposite to the contracting path to produce some large number of feature channels. In each convolutional layer, the feature map in the contracting path was concatenated with the corresponding features in the expansive path in order to provide high-resolution features. Similar to the FCN architecture, U-Net did not utilize fully connected layers; instead, it used the dense maps from the convolutional networks to segment the pixels of the images. Apart from the FCN technique, a deep convolutional encoder-decoder architecture for image segmentation was proposed by Badrinarayanan et al. in 2015 and was named SegNet [48, 49]. The SegNet architecture consisted of an encoder network that produces low-resolution feature maps, a corresponding decoder network that transforms the features into a full resolution of the input using a combination of convolution and upsampling layers, and a softmax pixel-wise classification layer. The encoder network used the VGG16 classification model, which involves 13 convolutional layers. However, the same layers were mirrored in the decoder network. The novelty of this segmentation method lies in the way that the decoder upsamples the lower-resolution feature maps. Each upsampling layer in the decoder network matched to the corresponding subsampling layer in the encoder network, preserving the locations of the feature maps in an input image during the upsampling process. The softmax classifier was utilized to generate predictions for the input pixels, which represented the segmentation map.

Noh et al. evolved a deep learning semantic segmentation technique called Deconvolution Network (DeconvNet) as an extension of FCN [50]. Analogous to the SegNet architecture, DeconvNet contained convolution networks that extracted representations from the input image and deconvolution networks that retrieved the feature maps into higher resolution as the dimension of the input image. DeconvNet was also a shape generator that produced a probability map of all input pixels and had a higher spatial resolution compared to that generated by FCN. Thus, the final output of the DeconvNet is the segmentation masks of the objects in the input images. Furthermore, an image segmentation technique via deep convolutional nets called DeepLab was introduced by Chen et al. [51, 52]. This method was proposed to provide a solution to some of the challenges faced in the semantic segmentation models. First, the authors replaced the last few subsampling layers in the network by Atrous convolution, which is a dilated convolution that extends the filter size and fills the hole locations with zeros to maintain the feature resolution. Second, they used the Atrous spatial pyramid pooling for multi-scale image representations. Then, they appended the DeepLab network with the fully connected conditional random field (CRF) to recover more precise boundaries of the segmented objects. Arguably, all of the aforementioned segmentation methods take the well-known classification networks and eliminate their FC-NN layers. Along with the encoder network, which extracts low-resolution feature maps, these methods build a corresponding decoder network to compensate for the loss of features and generate dense pixel-wise predictions. More details of these segmentation architectures are explained in a review paper [46].

10.2.4 PREVIOUS DEEP LEARNING WORKS ON SKIN LESION SEGMENTATION

Recently, these deep learning approaches such as FCN, U-Net, and SegNet have been stratified in the skin lesion segmentation field. In 2018, Li and Shen proposed a framework named the lesion indexing network (LIN) consisting of multi-scale fully convolutional residual networks (FCRN) and the lesion index calculation unit (LICU) [53]. Authors utilized the cropped center area of the skin lesion images from the ISIC 2017 challenge dataset. The performance of the LIN method was evaluated on the ISIC 2017 validation set: it achieved overall accuracy and Jaccard indices (JACs) of 95.0% and 75.3%, respectively. Bi et al. presented a cascaded multi-phase FCN followed by the parallel integration (PI) technique to segment lesion boundaries from dermoscopy images [54]. The PI technique was used for further refinement of the boundaries of the segmented lesions. They evaluated their segmentation approach utilizing two public datasets: the PH2 and ISIC 2016 skin lesion challenge datasets. Their method achieved overall segmentation accuracies of 94.24% and 95.51% and dice coefficient indices of 90.66% and 91.18% from the two datasets. Yu et al. built a two-stage approach for segmentation and classification of skin lesions using deep residual networks [55]. Their FCRN was ranked second in the segmentation challenge with an overall accuracy of 94.9% when using the ISIC 2016 challenge dataset.

Yuan et al. developed a skin lesion segmentation technique via deep FCN [56]. In order to handle the imbalance among lesion-tissue pixels, the authors extended the well-known FCN approach by employing the Jaccard distance as a loss function. Their technique achieved a dice index of 93.8% and an overall segmentation accuracy of 95.5% with the PH2 and ISIC 2016 datasets, respectively. Goyal and Yap presented a multi-class semantic segmentation via FCN, which was capable of segmenting three classes of skin lesions (i.e., melanoma, seborrheic keratoses [SK], and benign nevi) from the ISIC 2017 challenge dataset [57]. Their method achieved dice metrics of 65.3%, 55.7%, and 78.5% for melanoma, seborrheic keratosis, and benign lesions, respectively. Lin et al. proposed a comparison between two skin lesion segmentation methods, C-means clustering and U-Net-based histogram equalization [58]. They evaluated their methodologies utilizing the ISIC 2017 challenge dataset. The U-Net method obtained a dice coefficient measure of 77.0%, which significantly outperformed the clustering method's result of only 61.0%. In 2017, Yading Yuan developed a skin lesion segmentation utilizing deep convolutional-deconvolutional neural networks (CDNN) [59]. The CDNN model was trained with various color spaces of dermoscopy images using the ISIC 2017 dataset. CDNN approach was ranked first in the ISIC 2017 challenge, in which it obtained a Jaccard index (JAC) of 76.5%.

10.3   MATERIALS AND METHODS

10.3.1   DATASET

In this study, a well-known and public dermoscopy dataset, called the IEEE International Symposium on Biomedical Imaging (ISBI) 2017 segmentation challenge dataset, was used with the FrCN segmentation technique. This dataset is provided by the ISIC archive [60] and is available at [61]. The ISIC 2017 challenge dataset includes 8-bit RGB dermoscopy images with various image sizes ranging from 540×722 to $4,499 \times 6,748$ pixels. This dataset consists of 2,000 training images along with validation and testing datasets of 150 and 600 images, respectively. All of these dermoscopy images include skin lesions, which are labeled by expert dermatologist as benign nevi, seborrheic keratosis (SK), or melanoma. Table 10.1 summarizes the distribution of the dataset. Moreover, this dataset provided the original dermoscopy images paired with their ground-truth segmented lesion boundaries in the form of binary masks, which are manually annotated by expert dermatologists. Figure 10.2 illustrates an example of these dermoscopy images from the ISIC 2017 challenge dataset with their ground-truth segmentation masks.

10.3.2   DATA PREPROCESSING AND AUGMENTATION

A normalization process is applied to each channel of the RGB images to facilitate and increase the convergence speed of the segmentation method. Therefore, each channel is normalized in a range from 0 to 1. Due to the differences in the image

TABLE 10.1
Distribution of the ISIC 2017 Challenge Dataset

Data Type	Benign	SK	Melanoma	Total
Training Data	1,372	254	374	**2,000**
Validation Data	78	42	30	**150**
Test Data	393	90	117	**600**
Total	**1,843**	**386**	**521**	**2,750**

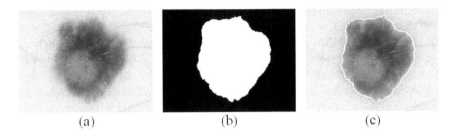

(a)   (b)   (c)

FIGURE 10.2   (a) and (b) are an exemplary pair of the original dermoscopy image and its segmentation mask, which was annotated by expert dermatologists. (c) illustrates the segmented lesion boundaries.

sizes of the ISIC 2017 dataset (i.e., the majority ratio of height to width of 3:4), we resized all of the dermoscopy images into 192×256 pixels utilizing bilinear interpolation according to the report in [56], where the optimal segmentation achievement was made with the image size of 192×256 among other different image sizes. In fact, deep learning methods require a large amount of data for appropriate training. However, a limit on the sizes of medical image datasets, especially a limit on reliable annotated ground-truths, is one of the challenges to adopting deep learning approaches. Due to this, data augmentation was applied to enlarge the training dataset. Augmentation is a process that generates augmented images utilizing the given original images using various transformation techniques such as rotation, scaling, and translation [62–64].

By taking the merit of the dermoscopy images (i.e., RGB color), the information from three channels—hue-saturation-value (HSV)—could be derived in addition to the original RGB images, producing more color space features. The augmented dataset includes the RGB and HSV dermoscopy images and their rotated images with angles of 0°, 90°, 180°, and 270°. Additionally, horizontal and vertical flipping was applied. Thus, a total of 32,000 [i.e., (2,000 RGB + 2,000 HSV) × 4 rotations × 2 flipping] images with their corresponding ground-truth binary masks were utilized to train the FrCN segmentation technique. Figure 10.3 shows a sample of the original RGB images in addition to the 15 augmented images. In fact, the data augmentation process reduces the overfitting problem of deep learning and improves the robustness of the deep network.

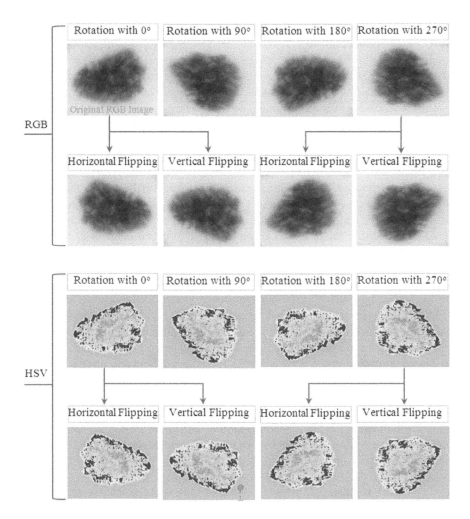

FIGURE 10.3 The augmentation processes for the training images. Top two rows represent RGB images, while the bottom two rows are for HSV images. From each original RGB image (top left corner), we generate a total of 16 different images including four rotations and four horizontal and vertical flippings on the original RGB and HSV images.

10.3.3 FULL-RESOLUTION CONVOLUTIONAL NETWORKS

In general, CNN contains two essential parts to obtain pixel-wise classification for segmentation: the first includes convolutional and subsampling layers and the second includes upsampling layers. Convolutional and subsampling layers represent the first part of these networks. The convolutional layers are in charge of extracting deep feature representations from the input image utilizing different filter types of various sizes. The subsampling layers are utilized to reduce the dimensionality of the extracted feature maps, but they also reduce the resolution of the feature maps [65, 66]. In image-wise classification tasks (i.e., image classification via CNNs), the pooling or subsampling layer does promote the robustness of the classifier, since it reduces overfitting, eliminates the redundancy of features, and minimizes computation time [45, 54]. This is because all the reduced attributes represent a single class label of the input image. However, in the pixel-wise segmentation tasks, the subsampling layers cause loss of spatial resolution in the features of input images.

The second part of the segmentation networks is the upsampling layers followed by the softmax classifier, which exploits the extracted features of each individual pixel to distinguish it as a tissue or a lesion pixel. Recent deep learning segmentation approaches have used complicated procedures to compensate for the missing features of particular pixels, due to the reduced image size, via upsampling, deconvolution, bilinear interpolation, decoding, or Atrous convolution, increasing the number of hyper-parameters [45, 47, 48, 50, 51]. Furthermore, these mechanisms require further PI or CRF operations to refine boundaries of the segmented objects [51, 54]. These deep learning segmentation networks suffer from reduced spatial resolution and loss of details. Because pixel-wise dependency is not adequately addressed throughout the previous deep learning segmentation methods such as FCN, U-Net, and SegNet, important information about some pixels is left out and is not easy to retrieve.

The full-resolution convolutional networks (FrCN) [12] is an end-to-end supervised deep network that is trained via mapping the entire input image to its corresponding ground-truth masks with no resolution loss, leading to better segmentation performance of skin lesion boundaries. In fact, the FrCN method is a resolution-preserving model that leads to high-level feature learning and improves the segmentation performance. The full-resolution feature maps of the input pixels are reserved by removing all the subsampling layers in the architecture. Figure 10.4 illustrates the architecture of the FrCN segmentation method. The FrCN allows each pixel in the input dermoscopy image to extract its own features utilizing the convolutional layers.

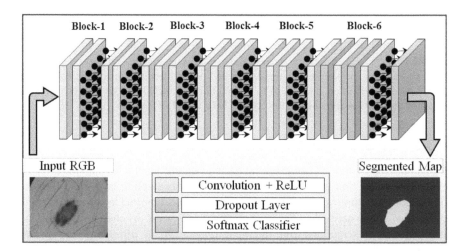

FIGURE 10.4 An illustration of the FrCN segmentation method. There are no subsampling or fully connected layers.

The FrCN architecture contains 16 convolutional layers, which are inspired by the VGG-16 network [67].

It has multiple stacks of convolutional layers with different sizes of feature maps, where each stack learns distinctive representations and causes an increase over the network complexity. The last three FC-NN layers in Block-6 were substituted with the convolutional layers to allow the network to process input images with arbitrary sizes as well as to generate pixel-wise maps of dense prediction with the same sizes as the input images. Details of the number of maps and the convolutional filter sizes in each layer are given in Table 10.2. The early layers of FrCN learn localization information, while the late layers learn subtle and fine feature representations of skin lesion boundaries.

A convolutional network is originally translation invariant, in which its main role is to extract different features from the entire dermoscopy images and then produce feature maps using convolution operations [68]. In the deep learning network, the filter kernels of the convolutional layers are represented by weights, which are automatically updated during the training process. The basic components of this convolutional network are the filter kernels W, convolution operator $*$, and activation function $\phi(\cdot)$. Therefore, the kth feature map of layer L is computed as follows:

$$F_L^k = \phi\left(W_L^k * F_{L-1}^k + b_L^k\right), \tag{10.1}$$

where b_L^k is the bias applied to each feature map for each layer.

A nonlinear activation process gets used via rectified linear units (ReLUs) instantly after each convolutional layer to provide nonlinearity to the network. A ReLU activation function is widely used rather than the traditional sigmoid and tanh functions due to its capability to alleviate the vanishing gradient problem and to

TABLE 10.2
Details of Architecture Layers of the FrCN Method

Block	Layer	Filter Size, Maps	Block	Layer	Filter Size, Maps
Input data	$192 \times 256 \times 3$	–	**Block-5**	Conv. 11	$3 \times 3, 512$
Block-1	Conv. 1	$3 \times 3, 64$		Conv. 12	$3 \times 3, 512$
	Conv. 2	$3 \times 3, 64$		Conv. 13	$3 \times 3, 512$
Block-2	Conv. 3	$3 \times 3, 128$	**Block-6**	Conv. 14	$7 \times 7, 4096$
	Conv. 4	$3 \times 3, 128$		Dropout 1	$p=0.5$
Block-3	Conv. 5	$3 \times 3, 256$		Conv. 15	$1 \times 1, 4096$
	Conv. 6	$3 \times 3, 256$		Dropout 2	$p=0.5$
	Conv. 7	$3 \times 3, 256$		Conv. 16	$1 \times 1, 2$
Block-4	Conv. 8	$3 \times 3, 512$		Softmax	–
	Conv. 9	$3 \times 3, 512$	**Output map**	$192 \times 256 \times 3$	–
	Conv. 10	$3 \times 3, 512$			

train the network with higher computational efficiency [69, 70]. The ReLU activation function is defined as follows:

$$\phi(x) = \max(0, x) = \begin{cases} x, & \text{if } x \geq 0 \\ 0, & \text{if } x < 0. \end{cases} \tag{10.2}$$

In addition, a dropout is used with $p = 0.5$ after the convolutional layers *Conv14* and *Conv15*, as shown in the FrCN architecture in Figure 10.4 and Table 10.2. Dropout is a process used in deep learning to tackle the overfitting problem of deep layers [71]. During training, it randomly eliminates some neural units with their connections to prevent overfitting to the training data. Hence, every unit is retrained with a particular probability, leading the outgoing weights of that unit to be multiplied by p at the test time.

Finally, a softmax classifier known as multinomial logistic regression is fed in the last layer of the FrCN architecture to classify each pixel in the dermoscopy image into two binary classes (i.e., lesion and non-lesion). This logistic regression function generates a map of predictions for each pixel, which involves the segmented map. As a result, the resolutions of the spatial output maps are same as that of the input data. Cross-entropy is used as a loss function, in which the overall loss H of each pixel is minimized throughout the training process as defined by

$$H = -(y \cdot \log(\hat{y}) + (1 - y)\log(1 - \hat{y})), \tag{10.3}$$

where y and $\hat{y}$ are the ground-truth delineation and predicted segmented map, respectively. The cross-entropy loss function is utilized when a deep convolutional network is applied for a pixel-wise recognition [45, 56].

10.3.4   Training and Testing

Biomedical deep learning applications still suffer from the lack of medical data [62]. One reason is the cost and complexity of the labeling procedure during data acquisition [72]. To overcome the challenge of insufficient training data in the field of skin lesion dermoscopy images, we exploited two mechanisms, data augmentation as described above and transfer learning [54, 57, 62, 72–75]. For transfer learning, we utilized the pre-training network weights from the trained VGG-16 [67] using the public ImageNet database [76]. Subsequently, the deep FrCN segmentation technique was fine-tuned or retrained using the augmented training data of the dermoscopy images. Regarding the optimization of the FrCN, a separate validation dataset consisting of 150 dermoscopy images was utilized. Therefore, the evaluations of the optimization network and the final network performance were performed independently utilizing the validation and test datasets, respectively. This procedure is called an unbiased double cross-validation scheme, and it is more commonly used and reliable than the trial-and-error process in the most recent deep learning methods [77, 78].

To ensure generalization of deep learning architectures and avoid overfitting to training data, deep learning networks must be trained, validated, and tested using separate datasets as shown in Figure 10.5. A training dataset (i.e., 72.7% of the whole dataset) is used to train the deep learning network with hyper-parameters, which are evaluated utilizing the validation dataset (5.5% of the whole dataset). The optimal deep learning model is selected according to its high efficiency on the validation set. Then, further evaluation is performed on the test dataset (21.8% of the whole dataset) to obtain the overall performance of the network [65, 79].

The objective of training deep learning approaches is to optimize the weight parameters in each layer. The optimization procedure of a single cycle performs as follows [38, 80]. First, the forward propagation pass sequentially computes the output in each layer utilizing the training data. In the last output layer, the error between the ground-truth and predicted labels gets computed using the loss function. To reduce the training error, back-propagation is proceeded through the network layers. Consequently, the training weights of FrCN get updated using the training data. Furthermore, the performance of FrCN is compared against the performance of the recent deep learning approaches such as FCN, U-Net, and SegNet utilizing the test subset of 600 dermoscopy images from the ISIC 2017 dataset. All the segmentation models were trained using the superior Adadelta optimization method (details in Section 10.3.6.2) with a batch size of 20. The learning rate was initially set as 0.2 and then reduced with automated updating throughout the training process. The training network tends to have convergence at about the 200th epoch.

The system implementations of all experiments were performed on a personal computer (PC) with the following specifications: a CPU of Intel® Core(TM) i7-6850 K @ 3.360 GHz with 16 GB RAM and a GPU of NVIDIA GeForce GTX 1080. This work was conducted with Python 2.7.14 on Ubuntu 16.04 OS using the Keras and Theano DL libraries [81, 82].

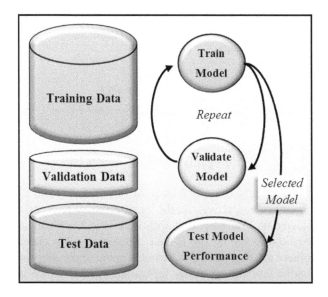

FIGURE 10.5 Functionality scheme of training, validation, and testing.

10.3.5 Evaluation Metrics

To quantitatively evaluate the efficiency and capability of the deep FrCN method on skin lesion boundary segmentation, the following statistical evaluation measures were utilized. Sensitivity (a.k.a., recall or true positive rate) measures the ratio of the skin lesion pixels that are correctly segmented, while the specificity (a.k.a., the true negative rate) indicates the proportion of the skin non-lesion pixels that are not correctly segmented [57, 83, 84]. The dice coefficient, also known as the F1-score, was also utilized to quantify the efficiency of the FrCN segmentation method. This measures the similarity between the segmented lesions and the delineated ground-truths [85, 86]. The Jaccard index (JAC) is an intersection over union of the ground-truth masks with the segmented lesions [87]. Accuracy is also provided to show the overall pixel-wise segmentation performance. The Matthew correlation coefficient (MCC) measures the correlation between the segmented and annotated skin lesion pixels and produces values in a range from −1 to +1 [87]. A higher MCC value implies a better segmentation performance. All of these metrics were computed in relation to the elements of the confusion matrix as follows.

$$\text{Sensitivity (SEN)} = \frac{\text{TP}}{\text{TP} + \text{FN}}, \tag{10.4}$$

$$\text{Specificity (SPE)} = \frac{\text{TN}}{\text{TN} + \text{FP}}, \tag{10.5}$$

$$\text{Dice Index (DIC)} = \frac{2 \cdot \text{TP}}{(2 \cdot \text{TP}) + \text{FP} + \text{FN}}, \tag{10.6}$$

$$\text{Jaccard Index (JAC)} = \frac{\text{TP}}{\text{TP} + \text{FN} + \text{FP}}, \tag{10.7}$$

$$\text{Accuracy (ACC)} = \frac{\text{TP} + \text{TN}}{\text{TP} + \text{FN} + \text{TN} + \text{FP}}, \tag{10.8}$$

$$\text{MCC} = \frac{\text{TP} \cdot \text{TN} - \text{FP} \cdot \text{FN}}{\sqrt{(\text{TP} + \text{FP})(\text{TP} + \text{FN})(\text{TN} + \text{FP})(\text{TN} + \text{FN})}}, \tag{10.9}$$

where TP and FP denote the true and false positives, while TN and FN indicate the true and false negatives, respectively. If the lesion pixels were segmented correctly, they were considered as TPs; otherwise, they were FNs. On the contrary, the non-lesion pixels were considered as TNs if their segmentation was classified correctly as non-lesion; otherwise, they were FPs. An illustration of the sensitivity and specificity concept with a dermoscopy image for classification is shown in Figure 10.6. Moreover, the curve of the receiver operator characteristic (ROC) with its area under the curve (AUC) was used for further segmentation evaluation.

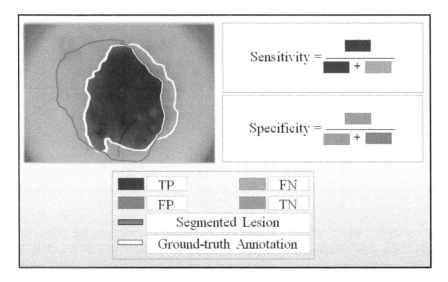

FIGURE 10.6 Concepts of sensitivity and specificity in terms of TP, FN, FP, and TN pixels. The white contour indicates the ground-truth of the lesion boundary annotation by a dermatologist, while the red one is an exemplary segmented boundary.

10.3.6 Key Component Optimization

This section presents the effect of some key components of the performance of the FrCN method over the skin lesion segmentation to develop a robust deep model capable of overcoming the overfitting and speeding up its convergence during the training process. These components involve data augmentation and network optimization. For each component, some experiments have been performed with different alternative settings, while the other components are fixed. The modified models of all these experiments are trained using the training data of the ISIC 2017 dataset, while the models are evaluated using only the validation data (i.e., 150 dermoscopy images) of the same dataset.

10.3.6.1 The Effect of Training Data

This section presents the effect of training the deep FrCN with different training dataset sizes (i.e., original versus augmented training datasets). The original training dataset consists of 2,000 dermoscopy images, while the augmented dataset contains 32,000 images. In general, the performance of deep learning approaches improves as the size of the training dataset increases. The effect of training data size on the FrCN segmentation method is shown in Table 10.3. The FrCN was first trained with the original training images (i.e., without augmentation), which provided a dice coefficient index of 79.59% on the validation dataset. Subsequently, the network was trained with the augmented training dataset, which yielded an improvement of 8.15% in terms of dice coefficient index. The results show an improvement in terms of Jaccard index in the case of the augmented dataset against the original, with 73.58% and 63.36%, respectively. Significantly, the augmented data also affects

TABLE 10.3

Segmentation Performance with Different Optimization Components

Component	Strategy	Performance Evaluation (%)					
		SEN	SPE	DIC	JAC	MCC	ACC
Data Augmentation	**Without Aug.**	73.40	92.07	79.59	63.36	72.88	88.81
	With Aug.	**81.20**	**94.15**	**87.74**	**73.58**	**82.89**	**91.89**
Optimizer	**SGD**	74.05	96.44	77.57	63.36	73.22	92.53
	Adam	78.60	**97.35**	82.23	69.83	78.81	**94.08**
	Adadelta	**81.20**	94.15	**87.74**	**73.58**	**82.89**	91.89

the sensitivity, which reflects the ability of the network to segment the skin lesion correctly, with an improvement of 7.80% as reported in Table 10.3. Due to this, the overall segmentation accuracy is increased from 88.81% to 91.89%. These results ensure that training deep learning models with a larger training data achieves better performance and provides more feasible and reliable models. The overfitting problem is clearly shown in Figure 10.7 (a), which presents how the loss function diverges over epochs on the validation dataset when the network is trained with the original training dataset without data augmentation. In contrast, the loss function converged rapidly in training the network with the larger augmented dataset.

10.3.6.2 The Effect of Network Optimizers

In this section, the Adadelta [88] optimization method was tested against Adam [89] and Stochastic Gradient Descent (SGD) [90] optimizers. The learning rate of the Adam and SGD optimizers was set as 0.003. Table 10.3 shows the statistical evaluation of these three optimizers with various measures. Adadelta showed the superior performance over the skin lesion segmentation against Adam and SGD methods,

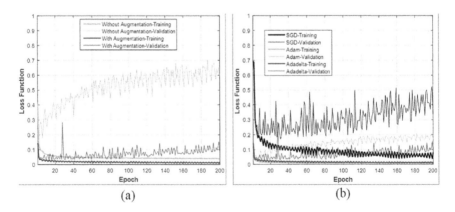

FIGURE 10.7 Optimization performance in terms of loss function. Comparison between the training and validation loss functions (a) with and without data augmentation, and (b) using Adadelta, Adam, and SGD optimization methods.

with overall Jaccard indices of 73.58%, 69.83%, and 63.36% on the validation dataset, respectively. Furthermore, Figure 10.7 (b) illustrates how the loss function declined during the training process on the training datasets. This figure provides an indication of the convergence speed of the network under training over epochs. Clearly, the training and validation curves of Adadelta are significantly better than those of Adam and SGD. In addition, these curves clarify the network performance on the validation data, in which a small variation between them could give the FrCN the capability to perform well with unseen test data. Although the loss function of the training data using the Adam optimizer decreases rapidly over epochs, the loss for the validation data diverges, leading reduced performance for skin lesion boundary segmentation compared to Adadelta. However, the SGD optimizer provides the worst performance with this dataset, as illustrated in Figure 10.7 (b).

10.4   RESULTS OF SKIN LESION SEGMENTATION

10.4.1   SEGMENTATION PERFORMANCE ON THE ISIC 2017 TEST DATASET

This section presents the skin lesion segmentation performance of the FrCN against the latest deep learning methods, utilizing 600 dermoscopy images from the ISIC 2017 test dataset. Quantitatively, Tables 10.4 and 10.5 summarize the segmentation performance results of the FrCN compared to FCN, U-Net, and SegNet. In this work, the segmentation of FCN was performed utilizing a stride of 32. Table 10.4 explores the performances in terms of specificity, sensitivity, and overall pixel-wise segmentation accuracy, while the measurements in Table 10.5 are reported in terms of the dice, Jaccard, and Matthew correlation indices for benign, SK, melanoma, and overall cases. The FrCN method outperformed the FCN, U-Net, and SegNet, with overall dice, Jaccard, and accuracy indices of 87.08%, 77.11%, and 94.03%, respectively. The highest specificity for overall skin lesion segmentation was achieved by U-Net, with 97.24%.

An analysis of the segmentation performance on the diagnostic categories of the skin lesions is also provided. Measurements of each diagnostic class including benign, seborrheic keratosis (SK), and melanoma cases are shown in Tables 10.4 and 10.5. The segmentation accuracies of the benign, SK, and melanoma cases in the ISIC 2017 test dataset for the FrCN were 95.62%, 91.29%, and 90.78%, respectively. Finally, the ROC curves with their AUCs of the FrCN versus FCN, U-Net, and SegNet are illustrated in Figure 10.8 separately for each diagnostic class. The overall AUCs of FCN, U-Net, SegNet, and FrCN were 88.32%, 82.19%, 87.71%, and 91.04%, respectively. For qualitative evaluation, Figure 10.9 (a), (b), and (c) present some exemplar segmentation results of the FrCN compared to the ground-truth contours for the benign, SK, and melanoma cases from the ISIC 2017 test dataset. The relevant bars in Figure 10.9 indicate the sensitivity, Jaccard index, and accuracy for each image. Obviously, the FrCN segmentation method provides promising results as compared against other methods.

10.4.2   NETWORK COMPUTATION

The computation time for the training process was about 17.5 hours over 200 epochs. However, the decoding (i.e., inference time) for a single dermoscopy

TABLE 10.4

Segmentation Performance (%) of the FrCN Compared to FCN, U-Net, and SegNet in Terms of Sensitivity, Specificity, and Accuracy for Benign, SK, Melanoma, and Overall Cases

Method	Benign Cases			SK Cases			Melanoma Cases			Overall		
	SEN	SPE	ACC	SEN	SPE	ACC	SEN	SPE	ACC	SEN	SPE	ACC
FCN	85.25	96.95	94.45	75.10	95.91	90.96	70.67	96.18	88.25	79.98	96.66	92.72
U-Net	76.76	97.26	92.89	43.81	97.64	84.83	58.71	96.81	84.98	67.15	97.24	90.14
SegNet	85.19	96.30	93.93	70.58	92.50	87.29	73.78	94.26	87.90	80.05	95.37	91.76
FrCN	88.95	97.44	95.62	82.37	94.08	91.29	78.91	96.04	90.78	85.40	96.69	94.03

TABLE 10.5

Segmentation Performance (%) of the FrCN Compared to FCN, U-Net, and SegNet in Terms of Dice, Jaccard, and MCC Indices for Benign, SK, Melanoma, and Overall Cases

Method	Benign Cases			SK Cases			Melanoma Cases			Overall		
	DIC	JAC	MCC	DIC	JAC	MCC	DIC	JAC	MCC	DIC	JAC	MCC
FCN	86.77	76.63	83.28	79.81	66.40	74.26	78.89	65.14	71.84	83.83	72.17	79.30
U-Net	82.16	69.72	78.05	57.88	40.73	53.89	70.82	54.83	63.71	76.27	61.64	71.23
SegNet	85.69	74.97	81.84	72.54	56.91	64.32	79.11	65.45	71.03	82.09	69.63	76.79
FrCN	89.68	81.28	86.90	81.83	69.25	76.11	84.02	72.44	77.90	87.08	77.11	83.22

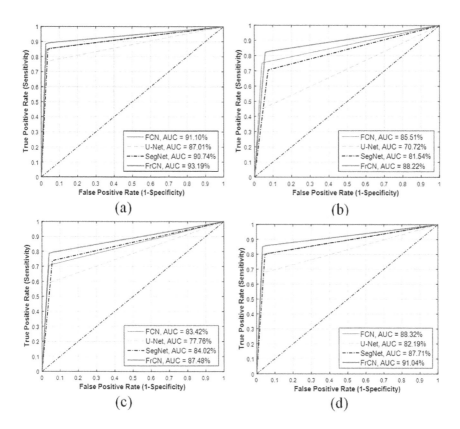

FIGURE 10.8   ROC curves of different deep learning methods for skin lesion boundary segmentation on the ISIC 2017 test dataset for (a) benign, (b) SK, (c) melanoma, and (d) overall clinical cases.

image took about 9.7 seconds. This should make the FrCN segmentation approach applicable for clinical practice. In addition, Table 10.6 shows the trainable parameters along with the training computation time per epoch and the test time per single dermoscopy image for each of the segmentation techniques. Clearly, the FrCN approach seems feasible for medical practices, since less than 10 seconds of inference time is required to segment the suspicious lesion from the dermoscopy image. The computational speed during the training process of the FrCN approach was faster than other segmentation techniques. Table 10.6 shows the speed of the FrCN approach in the training process time compared to the FCN model (i.e., from 651 to 315 seconds per epoch). Indeed, this is because FrCN was executed without the need for the downsampling processes, but keeping the same convolution operations. In summary, the deep FrCN segmentation method outperforms the most popular deep learning methods of FCN, U-Net, and SegNet on the ISIC 2017 datasets.

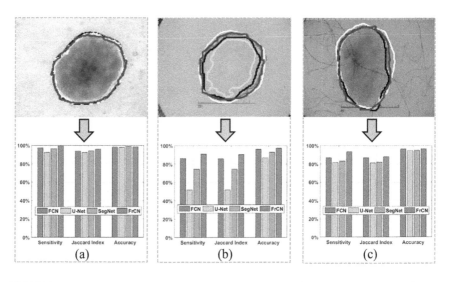

FIGURE 10.9   The lesion contours segmentation with statistical evaluation per each image of the FrCN against FCN, U-Net, and SegNet for (a) benign, (b) seborrheic keratosis, and (c) melanoma cases from the ISIC 2017 dataset. Segmentation results are depicted as follows: ground-truth, white; FrCN, blue; FCN, green; U-Net, red; and SegNet, black.

TABLE 10.6

Measurements in Seconds of the Training Time per Epoch and Test Time per Dermoscopy Image

Method	Trainable Parameters	Training Time/ Epoch	Test Time/ Single Image
FCN	134.3 M	651	10.6
U-Net	12.3 M	577	10.9
SegNet	11.5 M	543	11.13
FrCN	16.3 M	315	9.7

10.4.3   COMPARISONS OF SEGMENTATORS

To show the robustness and effectiveness of the FrCN segmentation method, comparisons with the latest research works in the literature are shown in Table 10.7. All of the segmentation models in this table were evaluated utilizing the overall diagnostic cases in the ISIC 2017 challenge dataset. In fact, 21 teams participated in the ISIC 2017 challenge. Table 10.7 provides a comparison with the top five models [59, 91–93] in the ISIC 2017 challenge. These models were ranked according to the segmentation performance of the Jaccard index. Clearly, the FrCN approach outperformed the top CDNN model [59] with a marginal increment of 0.61% in terms of the Jaccard index. In contrast, the deep ResNet method [92] showed a higher specificity of 98.50% for

TABLE 10.7

Performance (%) of the FrCN Method Compared to the Latest Studies in the Literature on Skin Lesion Segmentation

References	Ranked in the ISIC Challenge	SEN	SPE	DIC	JAC	ACC
Yuan et al. (CDNN) [59]	Top 1	82.50	97.50	84.90	76.50	93.40
Berseth [91]	Top 2	82.00	97.80	84.70	76.20	93.20
Bi et al. (ResNet) [92]	Top 3	80.20	**98.50**	84.40	76.00	93.40
Bi et al. (Multi-scale ResNet) [92]	Top 4	80.10	98.40	84.20	75.8	93.40
Menegola et al. [93]	Top 5	81.70	97.00	83.90	75.40	93.10
Lin et al. (U-Net) [58]	–	–	–	77.00	62.00	–
FrCN	–	**85.40**	96.69	**87.08**	**77.11**	**94.03**

the skin lesion boundary segmentation. However, the FrCN method provides promising results with fewer layers (i.e., only 16 layers) compared to the ResNet model, which employed a deeper network of 50 layers. Although the skin lesion U-Net-based histogram equalization technique proposed in [58] outperformed the clustering method, its segmentation performance was lower compared to the recent deep learning approach, with Jaccard and dice indices of 62.00% and 77.00%, respectively. The distribution performance of the segmented skin lesion boundaries of the ISIC 2017 test dataset (i.e., 600 images) in terms of Jaccard index via different segmentation networks is shown in Figure 10.10 and Table 10.7. This figure illustrates the boxplot performance of different models and also shows the median, lower, and upper Jaccard measures on the test dataset. Obviously, the FrCN method achieves a slightly higher median Jaccard value than the top method, with fewer counts below a Jaccard index of 0.4. With regard to the overall pixel-wise segmentation results of the skin lesion task, the FrCN method proves its feasibility and effectiveness compared to others

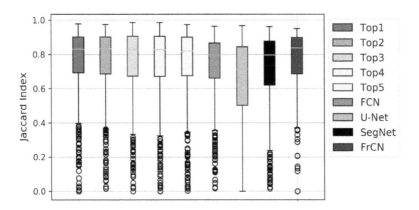

FIGURE 10.10 Boxplots of the Jaccard index for different segmentation networks on the ISIC 2017 challenge test dataset.

with a segmentation accuracy of 94.03%. The higher segmentation achievement of the FrCN model against others is due to FrCN directly learning the full-resolution features of each pixel through the convolutional layers of the network. It should be noted that although we utilized the fixed image size, which was a requirement for the conventional segmentation techniques, one can use any whole image size in FrCN without resizing. We conclude that the convolutional process in the segmentation task could extract and learn better prominent and specific features utilizing the full spatial resolution without the need for the downsampling process.

10.5 DISCUSSION

Automatic delineation of skin lesion boundaries is highly demanded as a prerequisite step for melanoma recognition. Accurate segmentation of skin lesions enables the classification CNNs to extract more specific and representative features from only segmented skin lesion areas instead of entire dermoscopy images. Generally, the overall segmentation performance of deep learning approaches improves as the size of the training dataset increases. The impact of various training dataset sizes is presented in [31, 62, 84], in which the results showed the significance of using larger training sets. In this work, the training datasets are augmented by utilizing two procedures. First, HSV images are generated from the original RGB images. Second, the training dermoscopy images are augmented by rotating the RGB and HSV images with four rotation angles. Furthermore, the images are flipped by applying horizontal and vertical flipping. The augmented training data should support the network to learn better features of the skin lesion characteristics. The results present the capability of the FrCN technique to segment skin lesions with accuracies higher than those obtained with the conventional methods. This is because the FrCN method is trained with the full spatial resolution of the input images. In our experiments, the segmentation performance of the FrCN technique was compared against three well-known segmentation approaches, namely FCN, U-Net, and SegNet, under the same test conditions. Quantitative evaluations with different indices of the recent deep learning methods are reported in Tables 10.4 and 10.5 for the ISIC 2017 test dataset. Significantly, the FrCN method outperformed other skin lesion segmentation approaches. The MCC index showed how the segmented skin lesions were correlated with the annotated ground-truths. The FrCN architecture obtained overall increments of 3.92%, 11.99%, and 6.43% in MCC as compared to FCN, U-Net, and SegNet, respectively. These promising segmentation results indicate the consistency and capability of the FrCN method.

In this work, further analysis was done to show the segmentation performance for clinical diagnosis cases. In the ISIC 2017 test dataset, the benign cases achieved a higher segmentation performance than the melanoma and SK cases, with Jaccard indices of 81.28%, 72.44%, and 69.25%, respectively, as shown in Table 10.4. This significant improvement for the benign cases is due to the larger existing percentage of the cases in the training dataset (i.e., 1,372/2,000 cases) and in the test dataset (i.e., 393/600), as numbered in the database distribution in Table 10.1. Note that due to the inequality of the numbers of each diagnostic type, the segmentation results of the overall indices shown in Tables 10.4 and 10.5 were not computed as direct averages of all cases, but were computed as percentages of their presence.

Figure 10.9 illustrates some examples of the segmentation results of the FrCN versus FCN, U-Net, and SegNet compared to the ground-truth contours. This figure shows the segmentation results of each clinical diagnostic case in the ISIC 2017 dataset. All approaches showed reasonable segmentation performance on the skin lesion cases such as the presented in Figure 10.9 (a). In contrast, FrCN achieved better segmentation when the dermoscopy image contained hair obstacles, as shown in Figure 10.9 (c). Also, Figure 10.9 (b) illustrates how the FrCN method segments one of the most challenging skin lesion cases with very low contrast.

In fact, the FrCN method is able to take an input image of any arbitrary size since there are no downsampling layers. However, there is a demand on the computational resources (i.e., enough memory), since the original dermoscopy images come with very large sizes (i.e., from 540×722 to $4,499 \times 6,748$). In this work, all images are resized in a preprocessing step for two reasons. One is to compare the conventional segmentation methods that require the fixed image size. The other is to take our computation resources into account.

10.6 CONCLUSIONS

In this chapter, we presented a full-resolution convolutional networks (FrCN) method for skin lesion segmentation. Unlike the previous well-known segmentation approaches, the FrCN method is able to utilize full spatial resolution features for each pixel of dermoscopy images, improving in the performance of pixel-wise segmentation. We have evaluated the FrCN method utilizing the ISIC 2017 challenge dataset. The results show that the FrCN method outperformed the recent FCN, U-Net, and SegNet techniques. In future work, a larger number of training dermoscopy images should be used to improve the segmentation performance of each class. Developing a computer-aided diagnostic system using segmented lesions is also necessary to distinguish between benign, melanoma, and normal skin lesions.

ACKNOWLEDGMENTS

This work was supported by the International Collaborative Research and Development Programme (funded by the Ministry of Trade, Industry and Energy [MOTIE, Korea]) (N0002252). This work was also supported by the National Research Foundation of Korea (NRF) grant funded by the Korean government (MEST) (NRF-2019R1A2C1003713).

BIBLIOGRAPHY

1. American Joint Committee on Cancer, "Melanoma of the skin," *Cancer Staging Manual*, pp. 209–220, Springer, New York, NY, 2002.
2. M. E. Celebi, H. A. Kingravi, B. Uddin, H. Lyatornid, Y. A. Aslandogan, W. V. Stoecker, and R. H. Moss, "A methodological approach to the classification of dermoscopy images," *Computerized Medical Imaging and Graphics*, vol. 31, no. 6, pp. 362–373, September, 2007.
3. R. L. Siegel, K. D. Miller, and A. Jemal, "Cancer statistics," *CA Cancer Journal for Clinicians*, vol. 68, no. 1, pp. 7–30, 2018.

4. American Cancer Society, "Cancer facts & figures 2018, American Cancer Society, Atlanta, 2018," Accessed [September 3, 2018]; https://www.cancer.org/cancer/melan oma-skin-cancer.html.

5. A. M. Noone, N. Howlader, M. Krapcho, D. Miller, A. Brest, M. Yu, J. Ruhl, Z. Tatalovich, A. Mariotto, D. R. Lewis, H. S. Chen, E. J. Feuer, and K. A. Cronin, "SEER cancer statistics review, 1975–2015," National Cancer Institute, 2018.

6. M. E. Vestergaard, P. Macaskill, P. E. Holt, and S. W. Menzies, "Dermoscopy compared with naked eye examination for the diagnosis of primary melanoma: A meta-analysis of studies performed in a clinical setting," *British Journal of Dermatology*, vol. 159, no. 3, pp. 669–676, 2008.

7. M. Silveira, J. C. Nascimento, J. S. Marques, A. R. Marçal, T. Mendonça, S. Yamauchi, J. Maeda, and J. Rozeira, "Comparison of segmentation methods for melanoma diagnosis in dermoscopy images," *IEEE Journal of Selected Topics in Signal Processing*, vol. 3, no. 1, pp. 35–45, 2009.

8. M. E. Celebi, H. Iyatomi, G. Schaefer, and W. V. Stoecker, "Lesion border detection in dermoscopy images," *Computerized Medical Imaging and Graphics*, vol. 33, no. 2, pp. 148–53, March, 2009.

9. H. Ganster, A. Pinz, R. Rohrer, E. Wildling, M. Binder, and H. Kittler, "Automated melanoma recognition," *IEEE Transactions on Medical Imaging*, vol. 20, no. 3, pp. 233–239, March, 2001.

10. E. Meskini, M. S. Helfroush, K. Kazemi, and M. Sepaskhah, "A new algorithm for skin lesion border detection in dermoscopy images," *Journal of Biomedical Physics and Engineering*, vol. 8, no. 1, pp. 117–126, 2018.

11. G. Schaefer, B. Krawczyk, M. E. Celebi, and H. Iyatomi, "An ensemble classification approach for melanoma diagnosis," *Memetic Computing*, vol. 6, no. 4, pp. 233–240, December, 2014.

12. M. A. Al-Masni, M. A. Al-antari, M. T. Choi, S. M. Han, and T. S. Kim, "Skin lesion segmentation in dermoscopy images via deep full resolution convolutional networks," *Computer Methods and Programs in Biomedicine*, vol. 162, pp. 221–231, August, 2018.

13. M. E. Yuksel, and M. Borlu, "Accurate segmentation of dermoscopic images by image thresholding based on type-2 fuzzy logic," *IEEE Transactions on Fuzzy Systems*, vol. 17, no. 4, pp. 976–982, August, 2009.

14. K. Mollersen, H. M. Kirchesch, T. G. Schopf, and F. Godtliebsen, "Unsupervised segmentation for digital dermoscopic images," *Skin Research and Technology*, vol. 16, no. 4, pp. 401–407, November, 2010.

15. M. E. Celebi, Q. Wen, S. Hwang, H. Iyatomi, and G. Schaefer, "Lesion border detection in dermoscopy images using ensembles of thresholding methods," *Skin Research and Technology*, vol. 19, no. 1, pp. E252–E258, 2013.

16. F. Peruch, F. Bogo, M. Bonazza, V. M. Cappelleri, and E. Peserico, "Simpler, faster, more accurate melanocytic lesion segmentation through MEDS," *IEEE Transactions on Biomedical Engineering*, vol. 61, no. 2, pp. 557–565, February, 2014.

17. H. Y. Zhou, G. Schaefer, A. H. Sadka, and M. E. Celebi, "Anisotropic mean shift based fuzzy c-means segmentation of dermoscopy images," *IEEE Journal of Selected Topics in Signal Processing*, vol. 3, no. 1, pp. 26–34, February, 2009.

18. S. Kockara, M. Mete, V. Yip, B. Lee, and K. Aydin, "A soft kinetic data structure for lesion border detection," *Bioinformatics*, vol. 26, no. 12, pp. i21–i28, June 15, 2010.

19. S. Suer, S. Kockara, and M. Mete, "An improved border detection in dermoscopy images for density based clustering," *BMC Bioinformatics*, vol. 12, no. 10, p. S12.

20. F. Y. Xie, and A. C. Bovik, "Automatic segmentation of dermoscopy images using self-generating neural networks seeded by genetic algorithm," *Pattern Recognition*, vol. 46, no. 3, pp. 1012–1019, March, 2013.

21. M. E. Celebi, H. A. Kingravi, H. Iyatomi, Y. A. Aslandogan, W. V. Stoecker, R. H. Moss, J. M. Malters, J. M. Grichnik, A. A. Marghoob, H. S. Rabinovitz, and S. W. Menzies, "Border detection in dermoscopy images using statistical region merging," *Skin Research and Technology*, vol. 14, no. 3, pp. 347–353, August, 2008.
22. Q. Abbas, M. E. Celebi, and I. F. Garcia, "Skin tumor area extraction using an improved dynamic programming approach," *Skin Research and Technology*, vol. 18, no. 2, pp. 133–142, May, 2012.
23. Q. Abbas, M. E. Celebi, I. F. Garcia, and M. Rashid, "Lesion border detection in dermoscopy images using dynamic programming," *Skin Research and Technology*, vol. 17, no. 1, pp. 91–100, February, 2011.
24. B. Erkol, R. H. Moss, R. J. Stanley, W. V. Stoecker, and E. Hvatum, "Automatic lesion boundary detection in dermoscopy images using gradient vector flow snakes," *Skin Research and Technology*, vol. 11, no. 1, pp. 17–26, February, 2005.
25. M. Mete, and N. M. Sirakov, "Lesion detection in demoscopy images with novel density-based and active contour approaches," *BMC Bioinformatics*, vol. 11, no. 6, p. S23, 2010.
26. A. R. Sadri, M. Zekri, S. Sadri, N. Gheissari, M. Mokhtari, and F. Kolahdouzan, "Segmentation of dermoscopy images using wavelet networks," *IEEE Transactions on Biomedical Engineering*, vol. 60, no. 4, pp. 1134–1141, 2013.
27. K. Korotkov, and R. Garcia, "Computerized analysis of pigmented skin lesions: A review," *Artificial Intelligence in Medicine*, vol. 56, no. 2, pp. 69–90, October, 2012.
28. M. E. Celebi, Q. Wen, H. Iyatomi, K. Shimizu, H. Zhou, and G. Schaefer, "A state-of-the-art survey on lesion border detection in dermoscopy images," *Dermoscopy Image Analysis*, pp. 97–129, CRC Press, 2015.
29. R. B. Oliveira, J. P. Papa, A. S. Pereira, and J. M. R. S. Tavares, "Computational methods for pigmented skin lesion classification in images: Review and future trends," *Neural Computing & Applications*, vol. 29, no. 3, pp. 613–636, February, 2018.
30. S. Pathan, K. G. Prabhu, and P. C. Siddalingaswamy, "Techniques and algorithms for computer aided diagnosis of pigmented skin lesions-A review," *Biomedical Signal Processing and Control*, vol. 39, pp. 237–262, 2018.
31. M. A. Al-Masni, M. A. Al-Antari, J. M. Park, G. Gi, T. Y. Kim, P. Rivera, E. Valarezo, M. T. Choi, S. M. Han, and T. S. Kim, "Simultaneous detection and classification of breast masses in digital mammograms via a deep learning YOLO-based CAD system," *Computer Methods and Programs in Biomedicine*, vol. 157, pp. 85–94, 2018.
32. G. Carneiro, J. Nascimento, and A. P. Bradley, "Automated analysis of unregistered multi-view mammograms with deep learning," *IEEE Transactions on Medical Imaging*, vol. 36, no. 11, pp. 2355–2365, 2017.
33. N. Dhungel, G. Carneiro, and A. P. Bradley, "A deep learning approach for the analysis of masses in mammograms with minimal user intervention," *Medical Image Analysis*, vol. 37, pp. 114–128, April, 2017.
34. G. Litjens, T. Kooi, B. E. Bejnordi, A. A. A. Setio, F. Ciompi, M. Ghafoorian, J. van der Laak, B. van Ginneken, and C. I. Sanchez, "A survey on deep learning in medical image analysis," *Medical Image Analysis*, vol. 42, pp. 60–88, December, 2017.
35. X. Zhao, Y. Wu, G. Song, Z. Li, Y. Zhang, and Y. Fan, "A deep learning model integrating FCNNs and CRFs for brain tumor segmentation," *Medical Image Analysis*, vol. 43, pp. 98–111, January, 2018.
36. E. Gibson, W. Li, C. Sudre, L. Fidon, D. I. Shakir, G. Wang, Z. Eaton-Rosen, R. Gray, T. Doel, Y. Hu, T. Whyntie, P. Nachev, M. Modat, D. C. Barratt, S. Ourselin, M. J. Cardoso, and T. Vercauteren, "NiftyNet: A deep-learning platform for medical imaging," *Computer Methods and Programs in Biomedicine*, vol. 158, pp. 113–122, May, 2018.

37. M. A. Al-Antari, M. A. Al-Masni, M. T. Choi, S. M. Han, and T. S. Kim, "A fully integrated computer-aided diagnosis system for digital X-ray mammograms via deep learning detection, segmentation, and classification," *International Journal of Medical Informatics*, vol. 117, pp. 44–54, September, 2018.

38. Y. LeCun, Y. Bengio, and G. Hinton, "Deep learning," *Nature*, vol. 521, no. 7553, pp. 436–44, May 28, 2015.

39. M. A. Al-Masni, M. A. Al-Antari, J. M. Park, G. Gi, T. Y. Kim, P. Rivera, E. Valarezo, S. M. Han, and T. S. Kim, "Detection and classification of the breast abnormalities in digital mammograms via regional Convolutional Neural Network," in *Proceedings of the 39th Annual International Conference of the IEEE Engineering in Medicine and Biology Society (EMBC)*, Jeju Island, Republic of Korea, 2017, pp. 1230–1233.

40. A. Krizhevsky, I. Sutskever, and G. E. Hinton, "ImageNet classification with deep convolutional neural networks," *Communications of the ACM*, vol. 60, no. 6, pp. 84–90, June, 2017.

41. D. Ciresan, A. Giusti, L. M. Gambardella, and J. Schmidhuber, "Deep neural networks segment neuronal membranes in electron microscopy images," in *Advances in Neural Information Processing Systems*, 2012, pp. 2843–2851.

42. C. Cernazanu-Glavan, and S. Holban, "Segmentation of bone structure in X-ray images using convolutional neural network," *Advances in Electrical and Computer Engineering*, vol. 13, no. 1, pp. 87–94, 2013.

43. M. Melinščak, P. Prentašić, and S. Lončarić, "Retinal vessel segmentation using deep neural networks," in *10th International Conference on Computer Vision Theory and Applications (VISAPP 2015)*, pp. 577–582, 2015.

44. J. Long, E. Shelhamer, and T. Darrell, "Fully convolutional networks for semantic segmentation," in *IEEE Conference on Computer Vision and Pattern Recognition (CVPR)*, 2015, pp. 3431–3440.

45. E. Shelhamer, J. Long, and T. Darrell, "Fully convolutional networks for semantic segmentation," *IEEE Transactions on Pattern Analysis and Machine Intelligence*, vol. 39, no. 4, pp. 640–651, April, 2017.

46. A. Garcia-Garcia, S. Orts-Escolano, S. Oprea, V. Villena-Martinez, and J. Garcia-Rodriguez, "A review on deep learning techniques applied to semantic segmentation," arXiv preprint arXiv:1704.06857, 2017.

47. O. Ronneberger, P. Fischer, and T. Brox, "U-net: Convolutional networks for biomedical image segmentation," in *International Conference on Medical Image Computing and Computer-Assisted Intervention*, 2015, pp. 234–241.

48. V. Badrinarayanan, A. Kendall, and R. Cipolla, "SegNet: A deep convolutional encoder-decoder architecture for image segmentation," *IEEE Transactions on Pattern Analysis and Machine Intelligence*, vol. 39, no. 12, pp. 2481–2495, December, 2017.

49. V. Badrinarayanan, A. Kendall, and R. Cipolla, "Segnet: A deep convolutional encoder-decoder architecture for image segmentation," arXiv preprint arXiv:1511.00561, 2015.

50. H. Noh, S. Hong, and B. Han, "Learning deconvolution network for semantic segmentation," in *Proceedings of the IEEE International Conference on Computer Vision*, 2015, pp. 1520–1528.

51. L. C. Chen, G. Papandreou, I. Kokkinos, K. Murphy, and A. L. Yuille, "DeepLab: Semantic image segmentation with deep convolutional nets, atrous convolution, and fully connected CRFs," *IEEE Transactions on Pattern Analysis and Machine Intelligence*, vol. 40, no. 4, pp. 834–848, April, 2018.

52. L. Chen, G. Papandreou, I. Kokkinos, K. Murphy, and A. L. Yuille, "Deeplab: Semantic image segmentation with deep convolutional nets, atrous convolution, and fully connected CRFs," CoRR abs/1606.00915, 2016.

53. Y. X. Li, and L. L. Shen, "Skin lesion analysis towards melanoma detection using deep learning network," *Sensors*, vol. 18, no. 2, February, 2018.

54. L. Bi, J. Kim, E. Ahn, A. Kumar, M. Fulham, and D. Feng, "Dermoscopic image segmentation via multistage fully convolutional networks," *IEEE Transactions on Biomedical Engineering*, vol. 64, no. 9, pp. 2065–2074, 2017.
55. L. Q. Yu, H. Chen, Q. Dou, J. Qin, and P. A. Heng, "Automated melanoma recognition in dermoscopy images via very deep residual networks," *IEEE Transactions on Medical Imaging*, vol. 36, no. 4, pp. 994–1004, 2017.
56. Y. D. Yuan, M. Chao, and Y. C. Lo, "Automatic skin lesion segmentation using deep fully convolutional networks with jaccard distance," *IEEE Transactions on Medical Imaging*, vol. 36, no. 9, pp. 1876–1886, 2017.
57. M. Goyal, and M. H. Yap, "Multi-class semantic segmentation of skin lesions via fully convolutional networks," arXiv preprint arXiv:1711.10449, 2017.
58. B. S. Lin, K. Michael, S. Kalra, and H. R. Tizhoosh, "Skin lesion segmentation: U-Nets versus clustering," in *2017 IEEE Symposium Series on Computational Intelligence (SSCI)*, 2017, pp. 1–7.
59. Y. Yuan, "Automatic skin lesion segmentation with fully convolutional-deconvolutional networks," arXiv preprint arXiv:1703.05165, 2017.
60. N. C. F. Codella, D. Gutman, M. E. Celebi, B. Helba, M. A. Marchetti, S. W. Dusza, A. Kalloo, K. Liopyris, N. Mishra, H. Kittler, and A. Halpern, "Skin lesion analysis toward melanoma detection: A challenge at the 2017 International Symposium on Biomedical Imaging (ISBI), hosted by the International Skin Imaging Collaboration (ISIC)," in *15th IEEE International Symposium on Biomedical Imaging (ISBI 2018)*, 2018, pp. 68–172.
61. International Skin Imaging Collaboration, "ISIC 2017: Skin lesion analysis towards melanoma detection," Accessed [October 19, 2018]; https://challenge.kitware.com/#challenges.
62. Z. Jiao, X. Gao, Y. Wang, and J. Li, "A deep feature based framework for breast masses classification," *Neurocomputing* vol. 197, pp. 221–231, 2016.
63. T. Kooi, G. Litjens, B. van Ginneken, A. Gubern-Merida, C. I. Sanch021, R. Mann, A. den Heeten, and N. Karssemeijer, "Large scale deep learning for computer aided detection of mammographic lesions," *Medical Image Analysis*, vol. 35, pp. 303–312, January, 2017.
64. H. R. Roth, L. Lu, J. M. Liu, J. H. Yao, A. Seff, K. Cherry, L. Kim, and R. M. Summers, "Improving computer-aided detection using convolutional neural networks and random view aggregation," *IEEE Transactions on Medical Imaging*, vol. 35, no. 5, pp. 1170–1181, May, 2016.
65. M. Lin, Q. Chen, and S. Yan, "Network in network," in rXiv preprint arXiv:1312.4400, pp. 1–10, 2013.
66. D. Scherer, A. Müller, and S. Behnke, "Evaluation of pooling operations in convolutional architectures for object recognition," in *Artificial Neural Networks–ICANN 2010*, 2010, pp. 92–101.
67. K. Simonyan, and A. Zisserman, "Very deep convolutional networks for large-scale image recognition," arXiv preprint arXiv:1409.1556, 2014.
68. Q. Guo, F. L. Wang, J. Lei, D. Tu, and G. H. Li, "Convolutional feature learning and Hybrid CNN-HMM for scene number recognition," *Neurocomputing*, vol. 184, pp. 78–90, April 5, 2016.
69. A. L. Maas, A. Y. Hannun, and A. Y. Ng., "Rectifier nonlinearities improve neural network acoustic models," in *Proceeding of 30th International Conference on Machine Learning (ICML)*, 2013, p. 3.
70. X. Glorot, A. Bordes, and Y. Bengio, "Deep sparse rectifier neural networks," in *Proceedings of the 14th International Conference on Artificial Intelligence and Statistics*, 2011, pp. 315–323.
71. N. Srivastava, G. Hinton, A. Krizhevsky, I. Sutskever, and R. Salakhutdinov, "Dropout: A simple way to prevent neural networks from overfitting," *Journal of Machine Learning Research*, vol. 15, no. 1, pp. 1929–1958, 2014.

72. S. Hoo-Chang, H. R. Roth, M. Gao, L. Lu, Z. Xu, I. Nogues, J. Yao, D. Mollura, and R. M. Summers, "Deep convolutional neural networks for computer-aided detection: CNN architectures, dataset characteristics and transfer learning," *IEEE Transactions on Medical Imaging*, vol. 35, no. 5, pp. 1285–1298, 2016.

73. Y. Bar, I. Diamant, L. Wolf, S. Lieberman, E. Konen, and H. Greenspan, "Chest pathology identification using deep feature selection with non-medical training," *Computer Methods in Biomechanics and Biomedical Engineering-Imaging and Visualization*, vol. 6, no. 3, pp. 259–263, 2018.

74. R. K. Samala, H. P. Chan, L. Hadjiiski, M. A. Helvie, J. Wei, and K. Cha, "Mass detection in digital breast tomosynthesis: Deep convolutional neural network with transfer learning from mammography," *Medical Physics*, vol. 43, no. 12, pp. 6654–6666, December, 2016.

75. J. Yosinski, J. Clune, Y. Bengio, and H. Lipson, "How transferable are features in deep neural networks?" in *Advances in Neural Information Processing Systems*, 2014, pp. 3320–3328.

76. O. Russakovsky, J. Deng, H. Su, J. Krause, S. Satheesh, S. Ma, Z. H. Huang, A. Karpathy, A. Khosla, M. Bernstein, A. C. Berg, and L. Fei-Fei, "ImageNet large scale visual recognition challenge," *International Journal of Computer Vision*, vol. 115, no. 3, pp. 211–252, December, 2015.

77. E. Szymanska, E. Saccenti, A. K. Smilde, and J. A. Westerhuis, "Double-check: validation of diagnostic statistics for PLS-DA models in metabolomics studies," *Metabolomics*, vol. 8, no. 1, pp. S3–S16, June, 2012.

78. S. Smit, M. J. van Breemen, H. C. J. Hoefsloot, A. K. Smilde, J. M. F. G. Aerts, and C. G. de Koster, "Assessing the statistical validity of proteomics based biomarkers," *Analytica Chimica Acta*, vol. 592, no. 2, pp. 210–217, June 5, 2007.

79. T. Hastie, R. Tibshirani, and J. Friedman, *The Elements of Statistical Learning; Data Mining, Inference and Prediction*, Second ed., Springer, New York, 2008.

80. S. Min, B. Lee, and S. Yoon, "Deep learning in bioinformatics," *Briefings in Bioinformatics*, vol. 18, no. 5, pp. 851–869, September, 2017.

81. Image Segmentation Keras, "Implementation of Segnet, FCN, UNet and other models in Keras," Accessed [October 23, 2018]; https://github.com/divamgupta/image-segmentation-keras.

82. Keras, "Keras: The python deep learning library," Accessed [January 8, 2019]; https://keras.io/.

83. M. Dong, X. Lu, Y. Ma, Y. Guo, Y. Ma, and K. Wang, "An efficient approach for automated mass segmentation and classification in mammograms," *Journal of Digital Imaging*, vol. 28, no. 5, pp. 613–25, 2015.

84. M. A. Al-antari, M. A. Al-masni, S. U. Park, J. Park, M. K. Metwally, Y. M. Kadah, S. M. Han, and T. S. Kim, "An automatic computer-aided diagnosis system for breast cancer in digital mammograms via deep belief network," *Journal of Medical and Biological Engineering*, vol. 38, no. 3, pp. 443–456, June, 2018.

85. D. Zikic, Y. Ioannou, M. Brown, and A. Criminisi, "Segmentation of brain tumor tissues with convolutional neural networks," in *MICCAI-BRATS*, 2014, pp. 36–39.

86. S. Pereira, A. Pinto, V. Alves, and C. A. Silva, "Brain tumor segmentation using convolutional neural networks in MRI images," *IEEE Transactions on Medical Imaging*, vol. 35, no. 5, pp. 1240–1251, May, 2016.

87. D. M. Powers, "Evaluation: From precision, recall and F-measure to ROC, informedness, markedness & correlation," *Journal of Machine Learning Technologies*, vol. 2, no. 1, pp. 37–63, 2011.

88. M. D. Zeiler, "ADADELTA: An adaptive learning rate method," arXiv preprint arXiv:1212.5701, 2012.

89. D. P. Kingma, and J. L. Ba, "Adam: A method for stochastic optimization," in *International Conference on Learning Representations (ICLR)*, arXiv preprint arXiv:1412.6980, 2015.
90. T. Tieleman, and G. Hinton, "Lecture 6.5-rmsprop: Divide the gradient by a running average of its recent magnitude," 4, http://www.cs.toronto.edu/~tijmen/csc321/slides/lecture_slides_lec6.pdf, 2012.
91. M. Berseth, "ISIC 2017-skin lesion analysis towards melanoma detection," arXiv preprint arXiv:1703.00523v1, 2017.
92. L. Bi, J. Kim, E. Ahn, and D. Feng, "Automatic skin lesion analysis using large-scale dermoscopy images and deep residual networks," arXiv preprint arXiv:1703.04197, 2017.
93. A. Menegola, J. Tavares, M. Fornaciali, L. T. Li, S. Avila, and E. Valle, "RECOD titans at ISIC challenge 2017," arXiv preprint arXiv:1703.04819, 2017.

11 Skin Melanoma Classification Using Deep Convolutional Neural Networks

Khalid M. Hosny, Mohamed A. Kassem, and Mohamed M. Foaud

CONTENTS

11.1 INTRODUCTION

Computer-supported skin lesion inspection and dermal investigation have been extended over the past decade to classify and model dermatological diseases and human skin [1]. Automated systems for analyzing and synthesizing human skin face different difficulties and complexities such as humidity and seasonal variation in temperature, geographically based differences in diseases, environmental factors, and hair presences [2]. Computerized assessment of skin lesions is an active research area. The main target of this field is to create systems able to analyze images of pigmented skin lesions for automated diagnosis of cancerous lesions or to assist dermatologists in this task [3]. The most important challenge in this context is classifying the color images of the skin lesions with highly accurate classifiers [4]. In terms of dermoscopic images, skin lesions fall into different categories based on the disease type, severity, and stage [5]. There are many algorithms used to extract skin lesion features [6]. The "ABDC" rule is the most known algorithm, where A, B, D, and C refer to lesion Asymmetry, irregularity of the Border, Color variation, and lesion Diameter, respectively. COLOR is the most prevalent content that can be easily visually observed when retrieving images. This is due to its value in isolating diseased skin from wholesome [7]. However, color-based features are highly influenced by the condition of the patient and his/her external environment such as variations in natural skin complexion, lighting, and camera specifications [8].

During the last decade, there have been many studies on classifying non-melanocytic and melanocytic skin lesions. These studies vary in using statistical methods, traditional feature extraction, and classification methods [11–15]. Recently, deep learning has been raised as a promising tool for feature extraction of images through different models. In these models, many layers of neurons have been used to fulfill different layers to extract the fine details of the image, where the new layer combines the characteristics of the previous one. This comprises the newest generic of such models, which are called convolutional neural networks (CNNs) [16]. Applying deep learning techniques in recent healthcare research area has had positive results, as suggested by Dubal et al. and Kumar et al. [17, 18]. There are many studies using different learning techniques, as can be found in Nasr-Esfahan et al., Premaladha and Ravichandran, and Kostopoulos et al. [19–22]. Various kinds of skin tumor have been found, like squamous cell carcinoma, basal cell carcinoma, and melanoma; the last of these is the most unpredictable. Jain et al. [23] suggested that detecting melanoma at an early stage is helpful for curing it. Precisely analyzing each patient sample is a tedious task for dermatologists; therefore, an efficient automated system with a high classification rate is necessary to assess the dangers associated with the given samples [24].

Martin et al. [25] showed that melanoma is a highly invasive and malignant tumor that can reach the bloodstream easily and cause metastasis in affected patients. Azulay [26] reported that one of the hardest challenges in dermatology is to distinguish malignant melanotic lesions (melanoma) from non-malignant melanotic lesions (nevus), since both diseases share morphological characteristics. Melanomas may also take on features that may not be common to their morphology, which can cause it to be confused with a completely different melanoma disease. Goodfellow et al. [16] concluded that CNNs are the best tool for recognizing patterns in digital images. In this chapter, we assessed the existing methods for skin lesion classification, especially deep learning-based methods. It was observed that existing deep learning-based methods encounter two major problems that negatively affect their performance and significantly reduce their classification rates. First, color images of skin lesions are associated with complicated backgrounds. Second, the available skin lesion datasets contain a limited number of images, while deep learning classification methods need a large number of images for good training.

These challenges motivated us to present a new method that significantly improves the skin cancer classification rate and outperforms all previous methods. High skin lesion detection and classification rates enable dermatologists and physicians to make the right decisions at an earlier stage, which generally improves patients' health and saves their lives. The proposed method is a pretrained deep convolutional neural network-based classification method. The well-known datasets of color images of skin, MED-NODE, Dermatology Information System (DermIS), & DermQuest, are used to prove the validity of the proposed method and evaluate its performance against the other existing methods. The obtained results clearly show that the proposed method achieves the best classification rate.

11.2   LITERATURE REVIEW

This section summarizes some of the state of the art. By the end of this section, a table is presented to compare all the listed methods. Bunte et al. [7] proposed a classifying melanoma system. In addition to first, second, and third moment orders, Bunte and his coauthors used correlograms, coherence vectors, and histograms for the representation of color image features. A binary mask has been used by Chang et al. [9] to perform principal component analysis (PCA) for the classification of conventional photographs. They achieved a classification rate of 82.30%. In order to improve the performance of their system, they extracted gray-level co-occurrence-based features and used these features in classifying different benign skin lesions. Their system achieves an accuracy of 90.86% [9]. Kundu et al. [10] used the local binary pattern (LBP) for feature extraction and developed another classification system that utilized multilayer perceptron, naïve Bayes, and a support vector machine (SVM) for skin disease classification. Amelard et al. [11] proposed extracting a set of morphological high-level intuitive features (HLIFs) to describe the amount of lesion border irregularity. They incorporated the HLIFs into a set of low-level features to get the more semantic meaning of the feature set and to allow the system to provide an intuitive rationale for the classification decision. Their system achieved the classification rate of 87.38%. An automated model for skin cancer classification has been developed by Karabulut and Ibrikci [12]. They used the convolutional neural networks (CNNs) and the SVM. Texture analysis of block difference of inverse probabilities and LBP have been used as a preprocessing step. They achieved a classification rate of 71.4%.

A computer-aided diagnosis (CAD) system has been proposed by Almaraz-Damian et al. [13], where the ABCD is rule used to extract the features and SVM is used in classification. Their system achieved a classification rate of 75.1%. Giotis et al. [14] proposed a system that extracts the regions of the lesion by using the descriptors of the texture and the color. Jafari et al. [15] describes a guided filtering method in which the noise of the input images is reduced. Then, the features of skin lesions are extracted using the ABCD rule. This method achieved a classification rate of 79%. Implementation of deep learning with clinical images has been proposed by Nasr-Esfahan et al. [19]. The accuracy of the system is increased by preprocessing to correlate the illumination and the segmentation of the region of interest (ROI). The proposed system sends the image to the CNN after segmentation for feature extraction. The system achieves a classification rate of 81%. Premaladha and Ravichandran [20] proposed a CAD system that combines supervised and deep learning algorithms in skin image classification. The contrast-limited adaptive histogram equalization technique (CLAHE) is used to enhance the input image; then, the median filter with the normalized Otsu's segmentation (NOS) method is used to separate the affected lesion from normal skin. This system achieved a classification rate of 90.12% with artificial neural networks (ANN) and a rate of 92.89% with deep learning neural networks (DLNN). Kostopoulos et al. [21] proposed a computer-based analysis of plain photography using a probabilistic neural network (PNN) to extract the features and decide if the lesion is melanocytic or melanoma. The system achieved a classification rate of 76.2%. Esteva et al. [22] classified the skin lesion using a pretrained CNN called Inception v3. They achieved accuracy of about 71.2% when they augmented the images in the dataset using rotations with random angles from 0° and 359° and vertical flipping. Table 11.1 summarizes and compares these state-of-the-art methods.

In Hosny et al. [42], transfer learning has been applied to a CNN model. To extract ROI from images, images have been segmented to discard the background. To overcome the limitation of low numbers of images, different methods of augmentation have been carried on the segmented images. They achieved 97.70% accuracy. In [43], the last layers are replaced in a pretrained CNN model, with different ways of augmentation, and fine-tuned; an accuracy of 98.61% is achieved.

11.3 BACKGROUND

During the last few years, deep neural networks have gained researchers' attention around the world as an alternative approach to extracting features. The core idea of deep neural networks is that these networks automatically select the most important features without performing feature extraction [31]. This mechanism is a powerful step in machine intelligence [33]. Through the following subsections, a brief description of the convolutional neural networks is given.

11.3.1 PREPROCESSING

In machine vision and image-processing applications, segmenting the ROI is a very crucial task. Image segmentation usually improves the accuracy of classification, since this allows the background of the original image to be ignored. We utilized the segmentation method of Basavaprasad and Hegad [39] to segment the ROIs from the color images of skin. In this method, a number of bins are used to coarsely represent the input color image. The spatial information is used by coarse representation from a histogram-based windowing process, and then the coarse data of the image is clustered by hierarchical clustering.

TABLE 11.1
Comparison of Literature Review

Method	Dataset	Enhancement	Segmentation	Image type	Classification method	Accuracy (%)
Bunte et al. [7]	Real-world dataset University of Groningen	Yes	No	RGB	Image retrieval	84
Chang et al. [9]	Collected by authors	No	No	Gray level	SVM	82.30
Kundu et al. [10]	Collected by authors			Gray level	MLP	94.28
Amelard et al. [11]	DermIS & DermQuest	Yes	Yes	RGB	SVM	87.38
Almaraz et al. [13]	DermIS & DermQuest	Yes	Yes	RGB	SVM	75.1
Karabulut et al. [12]	DermIS & DermQuest	Yes	Yes	Black & white	DCNN	71.4
Karabulut et al. [12]	DermIS & DermQuest	Yes	Yes	RGB	SVM	71.4
Premaladha et al. [20]	MED-NODE	Yes	Yes	Black & white	DLNN	92.89
Premaladha et al. [20]	MED-NODE	Yes	Yes	Black & white	ANN	90.12
Nasr-Esfahan et al. [19]	MED-NODE	Yes	Yes	RGB	CNN	81
Giotis et al. [14]	MED-NODE	Yes	Yes	RGB	Majority vote	81
M.H. Jafari et al. [15]	MED-NODE	Yes	Yes	RGB	SVM	79
Kostopoulos et al. [21]	MED-NODE	Yes	Yes	RGB	PNN	76.2
Esteva et al. [22]	ISIC 2017	No	Yes	RGB	DCNN	71.2
Hosny et al. [42]	ISIC 2017	No	Yes	RGB	DCNN	95.91
Hosny et al. [43]	PH2	No	Yes	RGB	DCNN	98.61

11.3.2 Convolutional Neural Networks

A convolutional neural network (CNN) consists of stacked layers in which a given layer has learned from the previous layer in a linear or nonlinear way [27]. Any CNN model contains building blocks consisting of the following:

1. Convolutional layer: Applies a convolution operation to the input image by passing the result to the next layer.
2. Pooling layer (also called subsampling or down-sampling): Reduces the dimensionality of each feature map but retains the most important information. Spatial pooling can be of different types: max, average, sum, etc.
3. Nonlinear rectified linear unit (RELU) layer: Element-wise operation (applied per pixel) that replaces all negative pixel values in the feature map with zero and that is connected to a regular multilayer neural network called a fully connected layer.
4. Connected layer: A traditional multilayer perceptron that uses a SoftMax activation function in the output layer and in which every neuron in the previous layer is connected to every neuron on the next layer.
5. Loss layer: Specifies how training penalizes the deviation between the predicted and true labels, being at the backend.

CNNs have been used successfully with significant improved performance in different applications such as natural language processing [28] and visual tasks [29]. Many deep convolution neural networks (DCNNs) are available, including LeNet, Alex-net, ZFNet, GoogLeNet, VGGNet, and Resnet [30, 31].

11.3.2.1 Alex-Net CNN Model

Alex-net was developed by Krizhevsky et al. [29] as a DCNN model. It is used for the ImageNet large-scale visual recognition challenge (ILSVRC) [30]. The input image of Alex-net must be resized to the specific size $227 \times 227 \times 3$, where the width (W) is 227, the height (H) is 227, and the depth (D) is 3. The depth number refers to the red, green, and blue channels. The input image is filtered by using the first convolutional layer with a number of kernels (K) equal to 96, with a filter (F) size of 11×11, and a stride (S) equal to 4 pixels. The stride is defined as the distance in the kernel map between responsive field centers of neighboring neurons. The size of the convolution layer output volume is computed using the mathematical form ($W-F+2P$)/$S+1$, where P refers to the number of padded pixels, here equal to zero. For input size of $227 \times 227 \times 3$, the width $W = 227$, $F = 11$, $S = 4$, $K = 96$, and the size of the convolution layer output volume is ($227-11+0$)/4 + 1 = 55. Then, the size of the convolution layer output volume is $55 \times 55 \times 96$.

The computational load for the second convolutional layer is distributed to two GPUs; therefore, the load for each GPU is $55 \times 55 \times 48$. The output of the first convolutional layer is the input to the second convolutional layer, which consists of max pooling followed by convolution. The max pooling is $55/2 \times 55/2 \times 96 \approx 27 \times 27 \times 96$. The input in this layer is filtered using 256 filters. The size of each filter is $5 \times 5 \times 48$

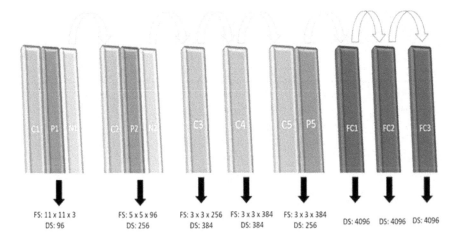

FS: 11 x 11 x 3   FS: 5 x 5 x 96   FS: 3 x 3 x 256   FS: 3 x 3 x 384   FS: 3 x 3 x 384
DS: 96   DS: 256   DS: 384   DS: 384   DS: 256   DS: 4096   DS: 4096   DS: 4096

FIGURE 11.1   Architecture of Alex-net. C, P, N, FC, FZ, and DS refer to convolutional layer, max pooling layer, local response normalization, fully connected layer, filter size, and depth size, respectively.

with a stride of 2 pixels. The output of this layer is $27 \times 27 \times 256$, which splits to two GPUs, where each GPU works with $27 \times 27 \times 128$.

The third, fourth, and fifth convolutional layers are created without any normalization and pooling layers. The third layer of the convolutional network is connected to the (pooled, normalized) outputs of the second convolutional layer with 384 kernels, each of size $3 \times 3 \times 192$. The fourth and fifth convolutional layers have 384 and 256 kernels of sizes $3 \times 3 \times 192$ and $3 \times 3 \times 128$, respectively. There are two fully connected layers containing 4096 neurons for each one. The output features of the fifth layer are the input of these two fully connected layers. Krizhevsky et al. [29] selected 1.2 million images from ImageNet [30] to construct 1000 classes to train the Alex-net. An illustration of the architecture of the Alex-net is displayed in Figure 11.1. Table 11.2 shows different layers with their associated parameters.

11.3.2.2   Resnet50

Resnet50 [44] is one of the deep residual networks (DRN). It is a very deep feedforward neural network (FFNN) with extra connections. A model of deep learning based on the residual learning passes the input from one layer to a far layer by escaping 2 to 5 layers; this method is called "skip connection". Some models like Alex-net try to find a solution by mapping some of the input to meet the same characteristics in the output, but there is an enforcement in the residual network to learn how to map some of the input to some of the output and input.

Resnet50 is a pretrained model consisting of 177 layers and containing a number of connections equaling 192×2 to connect layers with each other. The input layer restricts the size of the input images. The input must be resized to $224 \times 224 \times 3$, where the width is 224, the height is 224, and the depth is 3. The depth number refers to the color space red, green, and blue channels. Resnet50 layers consist of

TABLE 11.2

Different Layers and Their Associated Parameters

Layers	Input size	Output size	Filter size	Depth size	Stride size	Padding size
Conv#1+ RELU	$227 \times 227 \times 3$	$55 \times 55 \times 96$	11×11	96	4	
Max pooling	$55 \times 55 \times 96$	$27 \times 27 \times 96$	3×3		2	
Norm	$27 \times 27 \times 96$					
Conv#2 + RELU	$27 \times 27 \times 96$	$27 \times 27 \times 256$	5×5	256	1	2
Max pooling	$27 \times 27 \times 256$	$13 \times 13 \times 256$	3×3		2	
Norm	$13 \times 13 \times 256$					
Conv#3 + RELU	$13 \times 13 \times 256$	$13 \times 13 \times 384$	3×3	384	1	1
Conv#4 + RELU	$13 \times 13 \times 384$	$13 \times 13 \times 384$	3×3	384	1	1
Conv#5 + RELU	$13 \times 13 \times 384$	$13 \times 13 \times 256$	3×3	256	1	1
Max pooling	$13 \times 13 \times 256$	$6 \times 6 \times 256$	3×3		2	

convolutional, RELU, pooling, batch normalization, a number of additional layers to maintain the skipped connection fully connected, and a classification layer. The building block shown in Figure 11.2 is based on the following equation:

$$y = \mathcal{F}\left(x, \{W_i\}\right) + x \qquad (11.1)$$

where x and y are the input and the output vectors of the layers considered, $\mathcal{F}(x,\{W_i\})$ and represents the residual mapping to be learned. The dimensions of x and $\mathcal{F}$ must be equal.

Because Resnet50 contains 177 layers and 192×2 connections, a detailed part of Resnet50 has been shown in Figure 11.3. Furthermore, Table 11.3 shows different layers with their associated output size and the building block parameters.

11.3.3 TRANSFER KNOWLEDGE FOR DEEP NEURAL NETWORK

In traditional neural networks, utilization of these networks is restricted by the type of the trained data. In other words, the neural networks are not able to classify data if the tested and trained data are different. To overcome this problem, the transfer knowledge, which also called transfer learning methodology, is used with the DCNN. Transfer learning was developed as an alternative learning system to solve this problem [34]. In transfer learning, we must inquire about the knowledge that could be transferred. For example, Alex-net [29] and Resnet50 [44] were trained using the ImageNet [30] to classify images into 1000 classes. Transfer learning means the ability to use Alex-net and Resnet50 to classify different kinds of images with different training classes. To achieve this process, we need to define which knowledge or information could be transferred and which learning algorithms should be used in the knowledge transfer [35]. The relationship types of transfer learning and traditional machine learning are summarized in Table 11.4.

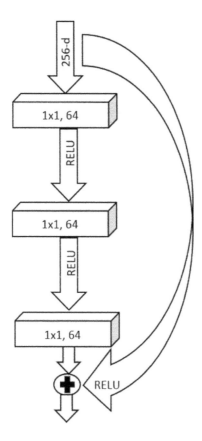

FIGURE 11.2 A "bottleneck" building block for Resnet50

11.4 METHODOLOGY

This section describes the steps of the proposed method to classify color-skin images. It is divided into three subsections. The augmentation process of the input color images of skin is discussed in the first subsection, which describes different augmentation methods that have been applied here. The implementation of the transfer learning methodology is described in the second subsection for Alex-net and Resnet50. The last subsection is devoted to describing the process of dropping out the last three layers for Alex-net and Resnet50 and how authors replaced these layers in order to adapt the process to the required task here.

11.4.1 THE AUGMENTATION PROCESS

DCNN required a large number of trained images to achieve good performance. Working with small datasets of medical images is the biggest challenge for the DCNN. Unfortunately, the available datasets of skin color images, such as MED-NODE [14], DermIS [36], and DermQuest [37], are small datasets. The MED-NODE dataset consists of 70 melanoma and 100 nevus images from the digital image archive

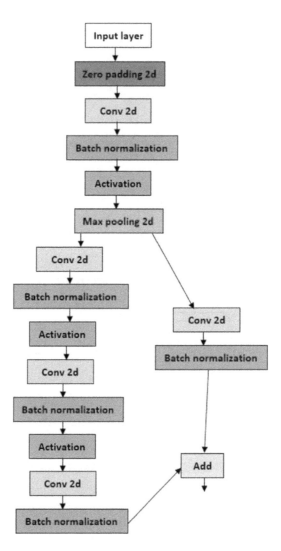

FIGURE 11.3   A detailed part of Resnet50.

of the Department of Dermatology, University Medical Center Groningen (UMCG) in the Netherlands. It is used for the development and testing of the system for skin cancer detection from macroscopic images.

The second dataset consists of 171 images of skin lesions. The images of this dataset were obtained using standard consumer-grade cameras in varying and unconstrained environmental conditions. These images, collected from the publicly available online databases DermIS [36] and DermQuest [37], are classified into 119 images of melanoma and 87 images of nevus; each image contains a single lesion of interest. The melanoma and nevus images are rotated with different angles ranging from 0° to 180° with fixed steps of 5° for each rotation. The number of images in

TABLE 11.3
Resnet50 Layers*

Layer name	Conv1	Conv2_x	Conv3_x	Conv4_x	Conv5_x	Others	FLOPs
Building blocks	7×7, stride 2	3×3, max pool, stide 2					
		$\begin{bmatrix} 1 \times 1,64 \\ 3 \times 3,64 \\ 1 \times 1,256 \end{bmatrix} \times 3$	$\begin{bmatrix} 1 \times 1,128 \\ 3 \times 3,128 \\ 1 \times 1,512 \end{bmatrix} \times 3$	$\begin{bmatrix} 1 \times 1,256 \\ 3 \times 3,256 \\ 1 \times 1,1024 \end{bmatrix} \times 3$	$\begin{bmatrix} 1 \times 1,512 \\ 3 \times 3,512 \\ 1 \times 1,2048 \end{bmatrix} \times 3$	average pool, 1000-d fc, SoftMax	3.8×10^9
Output size	112×112	56×56	28×28	14×14	7×7	1×1	FLOPs

* Brackets Refer to Building Blocks with the Numbers of Stacked Blocks.

TABLE 11.4
Relationship between Traditional Machine Learning and Various Transfer Learning Settings

	Traditional machine learning	Transfer learning	
		Unsupervised/inductive	Supervised/transudative
Domains of target & source	Same	Different but related	Different but related
Tasks of target & source	Same	Different but related	Same

the first dataset becomes (180/5+1) × 70 = 2590 melanoma images and (180/5+1) × 100 = 3700 nevus images.

Similarly, the number of images in the second dataset becomes (180/5+1) × 84 = 3108 melanoma images and (180/5+1) × 87 = 3219 nevus images. The original images are augmented by rotation with random angles ranging from 0° to 355° and translation with different translation parameters for every image. All dataset images were randomly divided into images for training (85% of the images) and images for testing samples of original, segmented, rotated, and translated images (15% of the images) as displayed in Figure 11.4. The three columns on the left are from DermIS & DermQuest, while those on the right are from MED-NODE.

11.4.2 Transfer Learning

To train a new DCNN model from scratch, it needs a huge amount of processing time in addition to a massive number of images like ImageNet. The weights of Alex-net and Resnet50 are initialized and fine-tuned to adapt weights according to

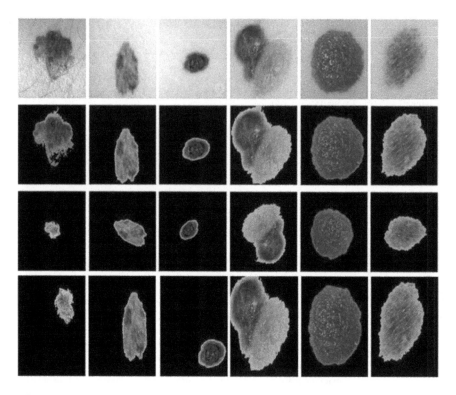

FIGURE 11.4   The three columns on the left were selected randomly from the DermIS & DermQuest datasets, and the three right columns are from MED-NODE. The first, second, third, and fourth rows show original, segmented, rotated, and translated images, respectively.

the datasets used in the proposed model. The weights will not change dramatically because of using a very small learning rate. The learning rate is used to update the weights of the convolutional layers. Stochastic gradient descent (SGD) algorithm has been used to update convolutional layer weights because it is computationally fast, in addition to processing a single training sample, which makes it easier to fit into memory. Unlike the convolutional layers, the weights of fully connected layers have been initialized randomly because we have changed the classification layer to multiclass SVM in Alex-net and to SoftMax in Resnet50.

11.4.3   DROPOUT LAYER

Overfitting is one of CNN's problems, particularly when using a small dataset for the learning model. A powerful tool used for CNN task regularization is the dropout of layers. Alex-net contains eight fully connected layers; the last fully connected layer, FC8, outputs 1000 class scores for different tasks. In this chapter, the last fully connected layer (FC8), the SoftMax layer, and the classification layer have been dropped out and replaced with a new fully connected layer (for two classes only: melanoma and nevus) and a multiclass SVM [38]. This process, clearly displayed in Figure 11.5,

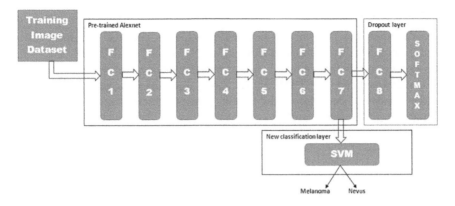

FIGURE 11.5   The architecture of Alex-net after transfer learning.

significantly increased the training speed and reduced the computational complexity of the DCNN. Also, it led to learning extra new robust features and finally reduced the negative effect of overfitting.

The same process of dropout layers has been applied with Resnet50 but with some differences. The last three layers, called FC100, FC100_softmax, and Classification layer_fc100, have been replaced with new layers to be fit with the required task here. The new layers are FC2, SoftMax, and Class Output. These layers are able to classify images into 2 classes (melanoma and nevus) instead of the 1000 classes in the original Resnet50. Figure 11.6 explains the modified Resnet50.

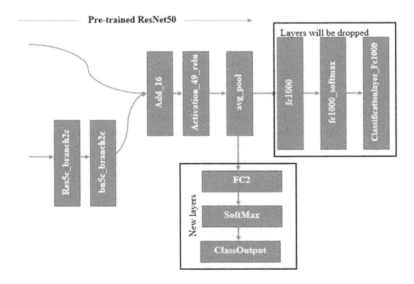

FIGURE 11.6   Resnet50 architecture after transfer learning.

11.5   EXPERIMENTAL RESULTS

The experiments are performed with an IBM-compatible computer equipped with a Core i5 processor. Two datasets with color images defined in RGB space are used. The datasets are divided into two classes, melanoma and nevus. Parallelization of the implemented code in CUDA and running it using the GPU allowed the use of a larger number of examples for the training process, which consequently provided a large gain in the generalization error of the models. The classification layer has been replaced by the multiclass SVM, where a fast-linear solver is compatible with the two classes of skin images: melanoma and nevus. Two types of experiments were performed. The first type was performed with high-quality microscopic images from the dataset [14], while the second type was performed with low-quality images acquired by the camera with unconstrained and varying environmental factors, which include noise such as hair and other factors.

Each experiment consisted of a number of runs where the batch size, the number of training epochs (a measure of the number of times to use all training vectors once to update the weights), and the initial learning rate were 10, 40, and 0.001, respectively; these values were fixed for all experiments. The images were randomly divided: 85% of the images were for training and 15% of the images were for testing. All images from the two datasets were preprocessed by segmenting the region of interest using the segmentation method [39]. The proposed model was evaluated using three metrics, namely, accuracy, sensitivity, and specificity [40], as follows:

$$\text{Accuracy} = \frac{t_p + t_n}{t_p + f_p + f_n + t_n} \qquad (11.2)$$

$$\text{Sensitivity} = \frac{t_p}{t_p + f_n} \qquad (11.3)$$

$$\text{Specificity} = \frac{t_n}{f_p + t_n} \qquad (11.4)$$

Where, t_p, f_p, f_n, and t_n refer to true positive, false positive, false negative, and true negative, respectively. The false-positive rate should be relatively small and the true negative rate should be relatively large, resulting in most points falling in the left part of the receiver operating characteristic (ROC) curve [41].

11.5.1   PERFORMANCE OF THE ALEX-NET

The modified architecture of Alex-net has been tested using two datasets: MED-NODE and DermIS [36] & DermQuest [37]. These datasets have been randomly divided into 85% for training and the remaining 15% for testing. A total of six experiments were conducted with proposed Alex-net using the first dataset, MED-NODE. The first one was performed using the original skin images of the dataset without any preprocessing. Because of noise like hair, background, and other factors, a segmentation process was carried out to extract ROI, keeping images in the same RGB color space. The segmented ROI images were used in the second experiments to measure the impact of noise reduction with Alex-net.

The segmented ROI images were augmented by rotating each image in the dataset by an angle ranging from $0°$ to $90°$ with a fixed step of $5°$. The numbers of melanoma and nevus images increased to $(90/5+1) \times 70 = 1330$ and $(90/5+1) \times 100 = 1900$, respectively. The third experiment was carried out to measure the performance of the modified Alex-net after the augmentation of the segmented images. The results of all experiments are summarized in Table 11.5. Comparing these results with the results of the first and second experiments clearly shows that increasing the number of training images results in a significant improvement in the evaluation metrics, accuracy, sensitivity, and specificity. This observation motivates us to conduct an additional experiment with an approximately doubled number of training images.

The fourth experiment was performed with the same classes and the same conditions except for the data augmentation. In this experiment, the segmented images were augmented by rotating each image in the dataset by an angle ranging from $0°$ to $180°$ with a fixed step of $5°$. Therefore, the numbers of melanoma and nevus images increased to $(180/5+1) \times 70 = 2590$ and $(180/5+1) \times 100 = 3700$, respectively. As shown in Table 11.5, the performance metrics were increased compared with the previous experiments. The fifth experiment was performed where each segmented image in the dataset was augmented by a rotation angle ranging from $0°$ to $355°$ with a fixed step of $5°$. The obtained result was improved compared with those from the third and fourth experiments. We used the same augmentation process by rotating each image with a fixed step angle of $5°$ from $0°$ to $90°$, $0°$ to $180°$, and $0°$ to $355°$ for the third, fourth, and fifth experiments, respectively.

That last experiment was done by using a combination of different augmentation approaches. Each segmented color image was translated with different translation parameters and rotated with a random rotation angle in the range between $0°$ and $355°$. The results of these six experiments clearly show that having more training images increases the ability of the proposed system to identify skin lesions of the type "melanoma" correctly (sensitivity) and its ability to identify skin lesions of type "nevus" correctly (specificity). To ensure this observation and the credibility of the proposed system and its ability to identify melanoma and nevus skin lesions correctly, another group of experiments was performed with more complicated colorskin images.

The second group of experiments was performed using noisy low-quality colorskin images. The dataset used in these experiments was created by a random selection of images from DermIS [36] and DermQuest [37]. As mentioned before, the images of the created dataset were segmented to isolate the ROI from the background to reduce the noise. The previous six experiments were repeated again using the DermIS & DermQuest dataset. The acquired results have proved that increasing the number of training images improved the evaluation metrics, accuracy, sensitivity, and specificity. Table 11.5 summarizes all obtained results of Alex-net using MED-NODE and the DermIS & DermQuest datasets.

TABLE 11.5
The Accuracy of the Proposed Model Using Alex-net

Experiment	MED-NODE dataset			DermIS & DermQuest dataset		
	Accuracy (%)	Sensitivity (%)	Specificity (%)	Accuracy (%)	Sensitivity (%)	Specificity (%)
First	78.11	97.85	58.37	73.08	61.54	84.62
Second	81.05	95.53	66.56	76.93	84.62	69.23
Third	91.65	97.59	85.71	85.95	80.83	91.06
Fourth	94.62	92.41	96.83	90.79	94.86	86.71
Fifth	95.42	95.42	96.03	92.37	92.29	92.46
Sixth	96.82	95.27	98.47	93.35	93.48	93.21

11.5.2   PERFORMANCE OF THE RESNET50

The modified Resnet50 architecture was tested using the MED-NODE dataset. The dataset was randomly divided into 85% for training and the remaining 15% for testing. Because Resnet50 is deeper than Alex-net, it has the ability to classify images better than Alex-net with a smaller number of images. Only three experiments were done with modified Resnet50 architecture. The first experiments were performed using the modified Resnet50 architecture and the original images from MED-NODE without segmentation and without augmentation. Images in the dataset were segmented to extract foreground containing important features (ROI) and discarding background, keeping segmented images in the same RGB color space. The segmented color images were used in the second experiment. The performance measures were greatly improved compared with using a dataset without segmentation. Finally, the segmented images in the dataset were augmented by rotating images from $0°$ to $90°$ with a fixed step angle of $5°$. To measure the impact of augmentation in performance measures, a third experiment was performed. Table 11.6 shows the performance of modified Resnet50 with the three experiments.

The same experiments have been repeated using noisy low-quality color-skin images from DermIS [36] and DermQuest [37]. As mentioned before to reduce the noise, the images of the dataset were segmented to isolate the ROI from the background. The previous three experiments were repeated using the DermIS & DermQuest dataset by dividing it into 85% for training and the remaining 15% for testing randomly. The acquired results proved that the modified Resnet50 architecture is better than the modified Alex-net architecture.

11.6   DISCUSSION

It is noted that the results obtained from the second dataset are smaller than the results obtained from the first dataset. This is normal and predictable according to the nature of the images in each dataset. The images of the first dataset are high-quality microscopic images acquired by specialized devices. On the other side, the images of the second dataset are low-quality images acquired by using a common camera in a noisy environment.

By comparing the performance measures of proposed Alex-net against proposed Resnet50 using different datasets, we found that the best performance measures were for the modified Resnet50. The performance of Resnet50 is the best because it contains many more layers than Alex-net. As discussed before, Resnet50 is enforced to learn how to map some of the input to some of the output and input while Alex-net tried to map some of the input to the output. Based on "skip connection", Resnet50 passes the input from one layer to a far layer by escaping two to five layers.

We could say that DCNN works well with any dataset of different resolutions. The performance increased with high-quality images, high resolution, and low levels of noise. The performance of the proposed methods was compared with the well-known existing methods [11–21, 42] and the comparison is summarized in Tables 11.7 and 11.8. The accuracy and the ROC curves were used to measure the performance of the proposed methods against the performance of the existing methods. For the

TABLE 11.6
The Accuracy of the Proposed Model Using Resnet50

Experiment	MED-NODE dataset			DermIS & DermQuest dataset		
	Accuracy (%)	Sensitivity (%)	Specificity (%)	Accuracy (%)	Sensitivity (%)	Specificity (%)
First	80	85.71	72.72	76.92	70.59	88.89
Second	92	100	83.33	92.31	86.67	100
Third	99.79	99.65	100	98.56	99.59	97.54

TABLE 11.7
Comparative Study Using the DermIS & DermQuest Dataset

Method	Preprocessing (enhancement)	Segmentation	Image type	Classification method	Accuracy (%)
Amelard [11]	Yes	Yes	RGB	SVM	87.38
Karabulut [12]	Yes	Yes	RGB	SVM	71.4
			Black & white	CNN	71.4
Almaraz [13]	Yes	Yes	RGB	SVM	75.1
Hosny [42]	No	Yes	RGB	DCNN	96.86
Proposed Alex-net	No	Yes	RGB	DCNN	93.35
Proposed Resnet50	**No**	**Yes**	**RGB**	**DRNN**	**98.56**

DermIS & DermQuest dataset, the performance of the proposed method was compared with the performance of the existing methods [11–13, 42]. The obtained results are listed in Table 11.7. All the mentioned methods in Table 11.7 used the same DermIS & DermQuest dataset. Amelard et al. [11] enhanced the images and detected the borders of skin lesions to segment important regions, and then they used the segmented images in the classification process and achieved an accuracy percentage of 87.38%. A similar approach was utilized in Karabulut and Ibrikci [12], where the input images were enhanced to reduce the noise and then the regions of interest were segmented in two ways. The first way is to keep the segmented images in RGB color space and the second way is to convert the segmented images to black-and-white images. The SVM and CNN were used and achieved the same accuracy, 71.4%, for the two methods. In Almaraz-Damian et al. [13], similar preprocessing steps were used. The obtained accuracy was marginally increased to 75.1%. In Hosny et al. [42], an automated process for skin cancer classification was proposed, where the authors applied transfer learning with Alex-net and replaced the classification layer with SoftMax. The achieved accuracy rate was 96.86%.

In order to improve the classification accuracy, the authors applied transfer learning with Alex-net and replaced the classification layer with the multiclass SVM. Unfortunately, this method achieved an accuracy rate of only 93.35%. It is clear that the achieved accuracy is lower than the achieved accuracy of the SoftMax method [42]. Therefore, utilizing an alternative CNN could be a fruitful solution for this problem. The authors applied transfer learning with Resnet50 to classify skin lesions using the same dataset where the proposed Resnet50-based classification method achieved a 98.56% accuracy rate and outperformed the Alex-net-based classification methods. The ROC curves of the proposed methods and the existing methods [11–13, 42] are plotted and displayed in Figure 11.7.

Another comparison was performed using the MED-NODE dataset; all methods mentioned in Table 11.8 use this dataset. The performance of the proposed methods was compared with that of the existing methods [14, 15, 19–21, 42]. Table 11.8 shows an overview of the conditions, image types, and classifiers used in the existing methods. It was noted that all of these methods applied two preprocessing

TABLE 11.8

Comparative Study for MED-NODE Dataset

Method	Preprocessing (enhancement)	Segmentation	Image type	Classifier	Accuracy (%)
Giotis [14]	Yes	Yes	RGB	Majority vote	81
M.H. Jafari [15]	Yes	Yes	RGB	SVM	79
Nasr-Esfahan [19]	Yes	Yes	RGB	CNN	81
Premaladha [20]	Yes	Yes	Black & white	ANN	90.12
				DLNN	92.89
Kostopoulos [21]	Yes	Yes	RGB	PNN	76.2
Hosny [42]	No	Yes	RGB	DCNN	97.70
Proposed Alex-net	No	Yes	RGB	DCNN	96.82
Proposed Resnet50	**No**	**Yes**	**RGB**	**DRNN**	**99.79**

steps. The first step was enhancement, while the second was segmentation to extract the region of interest. These methods used RGB color images, while the method of [20] converted these color images to black and white. Table 11.8 clearly shows that, on one hand, the performance of the proposed Alex-net method failed in terms of its classification rates compared with other methods [14, 15, 19–21, 42]. On the other hand, the proposed Resnet50-based classification method outperformed

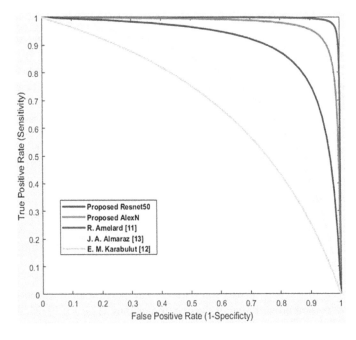

FIGURE 11.7 The ROC curves for the different methods with the DermIS & DermQuest dataset.

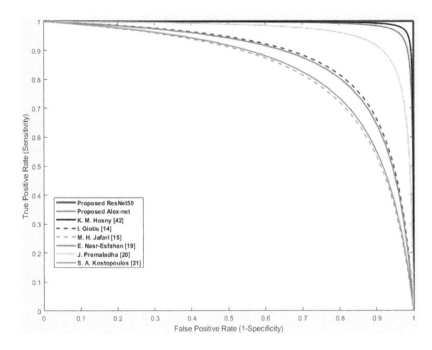

FIGURE 11.8   The ROC curves for the different methods with the MED-NODE dataset.

the existing methods [14, 15, 19–21, 42]. The ROC curves of the different methods are plotted and displayed in Figure 11.8. All results are consistent and clearly
show that the proposed methods are superior to all the existing methods [11–21].
From the results displayed in Tables 11.7 and 11.8 in addition to the ROC curves
in Figures 11.7 and 11.8, it was proved that the proposed method is superior to the
state of the art.

11.7   CONCLUSION

This chapter proposed two models using the theory of transfer learning for the Alex-
net and Resnet50 architecture to classify tumors as melanoma or not-melanoma.
The last three layers (fully connected, SoftMax, and classification layers) have been
dropped out and replaced with three new layers. In Alex-net, the new layers were
fully connected for two classes only, namely, a multiclass SVM and an output layer;
in Resnet50, these layers were fully connected for two different classes, namely,
SoftMax and a classification layer. This modification was done to create two classifications of skin images in addition to overcoming the overfitting problem. Two different datasets were used, one with high-quality images and the other with low-quality
images. These datasets were used to investigate the performance of the two proposed
models Alex-net and Resnet50 in a situation of using images of high and low quality.
Different ways of augmentation have been used to overcome the limitations of image
number in datasets. The proposed models have outperformed other models using

the same datasets. The sensitivity and specificity of the proposed models proved the credibility of the proposed model, because the increased sensitivity and specificity proved the truthfulness of the system.

BIBLIOGRAPHY

1. I. Maglogiannis, and C. N. Doukas, "Overview of advanced computer vision systems for skin lesions characterization", *Transactions on Information Technology in Biomedicine*, vol. 13, no. 5, pp. 721–733, 2009.
2. M. S. Arifin, M. G. Kibria, A. Firoze, M. A. Amini, and H. Yan, "Dermatological disease diagnosis using color-skin images", *International Conference on Machine Learning and Cybernetics*, vol. 5, pp. 1675–1680, 2012.
3. J. M. Gálvez, D. Castillo, L. J. Herrera, B. S. Román, O. Valenzuela, F. M. Ortuño, and I. Rojas, "Multiclass classification for skin cancer profiling based on the integration of heterogeneous gene expression series", *PLoS ONE*, vol. 13, no. 5, pp. 1–26, 2018.
4. A. Masood, A. A. Al-Jumaily, and T. Adnan, "Development of automated diagnostic system for skin cancer: Performance analysis of neural network learning algorithms for classification", *International Conference on Artificial Neural Networks*, pp. 837–844, 2014.
5. P. G. Cavalcanti, and J. Scharcanski, "Macroscopic pigmented skin lesion segmentation and its influence on lesion classification and diagnosis", *Color Medical Image Analysis*, vol. 6, pp. 15–39, 2013.
6. M. J. M. Vasconcelos, and L. Rosado, "No-reference blur assessment of dermatological images acquired via mobile devices", *Image and Signal Processing*, vol. 8509, pp. 350–357, 2014.
7. K. Bunte, M. Biehl, M. F. Jonkman, and N. Petkov, "Learning effective color features for content-based image retrieval in dermatology", *Pattern Recognition*, vol. 44, pp. 1892–1902, 2011.
8. S. V. Patwardhan, A. P. Dhawan, and P. A. Relue, "Classification of melanoma using tree structured wavelet transforms", *Computer Methods and Programs in Biomedicine*, vol. 72, no. 3, pp. 223–239, 2003.
9. W. Y. Chang, A. Huang, C. Y. Yang, C. H. Lee, Y. C. Chen, T. Y. Wu, and G. S. Chen, "Computer-aided diagnosis of skin lesions using conventional digital photography: A reliability and feasibility study", *PLoS ONE*, vol. 8, no. 11, pp. 1–9, 2013.
10. S. Kundu, N. Das, and M. Nasipuri, "Automatic detection of ringworm using Local Binary Pattern (LBP)", 2011, https://arxiv.org/abs/1103.0120.
11. R. Amelard, A. Wong, and D. A. Clausi, "Extracting morphological high-level intuitive features (HLIF) for enhancing skin lesion classification", *International Conference of the IEEE Engineering in Medicine and Biology Society*, pp. 4458–4461, 2012.
12. E. M. Karabulut, and T. Ibrikci, "Texture analysis of melanoma images for computer-aided diagnosis", *International Conference on Intelligent Computing, Computer Science and Information Systems (ICCSIS)*, vol. 2, pp. 26–29, 2016.
13. J. A. Almaraz-Damian, V. Ponomaryov, and E. R. Gonzalez, "Melanoma CADe based on ABCD rule and Haralick texture features", *International Kharkiv Symposium on Physics and Engineering of Microwaves, Millimeter and Submillimeter Waves (MSMW)*, pp. 1–4, 2016.
14. I. Giotis, N. Molders, S. Land, M. Biehl, M. F. Jonkman, and N. Petkov, "MED-NODE: A computer-assisted melanoma diagnosis system using non-dermoscopic images", *Expert Systems with Applications*, vol. 42, no. 19, pp. 6578–6585, 2015.
15. M. H. Jafari, S. Samavi, N. Karimi, S. M. R. Soroushmehr, K. Ward, and K. Najarian, "Automatic detection of melanoma using broad extraction of features from digital images", *International Conference of the IEEE Engineering in Medicine and Biology Society (EMBC)*, pp. 1357–1360, 2016.

16. I. Goodfellow, Y. Bengio, and A. Courville, *Deep Learning*. MIT Press, 2016.
17. P. Dubal, S. Bhatt, C. Joglekar, and S. Patil, "Skin cancer detection and classification", *International Conference on Electrical Engineering and Informatics (ICEEI)*, pp. 1–6, 2017.
18. D. Kumar, M. J. Shafiee, A. Chung, F. Khalvati, M. Haider, and A. Wong, "Discovery radiomics for computed tomography cancer detection", 2015, https://arxiv.org/abs/1509.00117.
19. E. Nasr-Esfahan, S. Samavi, N. Karimi, S. M. R. Soroushmehr, M. H. Jafari, K. Ward, and K. Najarian, "Melanoma detection by analysis of clinical images using convolutional neural network", *International Conference of the IEEE Engineering in Medicine and Biology Society (EMBC)*, pp. 1373–1376, 2016.
20. J. Premaladha, and K. S. Ravichandran, "Novel approaches for diagnosing melanoma skin lesions through supervised and deep learning algorithms", *Journal of Medical Systems*, vol. 40, no. 96, pp. 1–12, 2016.
21. S. A. Kostopoulos et al., "Adaptable pattern recognition system for discriminating Melanocytic Nevi from Malignant Melanomas using plain photography images from different image databases", *International Journal of Medical Informatics*, vol. 105, pp. 1–10, 2017.
22. A. Esteva, B. Kuprel, H. M. Blau, S. M. Swetter, J. Ko, R. A. Novoa, and S. Thrun, "Dermatologist-level classification of skin cancer with deep neural networks", *Nature*, vol. 542, no. 7639, pp. 115–118, 2017.
23. S. Jain, V. Jagtap, and N. Pise, "Computer aided melanoma skin cancer detection using image processing", *Procedia Computer Science*, vol. 48, pp. 735–740, 2015.
24. D. Gautam, and M. Ahmed, "Melanoma detection and classification using SVM based decision support system", *IEEE India Conference (INDICON)*, pp. 1–6, 2015.
25. Convolutional Neural Networks (CNNs / ConvNets), the Stanford CS class notes, Assignments, Spring 2017, http://cs231n.github.io/convolutional-networks/, Accessed: 18 August 2017.
26. R. D. Azulay, D. R. Azulay, and L. Azulay-Abulafia, *Dermatologia*. 6th edition. Rio de Janeiro: Guanabara Koogan, 2013.
27. Y. LeCun, L. Bottou, Y. Bengio, and P. Haffner, "Gradient-based learning applied to document recognition", *Proceedings of the IEEE*, vol. 86, no. 11, pp. 101–118, 1998.
28. S. Srinivas, R. K. Sarvadevabhatl, K. R. Mopur, N. Prabhu, S. S. S. Kruthiventi, and R. V. Babu, "A taxonomy of deep convolutional neural nets for computer vision", *Frontiers in Robotics and AI*, vol. 2, pp. 1–18, 2016.
29. A. Krizhevsky, I. Sutskever, and G. Hinton, "ImageNet classification with deep convolutional neural networks", *Advances in Neural Information Processing Systems*, vol. 25, no. 2, pp. 1097–1105, 2012.
30. O. Russakovsky et al., "ImageNet large scale visual recognition challenge", *International Journal of Computer Vision*, vol. 115, no. 3, pp. 211–252, 2015.
31. J. D. Prusa, and T. M. Khoshgoftaar, "Improving deep neural network design with new text data representations", *Journal of Big Data*, Springer, vol. 4, pp. 1–16, 2017.
32. T. A. Martin, A. J. Sanders, L. Ye, J. Lane, and W. G. jiang, "Cancer invasion and metastasis: Molecular and cellular perspective", Landes Bioscience, pp. 135–168, 2013.
33. J. Yosinski, J. Clune, Y. Bengio, and H. Lipson, "How transferable are features in deep neural networks?" *Advances in Neural Information Processing Systems*, vol. 2, pp. 3320–3328, 2014.
34. S. J. Pan, and Q. Yang, "A survey on transfer learning", *Transactions on Knowledge and Data Engineering*, vol. 22, no. 10, pp. 1345–1359, 2010.
35. K. Weiss, T. M. Khoshgoftaar, and D. Wang, "A survey of transfer learning", *Journal of Big Data*, Springer, vol. 3, no 9, pp. 1–15, 2016.
36. Dermatology Information System, 2012, http://www.dermis.net, Accessed: 16 August 2017.

37. DermQuest, 2012, http://www.dermquest.com, Accessed: 16 August 2017.

38. C. Chih-Chung, and L. Chih-Jen, "LIBSVM: A library for support vector machines", *Transactions on Intelligent Systems and Technology*, vol. 2, pp. 1–27, 2013.

39. B. Basavaprasad, and R. S. Hegad, "Color image segmentation using adaptive Growcut method", *Procedia Computer Science*, vol. 45, pp. 328–335, 2015.

40. M. Stojanovi et al., "Understanding sensitivity, specificity and predictive values", *Vojnosanitetski pregled*, vol. 71, no. 11, pp. 1062–1065, 2014.

41. T. Fawcett, "An introduction to ROC analysis", *Pattern Recognition Letter*, vol. 27, no. 8, pp. 861–874, 2006.

42. K. M. Hosny, M. A. Kassem, and M. M. Foaud, "Classification of skin lesions using transfer learning and augmentation with Alex-net", *PLoS ONE*, vol. 14, no. 5, pp. 1–17, 2019.

43. K. M. Hosny, M. A. Kassem, and M. M. Foaud, "Skin cancer classification using deep learning and transfer learning", *Cairo International Biomedical Engineering Conference (CIBEC)*, pp. 90–93, 2018.

44. K. He, X. Zhang, S. Ren, and J. Sun, "Deep residual learning for image recognition", *Conference on Computer Vision and Pattern Recognition (CVPR)*, pp. 770–778, 2016.

Index

Milton Keynes UK
Ingram Content Group UK Ltd.
UKHW031128141024
449569UK00006B/370